Student Solutions Manual

To Accompany

PHYSICS

Seventh Edition

John D. Cutnell

Southern Illinois University at Carbondale

Kenneth W. Johnson

Southern Illinois University at Carbondale

D0206941

WILEY

JOHN WILEY & SONS, INC.

Cover Photo: ©Ed Bock/Corbis Images

Copyright © 2007 John Wiley & Sons, Inc. All rights reserved. No part of this publication may be reproduced, stored in a retrieval system or transmitted in any form or by any means, electronic, mechanical, photocopying, recording, scanning, or otherwise, except as permitted under Sections 107 or 108 of the 1976 United States Copyright Act, without either the prior written permission of the Publisher, or authorization through payment of the appropriate per-copy fee to the Copyright Clearance Center, Inc., 222 Rosewood Drive, Danvers, MA 01923, (978)750-8400, fax (978)750-4470 or on the web at www.copyright.com. Requests to the Publisher for permission should be addressed to the Permissions Department, John Wiley & Sons, Inc., 111 River Street, Hoboken, NJ 07030-5774, (201)748-6011, fax (201)748-6008, or online at http://www.wiley.com/go/permissions.

To order books or for customer service please, call 1-800-CALL WILEY (225-5945).

ISBN-13 978- 0-471-66326-3
ISBN-10 0-471-63326-3

Printed in the United States of America

10 9 8 7 6 5 4 3 2 1

Printed and bound by Courier - Kendallville

PREFACE

This volume contains the complete solutions to those Conceptual Questions and Problems in the text that are marked with an **SSM** icon. These questions and problems are found at the end of each chapter in the text. There are about 100 such Conceptual Questions and about 600 such Problems.

The solutions are worked out with great care, and all the steps are included. Most of the solutions are comprised of two parts, a REASONING part followed by a SOLUTION part. In the REASONING part we explain what motivates our procedure for solving the problem, before any algebraic or numerical work is done. During the SOLUTION part, numerical calculations are performed, and the answer to the problem is obtained. Since the Conceptual Questions involve little or no mathematics, the REASONING and SOLUTION parts are combined into a single section.

We welcome any suggestions that you may have for improving the usefulness of these solutions. Please feel free to write us care of Physics Editor, Higher Education Division, John Wiley & Sons, Inc., 111 River Street, Hoboken, NJ 07030 or contact us at **www.wiley.com/college/cutnell**

CONTENTS

Chapter 1 INTRODUCTION AND MATHEMATICAL CONCEPTS
Conceptual Questions 1
Problems 3

Chapter 2 KINEMATICS IN ONE DIMENSION
Conceptual Questions 11
Problems 13

Chapter 3 KINEMATICS IN TWO DIMENSIONS
Conceptual Questions 24
Problems 26

Chapter 4 FORCES AND NEWTON'S LAWS OF MOTION
Conceptual Questions 41
Problems 42

Chapter 5 DYNAMICS OF UNIFORM CIRCULAR MOTION
Conceptual Questions 61
Problems 63

Chapter 6 WORK AND ENERGY
Conceptual Questions 71
Problems 72

Chapter 7 IMPULSE AND MOMENTUM
Conceptual Questions 83
Problems 85

Chapter 8 ROTATIONAL KINEMATICS
Conceptual Questions 97
Problems 98

Chapter 9 ROTATIONAL DYNAMICS
Conceptual Questions 106
Problems 107

Chapter 10 SIMPLE HARMONIC MOTION AND ELASTICITY
Conceptual Questions 119
Problems 120

Chapter 11 FLUIDS
Conceptual Questions 133
Problems 135

Chapter 12 TEMPERATURE AND HEAT
Conceptual Questions 149
Problems 150

Chapter 13 THE TRANSFER OF HEAT
Conceptual Questions 162
Problems 164

Chapter 14 THE IDEAL GAS LAW AND KINETIC THEORY
Conceptual Questions 170
Problems 171

Chapter 15 THERMODYNAMICS
Conceptual Questions 178
Problems 180

Chapter 16 WAVES AND SOUND
Conceptual Questions 194
Problems 195

Chapter 17 THE PRINCIPLE OF LINEAR SUPERPOSITION AND INTERFERENCE PHENOMENA
Conceptual Questions 210
Problems 212

Chapter 18 ELECTRIC FORCES AND ELECTRIC FIELDS
Conceptual Questions 222
Problems 224

Chapter 19 ELECTRIC POTENTIAL ENERGY AND THE ELECTRIC POTENTIAL
Conceptual Questions 238
Problems 240

Chapter 20 ELECTRIC CIRCUITS
Conceptual Questions 249
Problems 251

Chapter 21 MAGNETIC FORCES AND MAGNETIC FIELDS
Conceptual Questions 268
Problems 271

Chapter 22 ELECTROMAGNETIC INDUCTION
Conceptual Questions 280
Problems 281

Chapter 23 ALTERNATING CURRENT CIRCUITS
Conceptual Questions 290
Problems 293

Chapter 24 ELECTROMAGNETIC WAVES
Conceptual Questions 300
Problems 301

Chapter 25 THE REFLECTION OF LIGHT: MIRRORS
Conceptual Questions 308
Problems 309

Chapter 26 THE REFRACTION OF LIGHT: LENSES AND OPTICAL INSTRUMENTS
Conceptual Questions 315
Problems 317

Chapter 27 INTERFERENCE AND THE WAVE NATURE OF LIGHT
Conceptual Questions 333
Problems 335

Chapter 28 SPECIAL RELATIVITY
Conceptual Questions 344
Problems 346

Chapter 29 PARTICLES AND WAVES
Conceptual Questions 352
Problems 354

Chapter 30 THE NATURE OF THE ATOM
Conceptual Questions 360
Problems 361

Chapter 31 NUCLEAR PHYSICS AND RADIOACTIVITY
Conceptual Questions 369
Problems 371

Chapter 32 IONIZING RADIATION, NUCLEAR ENERGY, AND ELEMENTARY PARTICLES
Conceptual Questions 378
Problems 380

CHAPTER 1 | *INTRODUCTION AND MATHEMATICAL CONCEPTS*

CONCEPTUAL QUESTIONS

---

2. ***REASONING AND SOLUTION*** It is not always possible to add two numbers that have the same dimensions. In order to add any two physical quantities they must be expressed in the same *units*. Consider the two lengths: 1.00 m and 1.00 cm. Both quantities are lengths and, therefore, have the dimension [L]. Since the units are different, however, these two numbers cannot be added.

---

11. ***REASONING AND SOLUTION*** One can arrive back at the starting point after making eight consecutive displacements that add to zero only if one traverses three of the edges on one face and three edges on the opposite face (six displacements; the remaining two displacements occur in going from one opposite face to the other). For any particular starting point, there are four independent ways to traverse three edges on opposite faces. These are illustrated in the figures below. In (A) and (B), one traverses three consecutive edges before moving to the opposite face; in (C) and (D), one traverses two consecutive edges, moves to the opposite face, traverses three consecutive edges, then moves back to the original face and traverses the third edge on that face ending at the starting point.

(A)

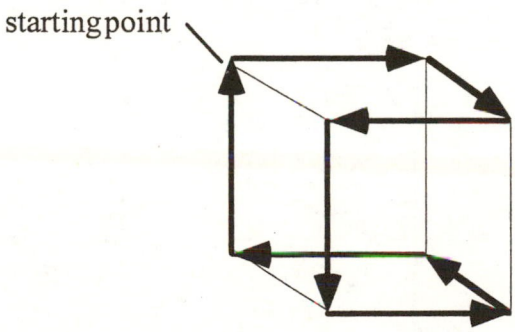

(C)

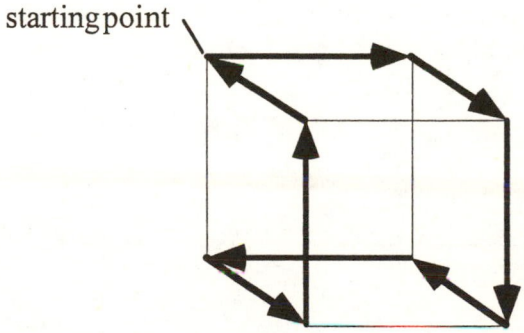

(B)

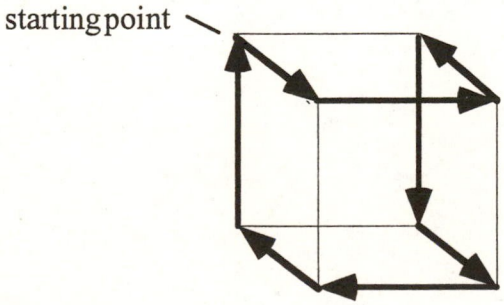

(D)

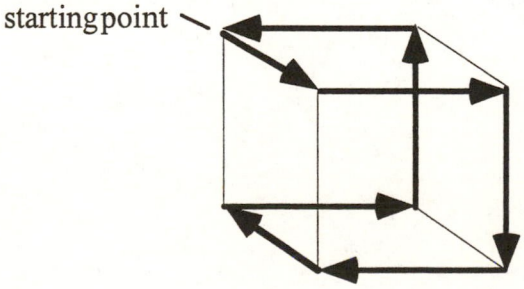

For any particular starting point, one can initially move in any one of three possible directions. Thus, there are a total of 4 (independent ways) × 3 (possible directions) or 12 ways to arrive back at the starting point that involve eight displacement vectors.

---

19. ***REASONING AND SOLUTION***

a. As the angle increases from 0° to 90°, the x component of the vector decreases in magnitude while the y component increases in magnitude. When the vector has rotated through an angle of 90°, the x component has zero magnitude and the magnitude of the y component is equal to the magnitude of the original vector (see figure below).

b. As the angle increases from 90° to 180°, the y component of the vector decreases in magnitude while the x component increases in magnitude. When the vector has rotated through an angle of 180°, the y component has zero magnitude and the magnitude of the x component is equal to the magnitude of the original vector.

c. As the angle increases from 180° to 270°, the x component of the vector decreases in magnitude while the y component increases in magnitude. When the vector has rotated through an angle of 270°, the x component has zero magnitude and the magnitude of the y component is equal to the magnitude of the original vector.

d. As the angle increases from 270° to 360°, the y component of the vector decreases in magnitude while the x component increases in magnitude. When the vector has rotated through an angle of 360°, the y component has zero magnitude and the magnitude of the x component is equal to the magnitude of the original vector.

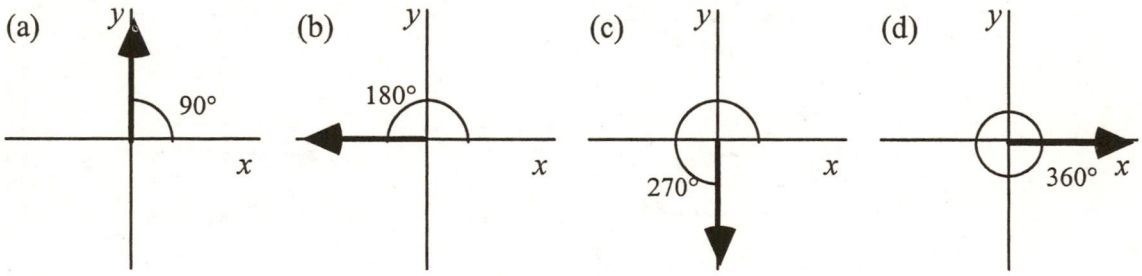

CHAPTER 1 | *INTRODUCTION AND MATHEMATICAL CONCEPTS*

PROBLEMS

---

1. ***REASONING AND SOLUTION*** We use the fact that 0.200 g = 1 carat and that, under the conditions stated, 1000 g has a weight of 2.205 lb to construct the two conversion factors: (0.200 g)/(1 carat) = 1 and (2.205 lb)/(1000 g) = 1. Then,

$$\left(3106\ \cancel{\text{carats}}\right)\left(\frac{0.200\ \cancel{\text{g}}}{1\ \cancel{\text{carat}}}\right)\left(\frac{2.205\ \text{lb}}{1000\ \cancel{\text{g}}}\right)=\boxed{1.37\ \text{lb}}$$

---

5. ***REASONING*** When converting between units, we write down the units explicitly in the calculations and treat them like any algebraic quantity. We construct the appropriate conversion factor (equal to unity) so that the final result has the desired units.

SOLUTION

a. Since 1.0×10^3 grams = 1.0 kilogram, it follows that the appropriate conversion factor is $(1.0\times10^3\ \text{g})/(1.0\ \text{kg}) = 1$. Therefore,

$$\left(5\times10^{-6}\ \cancel{\text{kg}}\right)\left(\frac{1.0\times10^3\ \text{g}}{1.0\ \cancel{\text{kg}}}\right)=\boxed{5\times10^{-3}\ \text{g}}$$

b. Since 1.0×10^3 milligrams = 1.0 gram,

$$\left(5\times10^{-3}\ \cancel{\text{g}}\right)\left(\frac{1.0\times10^3\ \text{mg}}{1.0\ \cancel{\text{g}}}\right)=\boxed{5\ \text{mg}}$$

c. Since 1.0×10^6 micrograms = 1.0 gram,

$$\left(5\times10^{-3}\ \cancel{\text{g}}\right)\left(\frac{1.0\times10^6\ \mu\text{g}}{1.0\ \cancel{\text{g}}}\right)=\boxed{5\times10^3\ \mu\text{g}}$$

---

9. ***REASONING*** The volume of water at a depth d beneath the rectangle is equal to the area of the rectangle multiplied by d. The area of the rectangle = (1.20 nautical miles) × (2.60 nautical miles) = 3.12 (nautical miles)^2. Since 6076 ft = 1 nautical mile and 0.3048 m = 1 ft, the conversion factor between nautical miles and meters is

$$\left(\frac{6076\ \cancel{\text{ft}}}{1\ \text{nautical mile}}\right)\left(\frac{0.3048\ \text{m}}{1\ \cancel{\text{ft}}}\right)=\frac{1.852\times10^{3}\ \text{m}}{1\ \text{nautical mile}}$$

SOLUTION The area of the rectangle of water in m^2 is, therefore,

$$\left[3.12\ \cancel{(\text{nautical miles})^2}\right]\left(\frac{1.852\times10^{3}\ \text{m}}{1\ \cancel{\text{nautical mile}}}\right)^2=1.07\times10^{7}\ \text{m}^2$$

Since 1 fathom = 6 ft, and 1 ft = 0.3048 m, the depth d in meters is

$$\left(16.0\ \cancel{\text{fathoms}}\right)\left(\frac{6\ \cancel{\text{ft}}}{1\ \cancel{\text{fathom}}}\right)\left(\frac{0.3048\ \text{m}}{1\ \cancel{\text{ft}}}\right)=2.93\times10^{1}\ \text{m}$$

The volume of water beneath the rectangle is

$$(1.07\times10^{7}\ \text{m}^2)\,(2.93\times10^{1}\ \text{m})=\boxed{3.13\times10^{8}\ \text{m}^3}$$

---

13. ***REASONING*** The shortest distance between the two towns is along the line that joins them. This distance, h, is the hypotenuse of a right triangle whose other sides are $h_o = 35.0$ km and $h_a = 72.0$ km, as shown in the figure below.

SOLUTION The angle θ is given by $\tan\theta = h_o / h_a$ so that

$$\theta=\tan^{-1}\left(\frac{35.0\ \text{km}}{72.0\ \text{km}}\right)=\boxed{25.9°\ \text{S of W}}$$

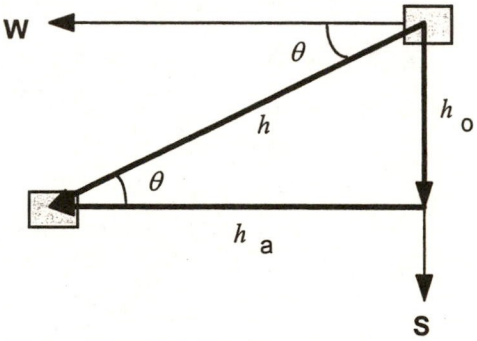

We can then use the Pythagorean theorem to find h.

$$h=\sqrt{h_o^2+h_a^2}=\sqrt{(35.0\ \text{km})^2+(72.0\ \text{km})^2}=\boxed{80.1\ \text{km}}$$

---

19. ***REASONING*** The drawing at the right shows the location of each deer A, B, and C. From the problem statement it follows that

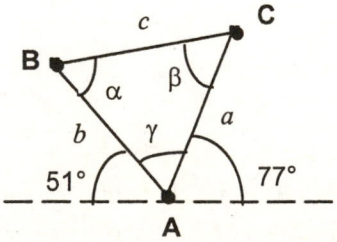

$$b = 62\ \text{m}$$

$$c = 95\ \text{m}$$

$$\gamma = 180° - 51° - 77° = 52°$$

Applying the law of cosines (given in Appendix E) to the geometry in the figure, we have

$$a^2 - 2ab\cos\gamma + (b^2 - c^2) = 0$$

which is an expression that is quadratic in a. It can be simplified to $Aa^2 + Ba + C = 0$, with

$$A = 1$$

$$B = -2b\cos\gamma = -2(62\text{ m})\cos 52° = -76\text{ m}$$

$$C = (b^2 - c^2) = (62\text{ m})^2 - (95\text{ m})^2 = -5181\text{ m}^2$$

This quadratic equation can be solved for the desired quantity a.

SOLUTION Suppressing units, we obtain from the quadratic formula

$$a = \frac{-(-76) \pm \sqrt{(-76)^2 - 4(1)(-5181)}}{2(1)} = 1.2 \times 10^2\text{ m} \quad \text{and} \quad -43\text{ m}$$

Discarding the negative root, which has no physical significance, we conclude that the distance between deer A and C is $\boxed{1.2 \times 10^2\text{ m}}$.

---

23. ***REASONING*** The performer walks out on the wire a distance d, and the vertical distance to the net is h. Since these two distances are perpendicular, the magnitude of the displacement is given by the Pythagorean theorem as $s = \sqrt{d^2 + h^2}$. Values for s and h are given, so we can solve this expression for the distance d. The angle that the performer's displacement makes below the horizontal can be found using trigonometry.

SOLUTION

a. Using the Pythagorean theorem, we find that

$$s = \sqrt{d^2 + h^2} \quad \text{or} \quad d = \sqrt{s^2 - h^2} = \sqrt{(26.7\text{ ft})^2 - (25.0\text{ ft})^2} = \boxed{9.4\text{ ft}}$$

b. The angle θ that the performer's displacement makes below the horizontal is given by

$$\tan\theta = \frac{h}{d} \quad \text{or} \quad \theta = \tan^{-1}\left(\frac{h}{d}\right) = \tan^{-1}\left(\frac{25.0\text{ ft}}{9.4\text{ ft}}\right) = \boxed{69°}$$

---

25. ***REASONING*** When a vector is multiplied by – 1, the magnitude of the vector remains the same, but the direction is reversed. Vector subtraction is carried out in the same manner as vector addition except that one of the vectors has been multiplied by –1.

SOLUTION Since both vectors point north, they are colinear. Therefore their magnitudes may be added by the rules of ordinary algebra.

a. Taking north as the positive direction, we have

$$\mathbf{A} - \mathbf{B} = \mathbf{A} + (-\mathbf{B}) = +2.43 \text{ km} + (-7.74 \text{ km}) = -5.31 \text{ km}$$

The minus sign in the answer for $\mathbf{A} - \mathbf{B}$ indicates that the direction is south so that

$$\boxed{\mathbf{A} - \mathbf{B} = 5.31 \text{ km, south}}$$

b. Similarly,

$$\mathbf{B} - \mathbf{A} = \mathbf{B} + (-\mathbf{A}) = +7.74 \text{ km} + (-2.43 \text{ km}) = +5.31 \text{ km}$$

The plus sign in the answer for $\mathbf{B} - \mathbf{A}$ indicates that the direction is north, so that

$$\boxed{\mathbf{B} - \mathbf{A} = 5.31 \text{ km, north}}$$

---

27. ***REASONING AND SOLUTION*** The single force needed to produce the same effect is equal to the resultant of the forces provided by the two ropes. The figure below shows the force vectors drawn to scale and arranged tail to head. The magnitude and direction of the resultant can be found by direct measurement using the scale factor shown in the figure.

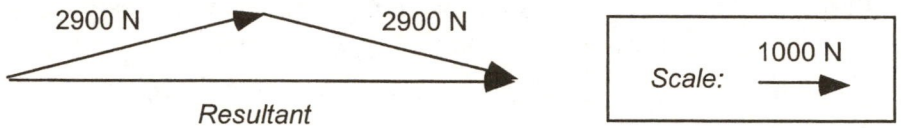

a. From the figure, the magnitude of the resultant is $\boxed{5600 \text{ N}}$

b. The single rope should be directed $\boxed{\text{along the dashed line}}$ in the text drawing.

---

31. ***REASONING AND SOLUTION*** In order to determine which vector has the largest x and y components, we calculate the magnitude of the x and y components explicitly and compare them. In the calculations, the symbol u denotes the units of the vectors.

$A_x = (100.0 \text{ u}) \cos 90.0° = 0.00 \text{ u}$ $\qquad$ $A_y = (100.0 \text{ u}) \sin 90.0° = 1.00 \times 10^2 \text{ u}$

$B_x = (200.0 \text{ u}) \cos 60.0° = 1.00 \times 10^2 \text{ u}$ $\qquad$ $B_y = (200.0 \text{ u}) \sin 60.0° = 173 \text{ u}$

$C_x = (150.0 \text{ u}) \cos 0.00° = 150.0 \text{ u}$ $\qquad$ $C_y = (150.0 \text{ u}) \sin 0.00° = 0.00 \text{ u}$

a. $\boxed{\mathbf{C} \text{ has the largest } x \text{ component.}}$

b. $\boxed{\mathbf{B} \text{ has the largest } y \text{ component.}}$

35. ***REASONING*** The x and y components of **r** are mutually perpendicular; therefore, the magnitude of **r** can be found using the Pythagorean theorem. The direction of **r** can be found using the definition of the tangent function.

SOLUTION According to the Pythagorean theorem, we have

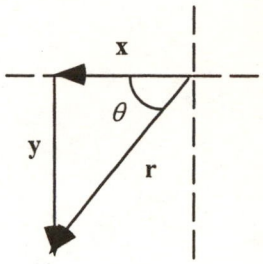

$$r = \sqrt{x^2 + y^2} = \sqrt{(-125\text{ m})^2 + (-184\text{ m})^2} = \boxed{222\text{ m}}$$

The angle θ is

$$\theta = \tan^{-1}\left(\frac{184\text{ m}}{125\text{ m}}\right) = \boxed{55.8°}$$

39. ***REASONING AND SOLUTION*** The force **F** can be first resolved into two components; the z component F_z and the projection onto the x-y plane, F_p as shown in the figure below on the left. According to that figure,

$$F_p = F \sin 54.0° = (475\text{ N}) \sin 54.0° = 384\text{ N}$$

The projection onto the x-y plane, F_p, can then be resolved into x and y components.

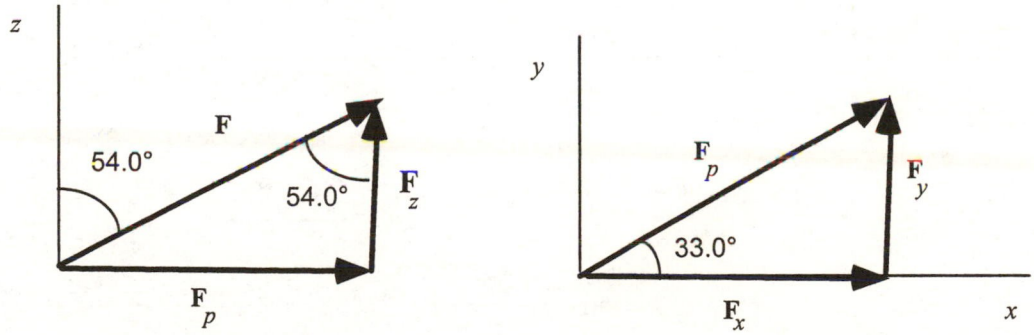

a. From the figure on the right,

$$F_x = F_p \cos 33.0° = (384\text{ N}) \cos 33.0° = \boxed{322\text{ N}}$$

b. Also from the figure on the right,

$$F_y = F_p \sin 33.0° = (384\text{ N}) \sin 33.0° = \boxed{209\text{ N}}$$

c. From the figure on the left,

$$F_z = F \cos 54.0° = (475\text{ N}) \cos 54.0° = \boxed{279\text{ N}}$$

41. ***REASONING*** The individual displacements of the golf ball, **A**, **B**, and **C** are shown in the figure. Their resultant, **R,** is the displacement that would have been needed to "hole the ball" on the very first putt. We will use the component method to find **R**.

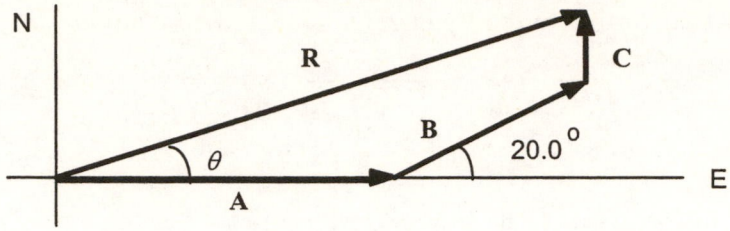

SOLUTION The components of each displacement vector are given in the table below.

Vector	***x* Components**	***y* Components**
A	(5.0 m) cos 0° = 5.0 m	(5.0 m) sin 0° = 0
B	(2.1 m) cos 20.0° = 2.0 m	(2.1 m) sin 20.0° = 0.72 m
C	(0.50 m) cos 90.0° = 0	(0.50 m) sin 90.0° = 0.50 m
R = **A** + **B** + **C**	7.0 m	1.22 m

The resultant vector **R** has magnitude

$$R = \sqrt{(7.0\text{ m})^2 + (1.22\text{ m})^2} = \boxed{7.1\text{ m}}$$

and the angle θ is

$$\theta = \tan^{-1}\left(\frac{1.22\text{ m}}{7.0\text{ m}}\right) = 9.9°$$

Thus, the required direction is $\boxed{\text{9.9° north of east}}$.

47. ***REASONING*** Since the finish line is coincident with the starting line, the net displacement of the sailboat is zero. Hence the sum of the components of the displacement vectors of the individual legs must be zero. In the drawing in the text, the directions to the right and upward are taken as positive.

SOLUTION In the horizontal direction $R_h = A_h + B_h + C_h + D_h = 0$

$$R_h = (3.20\text{ km})\cos 40.0° - (5.10\text{ km})\cos 35.0° - (4.80\text{ km})\cos 23.0° + D\cos\theta = 0$$

$$D\cos\theta = 6.14\text{ km}. \qquad (1)$$

In the vertical direction $R_v = A_v + B_v + C_v + D_v = 0$.

$$R_v = (3.20\text{ km})\sin 40.0° + (5.10\text{ km})\sin 35.0° - (4.80\text{ km})\sin 23.0° - D\sin\theta = 0.$$

$$D\sin\theta = 3.11\text{ km} \qquad (2)$$

Dividing (2) by (1) gives

$$\tan\theta = (3.11\text{ km})/(6.14\text{ km}) \quad \text{or} \quad \theta = \boxed{26.9°}$$

Solving (1) gives

$$D = (6.14\text{ km})/\cos 26.9° = \boxed{6.88\text{ km}}$$

---

53. ***REASONING*** The ostrich's velocity vector **v** and the desired components are shown in the figure at the right. The components of the velocity in the directions due west and due north are $\mathbf{v}_W$ and $\mathbf{v}_N$, respectively. The sine and cosine functions can be used to find the components.

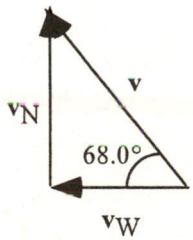

SOLUTION

a. According to the definition of the sine function, we have for the vectors in the figure

$$\sin\theta = \frac{v_N}{v} \quad \text{or} \quad v_N = v\sin\theta = (17.0\text{ m/s})\sin 68° = \boxed{15.8\text{ m/s}}$$

b. Similarly,

$$\cos\theta = \frac{v_W}{v} \quad \text{or} \quad v_W = v\cos\theta = (17.0\text{ m/s})\cos 68.0° = \boxed{6.37\text{ m/s}}$$

---

57. ***REASONING AND SOLUTION*** A single rope must supply the resultant of the two forces. Since the forces are perpendicular, the magnitude of the resultant can be found from the Pythagorean theorem.

a. Applying the Pythagorean theorem,

$$F = \sqrt{(475\text{ N})^2 + (315\text{ N})^2} = \boxed{5.70 \times 10^2\text{ N}}$$

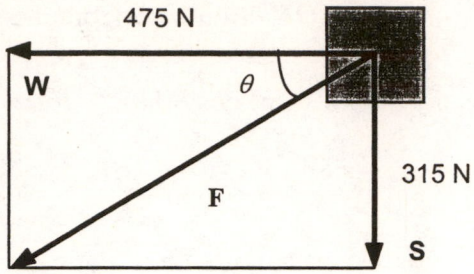

b. The angle θ that the resultant makes with the westward direction is

$$\theta = \tan^{-1}\left(\frac{315\text{ N}}{475\text{ N}}\right) = 33.6°$$

Thus, the rope must make an angle of $\boxed{33.6°\text{ south of west}}$.

---

59. ***REASONING AND SOLUTION*** We take due north to be the direction of the $+y$ axis. Vectors **A** and **B** are the components of the resultant, **C**. The angle that **C** makes with the x axis is then $\theta = \tan^{-1}(B/A)$. The symbol u denotes the units of the vectors.

a. Solving for B gives

$$B = A\tan\theta = (6.00\text{ u})\tan 60.0° = \boxed{10.4\text{ u}}$$

b. The magnitude of **C** is

$$C = \sqrt{A^2 + B^2} = \sqrt{(6.00\text{ u})^2 + (10.4\text{ u})^2} = \boxed{12.0\text{ u}}$$

CHAPTER 2 | *KINEMATICS IN ONE DIMENSION*

CONCEPTUAL QUESTIONS

---

4. ***REASONING AND SOLUTION*** Consider the four traffic lights **1**, **2**, **3** and **4** shown below. Let the distance between lights **1** and **2** be x_{12}, the distance between lights **2** and **3** be x_{23}, and the distance between lights **3** and **4** be x_{34}.

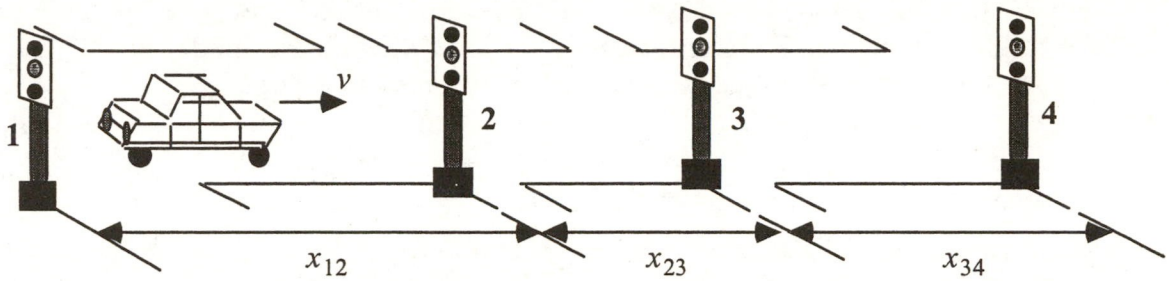

The lights can be timed so that if a car travels with a constant speed v, red lights can be avoided in the following way. Suppose that at time $t = 0$ s, light **1** turns green while the rest are red. Light **2** must then turn green in a time t_{12}, where $t_{12} = x_{12}/v$. Light **3** must turn green in a time t_{23} after light **2** turns green, where $t_{23} = x_{23}/v$. Likewise, light **4** must turn green in a time t_{34} after light **3** turns green, where $t_{34} = x_{34}/v$. Note that the timing of traffic lights is more complicated than indicated here when groups of cars are stopped at light **1.** Then the acceleration of the cars, the reaction time of the drivers, and other factors must be considered.

---

10. ***REASONING AND SOLUTION*** An object moving with a constant acceleration will slow down if the acceleration vector points in the opposite direction to the velocity vector; however, if the acceleration remains constant, the object will never come to a permanent halt. As time increases, the magnitude of the velocity will get smaller and smaller. At some time, the velocity will be instantaneously zero. If the acceleration is constant, however, the velocity vector will continue to change at the same rate. An instant after the velocity is zero, the magnitude of the velocity will begin increasing in the same direction as the acceleration. As time increases, the velocity of the object will then increase in the same direction as the acceleration. In other words, if the acceleration truly remains constant, the object will slow down, stop *for an instant*, reverse direction and then speed up.

---

14. ***REASONING AND SOLUTION*** The magnitude of the muzzle velocity of the bullet can be found (to a very good approximation) by solving Equation 2.9, $v^2 = v_0^2 + 2ax$, with $v_0 = 0$ m/s; that is

$$v = \sqrt{2ax}$$

where a is the acceleration of the bullet and x is the distance traveled by the bullet before it leaves the barrel of the gun (i.e., the length of the barrel).

Since the muzzle velocity of the rifle with the shorter barrel is greater than the muzzle velocity of the rifle with the longer barrel, the product ax must be greater for the bullet in the rifle with the shorter barrel. But x is smaller for the rifle with the shorter barrel, thus the *acceleration of the bullet must be larger in the rifle with the shorter barrel.*

CHAPTER 2 | *KINEMATICS IN ONE DIMENSION*

PROBLEMS

---

1. ***REASONING AND SOLUTION***
a. The total distance traveled is found by adding the distances traveled during each segment of the trip.

$$6.9 \text{ km} + 1.8 \text{ km} + 3.7 \text{ km} = \boxed{12.4 \text{ km}}$$

b. All three segments of the trip lie along the east-west line. Taking east as the positive direction, the individual displacements can then be added to yield the resultant displacement.

$$6.9 \text{ km} + (-1.8 \text{ km}) + 3.7 \text{ km} = +8.8 \text{ km}$$

The displacement is positive, indicating that it points due east. Therefore,

$$\text{Displacement of the whale} = \boxed{8.8 \text{ km, due east}}$$

---

3. ***REASONING AND SOLUTION*** The moving plane can be seen through the window of the stationary plane for the time that it takes the moving plane to travel its own length. To find the time, we write Equation 2.1 as

$$\text{Elapsed time} = \frac{\text{Distance}}{\text{Average speed}} = \frac{36 \text{ m}}{45 \text{ m/s}} = \boxed{0.80 \text{ s}}$$

---

9. ***REASONING AND SOLUTION***
a. The total displacement traveled by the bicyclist for the entire trip is equal to the sum of the displacements traveled during each part of the trip. The displacement traveled during each part of the trip is given by Equation 2.2: $\Delta x = \bar{v}\Delta t$. Therefore,

$$\Delta x_1 = (7.2 \text{ m/s})(22 \text{ min})\left(\frac{60 \text{ s}}{1 \text{ min}}\right) = 9500 \text{ m}$$

$$\Delta x_2 = (5.1 \text{ m/s})(36 \text{ min})\left(\frac{60 \text{ s}}{1 \text{ min}}\right) = 11\,000 \text{ m}$$

$$\Delta x_3 = (13 \text{ m/s})(8.0 \text{ min})\left(\frac{60 \text{ s}}{1 \text{ min}}\right) = 6200 \text{ m}$$

The total displacement traveled by the bicyclist during the entire trip is then

$$\Delta x = 9500 \text{ m} + 11\,000 \text{ m} + 6200 \text{ m} = \boxed{2.67\times10^4 \text{ m}}$$

b. The average velocity can be found from Equation 2.2.

$$\bar{v} = \frac{\Delta x}{\Delta t} = \frac{2.67\times10^4 \text{ m}}{(22 \text{ min} + 36 \text{ min} + 8.0 \text{ min})}\left(\frac{1 \text{ min}}{60 \text{ s}}\right) = \boxed{6.74 \text{ m/s, due north}}$$

---

13. ***REASONING*** Since the velocity and acceleration of the motorcycle point in the same direction, their numerical values will have the same algebraic sign. For convenience, we will choose them to be positive. The velocity, acceleration, and the time are related by Equation 2.4: $v = v_0 + at$.

SOLUTION

a. Solving Equation 2.4 for t we have

$$t = \frac{v - v_0}{a} = \frac{(+31 \text{ m/s}) - (+21 \text{ m/s})}{+2.5 \text{ m/s}^2} = \boxed{4.0 \text{ s}}$$

b. Similarly,

$$t = \frac{v - v_0}{a} = \frac{(+61 \text{ m/s}) - (+51 \text{ m/s})}{+2.5 \text{ m/s}^2} = \boxed{4.0 \text{ s}}$$

---

15. ***REASONING AND SOLUTION*** The initial velocity of the runner can be found by solving Equation 2.4 ($v = v_0 + at$) for v_0. Taking west as the positive direction, we have

$$v_0 = v - at = (+4.15 \text{ m/s}) - (+0.640 \text{ m/s}^2)(1.50 \text{ s}) = +3.19 \text{ m/s}$$

Therefore, the initial velocity of the runner is $\boxed{3.19 \text{ m/s, due west}}$.

---

21. ***REASONING*** The average acceleration is defined by Equation 2.4 as the change in velocity divided by the elapsed time. We can find the elapsed time from this relation because the acceleration and the change in velocity are given.

SOLUTION

a. The time Δt that it takes for the VW Beetle to change its velocity by an amount $\Delta v = v - v_0$ is (and noting that 0.4470 m/s = 1 mi/h)

$$\Delta t = \frac{v - v_0}{a} = \frac{(60.0\ \text{mi/h})\left(\frac{0.4470\ \text{m/s}}{1\ \text{mi/h}}\right) - 0\ \text{m/s}}{2.35\ \text{m/s}^2} = \boxed{11.4\ \text{s}}$$

b. From Equation 2.4, the acceleration (in m/s^2) of the dragster is

$$a = \frac{v - v_0}{t - t_0} = \frac{(60.0\ \text{mi/h})\left(\frac{0.4470\ \text{m/s}}{1\ \text{mi/h}}\right) - 0\ \text{m/s}}{0.600\ \text{s} - 0\ \text{s}} = \boxed{44.7\ \text{m/s}^2}$$

---

25. ***REASONING AND SOLUTION***

a. The magnitude of the acceleration can be found from Equation 2.4 ($v = v_0 + at$) as

$$a = \frac{v - v_0}{t} = \frac{3.0\ \text{m/s} - 0\ \text{m/s}}{2.0\ \text{s}} = \boxed{1.5\ \text{m/s}^2}$$

b. Similarly the magnitude of the acceleration of the car is

$$a = \frac{v - v_0}{t} = \frac{41.0\ \text{m/s} - 38.0\ \text{m/s}}{2.0\ \text{s}} = \boxed{1.5\ \text{m/s}^2}$$

c. Assuming that the acceleration is constant, the displacement covered by the car can be found from Equation 2.9 ($v^2 = v_0^2 + 2ax$):

$$x = \frac{v^2 - v_0^2}{2a} = \frac{(41.0\ \text{m/s})^2 - (38.0\ \text{m/s})^2}{2(1.5\ \text{m/s}^2)} = 79\ \text{m}$$

Similarly, the displacement traveled by the jogger is

$$x = \frac{v^2 - v_0^2}{2a} = \frac{(3.0\ \text{m/s})^2 - (0\ \text{m/s})^2}{2(1.5\ \text{m/s}^2)} = 3.0\ \text{m}$$

Therefore, the car travels 79 m – 3.0 m = $\boxed{76\ \text{m}}$ further than the jogger.

---

27. ***REASONING*** Since the belt is moving with constant velocity, the displacement ($x_0 = 0$ m) covered by the belt in a time t_{belt} is giving by Equation 2.2 (with x_0 assumed to be zero) as

$$x = v_{\text{belt}} t_{\text{belt}} \qquad (1)$$

Since Clifford moves with constant acceleration, the displacement covered by Clifford in a time t_{Cliff} is, from Equation 2.8,

$$x = v_0 t_{\text{Cliff}} + \tfrac{1}{2} a t_{\text{Cliff}}^2 = \tfrac{1}{2} a t_{\text{Cliff}}^2 \qquad (2)$$

The speed v_{belt} with which the belt of the ramp is moving can be found by eliminating x between Equations (1) and (2).

SOLUTION Equating the right hand sides of Equations (1) and (2), and noting that $t_{\text{Cliff}} = \frac{1}{4} t_{\text{belt}}$, we have

$$v_{\text{belt}} t_{\text{belt}} = \tfrac{1}{2} a \left(\tfrac{1}{4} t_{\text{belt}} \right)^2$$

$$v_{\text{belt}} = \tfrac{1}{32} a t_{\text{belt}} = \tfrac{1}{32} (0.37 \text{ m/s}^2)(64 \text{ s}) = \boxed{0.74 \text{ m/s}}$$

---

35. ***REASONING*** As the train passes through the crossing, its motion is described by Equations 2.4 ($v = v_0 + at$) and 2.7 $\left[x = \frac{1}{2}(v + v_0)t \right]$, which can be rearranged to give

$$v - v_0 = at \quad \text{and} \quad v + v_0 = \frac{2x}{t}$$

These can be solved simultaneously to obtain the speed v when the train reaches the end of the crossing. Once v is known, Equation 2.4 can be used to find the time required for the train to reach a speed of 32 m/s.

SOLUTION Adding the above equations and solving for v, we obtain

$$v = \tfrac{1}{2}\left(at + \frac{2x}{t} \right) = \tfrac{1}{2}\left[(1.6 \text{ m/s}^2)(2.4 \text{ s}) + \frac{2(20.0 \text{ m})}{2.4 \text{ s}} \right] = 1.0 \times 10^1 \text{ m/s}$$

The motion from the end of the crossing until the locomotive reaches a speed of 32 m/s requires a time

$$t = \frac{v - v_0}{a} = \frac{32 \text{ m/s} - 1.0 \times 10^1 \text{ m/s}}{1.6 \text{ m/s}^2} = \boxed{14 \text{ s}}$$

---

37. ***REASONING AND SOLUTION*** When air resistance is neglected, free fall conditions are applicable. The final speed can be found from Equation 2.9;

$$v^2 = v_0^2 + 2ay$$

where v_0 is zero since the stunt man falls from rest. If the origin is chosen at the top of the hotel and the upward direction is positive, then the displacement is $y = -99.4$ m. Solving for v, we have

$$v = -\sqrt{2ay} = -\sqrt{2(-9.80\text{ m/s}^2)(-99.4\text{ m})} = -44.1\text{ m/s}$$

The speed at impact is the magnitude of this result or $\boxed{44.1\text{ m/s}}$.

---

41. ***REASONING AND SOLUTION*** Equation 2.8 can be used to determine the displacement that the ball covers as it falls halfway to the ground. Since the ball falls from rest, its initial velocity is zero. Taking down to be the negative direction, we have

$$y = v_0 t + \tfrac{1}{2}at^2 = \tfrac{1}{2}at^2 = \tfrac{1}{2}(-9.80\text{ m/s}^2)(1.2\text{ s})^2 = -7.1\text{ m}$$

In falling all the way to the ground, the ball has a displacement of $y = -14.2$ m. Solving Equation 2.8 with this displacement then yields the time

$$t = \sqrt{\frac{2y}{a}} = \sqrt{\frac{2(-14.2\text{ m})}{-9.80\text{ m/s}^2}} = \boxed{1.7\text{ s}}$$

---

45. ***REASONING AND SOLUTION*** Since the balloon is released from rest, its initial velocity is zero. The time required to fall through a vertical displacement y can be found from Equation 2.8 $\left(y = v_0 t + \tfrac{1}{2}at^2\right)$ with $v_0 = 0$ m/s. Assuming upward to be the positive direction, we find

$$t = \sqrt{\frac{2y}{a}} = \sqrt{\frac{2(-6.0\text{ m})}{-9.80\text{ m/s}^2}} = \boxed{1.1\text{ s}}$$

---

51. ***REASONING AND SOLUTION*** The stone will reach the water (and hence the log) after falling for a time t, where t can be determined from Equation 2.8: $y = v_0 t + \tfrac{1}{2}at^2$. Since the stone is dropped from rest, $v_0 = 0$ m/s. Assuming that downward is positive and solving for t, we have

$$t = \sqrt{\frac{2y}{a}} = \sqrt{\frac{2(75\text{ m})}{9.80\text{ m/s}^2}} = 3.9\text{ s}$$

During that time, the displacement of the log can be found from Equation 2.8. Since the log moves with constant velocity, $a = 0\text{ m/s}^2$, and v_0 is equal to the velocity of the log.

$$x = v_0 t = (5.0\text{ m/s})(3.9\text{ s}) = 2.0\times10^1\text{ m}$$

Therefore, the horizontal distance between the log and the bridge when the stone is released is $\boxed{2.0\times10^1\text{ m}}$.

---

53. ***REASONING*** Once the man sees the block, the man must get out of the way in the time it takes for the block to fall through an additional 12.0 m. The velocity of the block at the instant that the man looks up can be determined from Equation 2.9. Once the velocity is known at that instant, Equation 2.8 can be used to find the time required for the block to fall through the additional distance.

SOLUTION When the man first notices the block, it is 14.0 m above the ground and its displacement from the starting point is $y = 14.0\text{ m} - 53.0\text{ m}$. Its velocity is given by Equation 2.9 $\left(v^2 = v_0^2 + 2ay\right)$. Since the block is moving down, its velocity has a negative value,

$$v = -\sqrt{v_0 + 2ay} = -\sqrt{(0\text{ m/s})^2 + 2(-9.80\text{ m/s}^2)(14.0\text{ m} - 53.0\text{ m})} = -27.7\text{ m/s}$$

The block then falls the additional 12.0 m to the level of the man's head in a time t which satisfies Equation 2.8:

$$y = v_0 t + \tfrac{1}{2}at^2$$

where $y = -12.0\text{ m}$ and $v_0 = -27.7\text{ m/s}$. Thus, t is the solution to the quadratic equation

$$4.90t^2 + 27.7t - 12.0 = 0$$

where the units have been suppressed for brevity. From the quadratic formula, we obtain

$$t = \frac{-27.7 \pm \sqrt{(27.7)^2 - 4(4.90)(-12.0)}}{2(4.90)} = 0.40\text{ s} \quad \text{or} \quad -6.1\text{ s}$$

The negative solution can be rejected as nonphysical, and the time it takes for the block to reach the level of the man is $\boxed{0.40\text{ s}}$.

---

59. ***REASONING*** In order to construct the graph, the time for each segment of the trip must be determined.

SOLUTION From the definition of average velocity

$$\Delta t = \frac{\Delta x}{\bar{v}}$$

Therefore,

$$\Delta t_1 = \left(\frac{10.0 \text{ km}}{15.0 \text{ km/h}}\right)\left(\frac{60 \text{ min}}{1.0 \text{ h}}\right) = 40 \text{ min}$$

$$\Delta t_2 = \left(\frac{15.0 \text{ km}}{10.0 \text{ km/h}}\right)\left(\frac{60 \text{ min}}{1.0 \text{ h}}\right) = 90 \text{ min}$$

Thus, the second time interval ends 40 min + 90 min = 130 min after the trip begins.

$$\Delta t_3 = \left(\frac{15.0 \text{ km}}{5.0 \text{ km/h}}\right)\left(\frac{60 \text{ min}}{1.0 \text{ h}}\right) = 180 \text{ min}$$

and the third time interval ends 130 min + 180 min = 310 min after the trip begins.

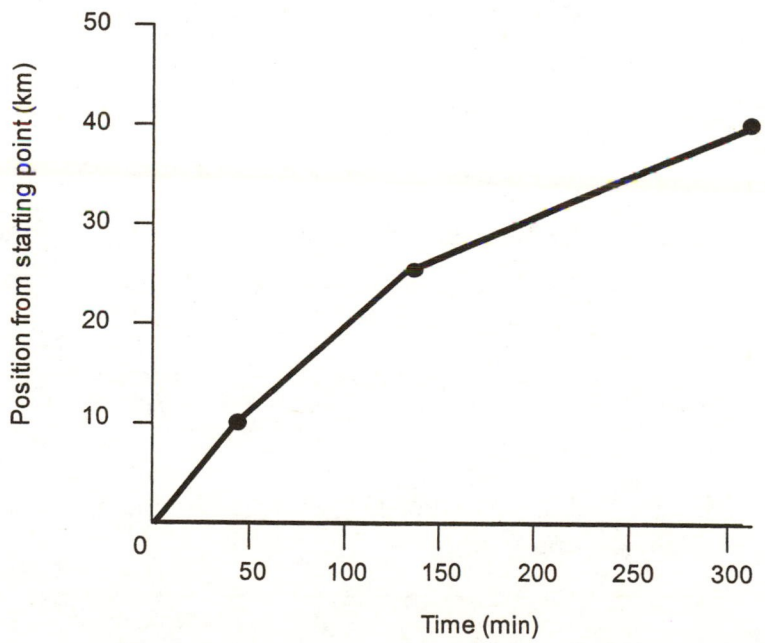

Note that the slope of each segment of the graph gives the average velocity during that interval.

61. ***REASONING*** The average acceleration is given by Equation 2.4 as $\bar{a} = (v_C - v_A)/\Delta t$. The velocities v_A and v_C can be found from the slopes of the position-time graph for segments A and C.

SOLUTION The average velocities in the segments A and C are

$$v_A = \frac{24\text{ km} - 0\text{ km}}{1.0\text{ h} - 0\text{ h}} = 24\text{ km/h}$$

$$v_C = \frac{27\text{ km} - 33\text{ km}}{3.5\text{ h} - 2.2\text{ h}} = -5\text{ km/h}$$

From the definition of average acceleration,

$$\bar{a} = \frac{\Delta v}{\Delta t} = \frac{v_C - v_A}{\Delta t} = \frac{(-5\text{ km/h}) - (24\text{ km/h})}{3.5\text{ h} - 0\text{ h}} = \boxed{-8.3\text{ km/h}^2}$$

---

65. ***REASONING AND SOLUTION*** The speed of the penny as it hits the ground can be determined from Equation 2.9: $v^2 = v_0^2 + 2ay$. Since the penny is dropped from rest, $v_0 = 0\text{ m/s}$. Solving for v, with downward taken as the positive direction, we have

$$v = \sqrt{2(9.80\text{ m/s}^2)(427\text{ m})} = \boxed{91.5\text{ m/s}}$$

---

67. ***REASONING AND SOLUTION*** The average acceleration of the plane can be found by solving Equation 2.9 $\left(v^2 = v_0^2 + 2ax\right)$ for a. Taking the direction of motion as positive, we have

$$a = \frac{v^2 - v_0^2}{2x} = \frac{(+6.1\text{ m/s})^2 - (+69\text{ m/s})^2}{2(+750\text{ m})} = \boxed{-3.1\text{ m/s}^2}$$

The minus sign indicates that the direction of the acceleration is opposite to the direction of motion, and the plane is slowing down.

---

73. ***REASONING*** Since the woman runs for a known distance at a known constant speed, we can find the time it takes for her to reach the water from Equation 2.1. We can then use Equation 2.1 to determine the total distance traveled by the dog in this time.

SOLUTION The time required for the woman to reach the water is

$$\text{Elapsed time} = \frac{d_{\text{woman}}}{v_{\text{woman}}} = \left(\frac{4.0\text{ km}}{2.5\text{ m/s}}\right)\left(\frac{1000\text{ m}}{1.0\text{ km}}\right) = 1600\text{ s}$$

In 1600 s, the dog travels a total distance of

$$d_{\text{dog}} = v_{\text{dog}}t = (4.5\text{ m/s})(1600\text{ s}) = \boxed{7.2\times10^3\text{ m}}$$

---

75. ***REASONING AND SOLUTION*** The stone requires a time, t_1, to reach the bottom of the hole, a distance y below the ground. Assuming downward to be the positive direction, the variables are related by Equation 2.8 with $v_0 = 0\text{ m/s}$:

$$y = \tfrac{1}{2}at_1^2 \qquad (1)$$

The sound travels the distance y from the bottom to the top of the hole in a time t_2. Since the sound does not experience any acceleration, the variables y and t_2 are related by Equation 2.8 with $a = 0\text{ m/s}^2$ and v_{sound} denoting the speed of sound:

$$y = v_{\text{sound}}\,t_2 \qquad (2)$$

Equating the right hand sides of Equations (1) and (2) and using the fact that the total elapsed time is $t = t_1 + t_2$, we have

$$\tfrac{1}{2}at_1^2 = v_{\text{sound}}t_2 \qquad \text{or} \qquad \tfrac{1}{2}at_1^2 = v_{\text{sound}}(t - t_1)$$

Rearranging gives

$$\tfrac{1}{2}at_1^2 + v_{\text{sound}}t_1 - v_{\text{sound}}t = 0$$

Substituting values and suppressing units for brevity, we obtain the following quadratic equation for t_1:

$$4.90t_1^2 + 343t_1 - 514 = 0$$

From the quadratic formula, we obtain

$$t_1 = \frac{-343 \pm \sqrt{(343)^2 - 4(4.90)(-514)}}{2(4.90)} = 1.47\text{ s} \quad \text{or} \quad -71.5\text{ s}$$

The negative time corresponds to a nonphysical result and is rejected. The depth of the hole is then found using Equation 2.8 with the value of t_1 obtained above:

$$y = v_0 t_1 + \tfrac{1}{2} a t_1^2 = (0 \text{ m/s})(1.47 \text{ s}) + \tfrac{1}{2}(9.80 \text{ m/s}^2)(1.47 \text{ s})^2 = \boxed{10.6 \text{ m}}$$

---

79. ***REASONING*** Since the car is moving with a constant velocity, the displacement of the car in a time t can be found from Equation 2.8 with $a = 0$ m/s^2 and v_0 equal to the velocity of the car: $x_{\text{car}} = v_{\text{car}} t$. Since the train starts from rest with a constant acceleration, the displacement of the train in a time t is given by Equation 2.8 with $v_0 = 0$ m/s:

$$x_{\text{train}} = \tfrac{1}{2} a_{\text{train}} t^2$$

At a time t_1, when the car just reaches the front of the train, $x_{\text{car}} = L_{\text{train}} + x_{\text{train}}$, where L_{train} is the length of the train. Thus, at time t_1,

$$v_{\text{car}} t_1 = L_{\text{train}} + \tfrac{1}{2} a_{\text{train}} t_1^2 \qquad (1)$$

At a time t_2, when the car is again at the rear of the train, $x_{\text{car}} = x_{\text{train}}$. Thus, at time t_2

$$v_{\text{car}} t_2 = \tfrac{1}{2} a_{\text{train}} t_2^2 \qquad (2)$$

Equations (1) and (2) can be solved simultaneously for the speed of the car v_{car} and the acceleration of the train a_{train}.

SOLUTION

a. Solving Equation (2) for a_{train} we have

$$a_{\text{train}} = \frac{2 v_{\text{car}}}{t_2} \qquad (3)$$

Substituting this expression for a_{train} into Equation (1) and solving for v_{car}, we have

$$v_{\text{car}} = \frac{L_{\text{train}}}{t_1 \left(1 - \dfrac{t_1}{t_2}\right)} = \frac{92 \text{ m}}{(14 \text{ s})\left(1 - \dfrac{14 \text{ s}}{28 \text{ s}}\right)} = \boxed{13 \text{ m/s}}$$

b. Direct substitution into Equation (3) gives the acceleration of the train:

$$a_{\text{train}} = \frac{2v_{\text{car}}}{t_2} = \frac{2\ (13\ \text{m/s})}{28\ \text{s}} = \boxed{0.93\ \text{m/s}^2}$$

CHAPTER 3 | *KINEMATICS IN TWO DIMENSIONS*

CONCEPTUAL QUESTIONS

---

2. ***REASONING AND SOLUTION*** An object thrown upward at an angle θ will follow the trajectory shown below. Its acceleration is that due to gravity, and, therefore, always points downward. The acceleration is denoted by a_y in the figure. In general, the velocity of the object has two components, v_x and v_y. Since $a_x = 0$, v_x always equals its initial value. The magnitude of the y component of the velocity, v_y, decreases as the object rises, drops to zero at the highest point, and then increases as the object falls downward.

a. Since $v_y = 0$ when the object is at its highest point, the velocity of the object points only in the x direction. As suggested in the figure below, the acceleration will be perpendicular to the velocity when the object is at its highest point and $v_y = 0$.

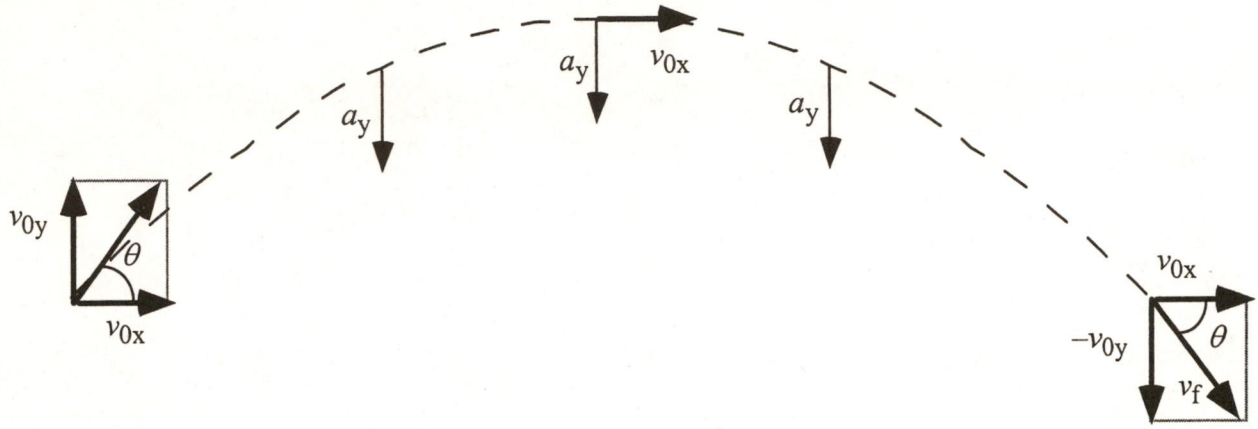

b. In order for the velocity and acceleration to be parallel, the x component of the velocity would have to drop to zero. However, v_x always remains equal to its initial value; therefore, the velocity and the acceleration can never be parallel.

---

9. ***REASONING AND SOLUTION*** Since the launch speed of projectile A is twice that of B, it follows that

$$(v_{0x})_A = 2(v_{0x})_B \quad \text{and} \quad (v_{0y})_A = 2(v_{0y})_B$$

As seen from the result of Example 6 in the text, the maximum height attained by either projectile is directly proportional to the square of v_{0y}^2; therefore, the ratio of the maximum heights is $H_A / H_B = (2)^2 = 4$. Example 8 in the text shows that the range of either projectile is directly proportional to the product of v_{0x} and t. Example 7 in the text shows that t is proportional to v_{0y}; thus,

$$\frac{R_A}{R_B} = \frac{(v_{0x})_A t_A}{(v_{0x})_B t_B} = \frac{(v_{0x})_A (v_{0y})_A}{(v_{0x})_B (v_{0y})_B} = \frac{[2(v_{0x})_B][2(v_{0y})_B]}{(v_{0x})_B (v_{0y})_B} = 4$$

This ratio is apparent in the following drawing.

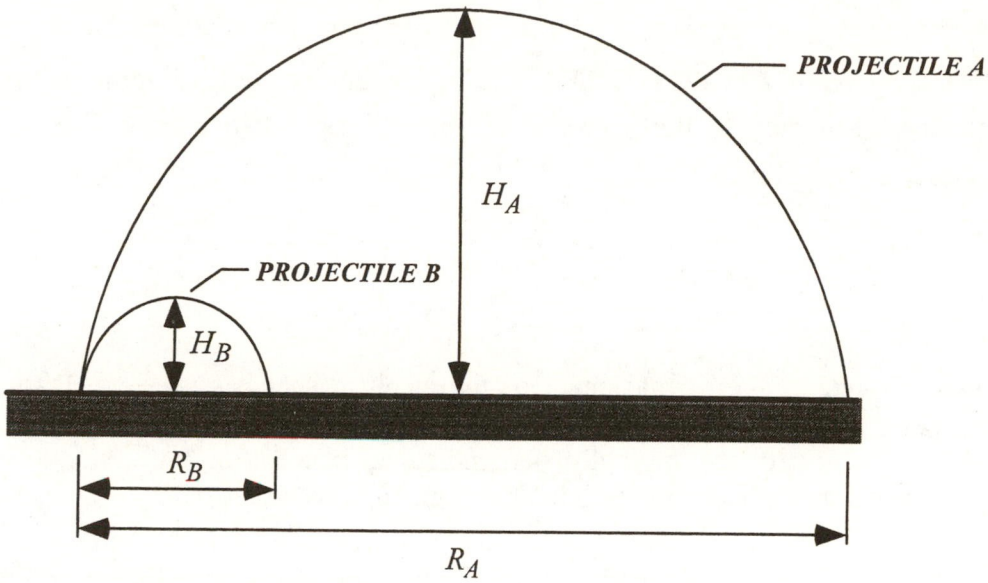

---

15. ***REASONING AND SOLUTION***

a. Since the child is at rest with respect to the floor of the RV and with respect to the cannon, there is no relative motion between the child, the floor of the RV and the cannon. Therefore, the range of the marble toward the front is the same as the range toward the rear from the point of view of the child.

b. The velocity of the marble relative to an observer standing still on the ground is $\mathbf{v}_{MG} = \mathbf{v}_{MV} + \mathbf{v}_{VG}$, where $\mathbf{v}_{MV}$ is the velocity of the marble relative to the vehicle and $\mathbf{v}_{VG}$ is the velocity of the vehicle relative to the ground. Suppose that the RV moves in the positive direction; then the quantity $\mathbf{v}_{VG}$ is positive. When the marble is shot toward the front of the RV, the x component of $\mathbf{v}_{MV}$ will also be a positive quantity. Alternatively, when the marble is shot toward the back of the RV, the x component of $\mathbf{v}_{MV}$ is a negative quantity. Thus, the magnitude of the x component of $\mathbf{v}_{MG}$ is smaller when the marble is shot toward the rear than when it is shot toward the front. The time that the marble is in the air is determined by the vertical motion which is the same in both cases. Since the magnitude of the x component of $\mathbf{v}_{MG}$ is smaller when the marble is shot toward the rear, it will cover less distance in the time it takes to hit the floor of the RV compared to when the marble is shot toward the front. Therefore, from the point of view of an observer on the ground, the range of the marble is greater when the marble is shot toward the front.

CHAPTER 3 | *KINEMATICS IN TWO DIMENSIONS*

PROBLEMS

---

1. ***REASONING AND SOLUTION*** The horizontal and vertical components of the plane's velocity are related to the speed of the plane by the Pythagorean theorem: $v^2 = v_h^2 + v_v^2$. Solving for v_h we have

$$v_h = \sqrt{v^2 - v_v^2} = \sqrt{(245\text{ m/s})^2 - (40.6\text{ m/s})^2} = \boxed{242\text{ m/s}}$$

---

5. ***REASONING*** The displacement of the elephant seal has two components; 460 m due east and 750 m downward. These components are mutually perpendicular; hence, the Pythagorean theorem can be used to determine their resultant.

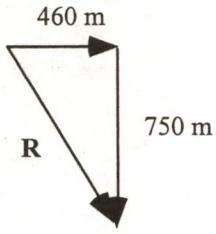

SOLUTION From the Pythagorean theorem,

$$R^2 = (460\text{ m})^2 + (750\text{ m})^2$$

Therefore,

$$R = \sqrt{(460\text{ m})^2 + (750\text{ m})^2} = \boxed{8.8 \times 10^2\text{ m}}$$

---

9. ***REASONING AND SOLUTION*** The escalator is the hypotenuse of a right triangle formed by the lower floor and the vertical distance between floors, as shown in the figure. The angle θ at which the escalator is inclined above the horizontal is related to the length L of the escalator and the vertical distance between the floors by the sine function:

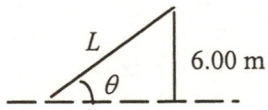

$$\sin\theta = \frac{6.00\text{ m}}{L} \qquad (1)$$

The length L of the escalator can be found from the right triangle formed by the components of the shopper's displacement up the escalator and to the right of the escalator.

From the Pythagorean theorem, we have

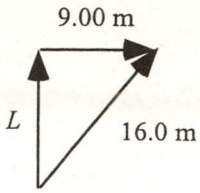

$$L^2 + (9.00\text{ m})^2 = (16.0\text{ m})^2$$

so that

$$L = \sqrt{(16.0\text{ m})^2 - (9.00\text{ m})^2} = 13.2\text{ m}$$

From Equation (1) we have

$$\theta = \sin^{-1}\left(\frac{6.00\text{ m}}{13.2\text{ m}}\right) = \boxed{27.0°}$$

---

13. ***REASONING AND SOLUTION*** As shown in Example 3, the time required for the package to hit the ground is given by $t = \sqrt{2y/a_y}$ and is independent of the plane's horizontal velocity. Thus, the time needed for the package to hit the ground is still $\boxed{14.6\text{ s}}$.

---

17. ***REASONING*** The upward direction is chosen as positive. Since the ballast bag is released from rest relative to the balloon, its initial velocity relative to the ground is equal to the velocity of the balloon relative to the ground, so that $v_{0y} = 3.0\text{ m/s}$. Time required for the ballast to reach the ground can be found by solving Equation 3.5b for t.

SOLUTION Using Equation 3.5b, we have

$$\tfrac{1}{2}a_y t^2 + v_{0y}t - y = 0 \quad \text{or} \quad \tfrac{1}{2}(-9.80\text{ m/s}^2)t^2 + (3.0\text{ m/s})t - (-9.5\text{ m}) = 0$$

This equation is quadratic in t, and t may be found from the quadratic formula. Using the quadratic formula, suppressing the units, and discarding the negative root, we find

$$t = \frac{-3.0 \pm \sqrt{(3.0)^2 - 4(-4.90)(9.5)}}{2(-4.90)} = \boxed{1.7\text{ s}}$$

---

21. ***REASONING*** Since the magnitude of the velocity of the fuel tank is given by $v = \sqrt{v_x^2 + v_y^2}$, it is necessary to know the velocity components v_x and v_y just before impact. At the instant of release, the empty fuel tank has the same velocity as that of the plane. Therefore, the magnitudes of the initial velocity components of the fuel tank are given by $v_{0x} = v_0 \cos\theta$ and $v_{0y} = v_0 \sin\theta$, where v_0 is the speed of the plane at the instant of release. Since the x motion has zero acceleration, the x component of the velocity of the plane remains equal to v_{0x} for all later times while the tank is airborne. The y component of

the velocity of the tank after it has undergone a vertical displacement y is given by Equation 3.6b.

SOLUTION

a. Taking up as the positive direction, the velocity components of the fuel tank just before it hits the ground are

$$v_x = v_{0x} = v\cos\theta = (135 \text{ m/s}) \cos 15° = 1.30\times10^2 \text{ m/s}$$

From Equation 3.6b, we have

$$v_y = -\sqrt{v_{0y}^2 + 2a_y y} = -\sqrt{(v_0 \sin\theta)^2 + 2a_y y}$$

$$= -\sqrt{\left[(135 \text{ m/s}) \sin 15.0°\right]^2 + \left[2(-9.80 \text{ m/s}^2)(-2.00\times10^3 \text{ m})\right]} = -201 \text{ m/s}$$

Therefore, the magnitude of the velocity of the fuel tank just before impact is

$$v = \sqrt{v_x^2 + v_y^2} = \sqrt{(1.30\times10^2 \text{ m/s})^2 + (201 \text{ m/s})^2} = \boxed{239 \text{ m/s}}$$

The velocity vector just before impact is inclined at an angle ϕ above the horizontal. This angle is

$$\phi = \tan^{-1}\left(\frac{201 \text{ m/s}}{1.30\times10^2 \text{ m/s}}\right) = \boxed{57.1°}$$

b. As shown in Conceptual Example 10, once the fuel tank in part a rises and falls to the same altitude at which it was released, its motion is identical to the fuel tank in part b. Therefore, the velocity of the fuel tank in part b just before impact is $\boxed{\text{239 m/s at an angle of 57.1° above the horizontal}}$.

---

25. ***REASONING*** The speed of the fish at any time t is given by $v = \sqrt{v_x^2 + v_y^2}$, where v_x and v_y are the x and y components of the velocity at that instant. Since the horizontal motion of the fish has zero acceleration, $v_x = v_{0x}$ for all times t. Since the fish is dropped by the eagle, v_{0x} is equal to the horizontal speed of the eagle and $v_{0y} = 0$. The y component of the velocity of the fish for any time t is given by Equation 3.3b with $v_{0y} = 0$. Thus, the speed at any time t is given by $v = \sqrt{v_{0x}^2 + (a_y t)^2}$.

SOLUTION

a. The initial speed of the fish is $v_0 = \sqrt{v_{0x}^2 + v_{0y}^2} = \sqrt{v_{0x}^2 + 0^2} = v_{0x}$. When the fish's speed doubles, $v = 2v_{0x}$. Therefore,

$$2v_{0x} = \sqrt{v_{0x}^2 + (a_y t)^2} \qquad \text{or} \qquad 4v_{0x}^2 = v_{0x}^2 + (a_y t)^2$$

Assuming that downward is positive and solving for t, we have

$$t = \sqrt{3}\frac{v_{0x}}{a_y} = \sqrt{3}\left(\frac{6.0\text{ m/s}}{9.80\text{ m/s}^2}\right) = \boxed{1.1\text{ s}}$$

b. When the fish's speed doubles again, $v = 4v_{0x}$. Therefore,

$$4v_{0x} = \sqrt{v_{0x}^2 + (a_y t)^2} \qquad \text{or} \qquad 16v_{0x}^2 = v_{0x}^2 + (a_y t)^2$$

Solving for t, we have

$$t = \sqrt{15}\frac{v_{0x}}{a_y} = \sqrt{15}\left(\frac{6.0\text{ m/s}}{9.80\text{ m/s}^2}\right) = 2.37\text{ s}$$

Therefore, the additional time for the speed to double again is $(2.4\text{ s}) - (1.1\text{ s}) = \boxed{1.3\text{ s}}$.

---

29. ***REASONING*** Once the diver is airborne, he moves in the x direction with constant velocity while his motion in the y direction is accelerated (at the acceleration due to gravity). Therefore, the magnitude of the x component of his velocity remains constant at 1.20 m/s for all times t. The magnitude of the y component of the diver's velocity after he has fallen through a vertical displacement y can be determined from Equation 3.6b: $v_y^2 = v_{0y}^2 + 2a_y y$. Since the diver runs off the platform horizontally, $v_{0y} = 0$. Once the x and y components of the velocity are known for a particular vertical displacement y, the speed of the diver can be obtained from $v = \sqrt{v_x^2 + v_y^2}$.

SOLUTION For convenience, we will take downward as the positive y direction. After the diver has fallen 10.0 m, the y component of his velocity is, from Equation 3.6b,

$$v_y = \sqrt{v_{0y}^2 + 2a_y y} = \sqrt{0^2 + 2(9.80\text{ m/s}^2)(10.0\text{ m})} = 14.0\text{ m/s}$$

Therefore,

$$v = \sqrt{v_x^2 + v_y^2} = \sqrt{(1.20\text{ m/s})^2 + (14.0\text{ m/s})^2} = \boxed{14.1\text{ m/s}}$$

---

33. ***REASONING AND SOLUTION*** The horizontal displacement of the jumper is given by Equation 3.5a with $a_x = 0$: $x = v_{0x}t = (v_0 \cos\theta)t$. The vertical component of the jumper's velocity at any time t is given by Equation 3.3b: $v_y = v_{0y} + a_y t$. At the instant that the jumper lands, $v_y = -v_{0y}$. Therefore, the vertical component of the jumper's velocity is

$$-v_{0y} = v_{0y} + a_y t$$

Solving for t and assuming that upward is the positive direction gives

$$t = \frac{-2v_{0y}}{a_y} = \frac{-2v_0 \sin\theta}{a_y}$$

Substituting this expression for t into the expression for x gives

$$x = (v_0 \cos\theta)t = v_0 \cos\theta \left(\frac{-2v_0 \sin\theta}{a_y} \right)$$

or

$$x = \frac{-2v_0^2 \sin\theta \cos\theta}{a_y}$$

Solving for v_0 gives

$$v_0 = \sqrt{\frac{-x\,a_y}{2\cos\theta \sin\theta}} = \sqrt{\frac{-(8.7\text{ m})(-9.80\text{ m/s}^2)}{2\cos 23° \sin 23°}} = \boxed{11\text{ m/s}}$$

---

37. ***REASONING*** The angle θ can be found from

$$\theta = \tan^{-1}\left(\frac{2400\text{ m}}{x}\right) \qquad (1)$$

where x is the horizontal displacement of the flare. Since $a_x = 0$, it follows that $x = (v_0 \cos 30.0°)t$. The flight time t is determined by the vertical motion. In particular, the time t can be found from Equation 3.5b. Once the time is known, x can be calculated.

SOLUTION From Equation 3.5b, assuming upward is the positive direction, we have

$$y = -(v_0 \sin 30.0°)t + \tfrac{1}{2}a_y t^2$$

which can be rearranged to give the following equation that is quadratic in t:

$$\tfrac{1}{2}a_y t^2 - (v_0 \sin\ 30.0°)t - y = 0$$

Using $y = -2400$ m and $a_y = -9.80$ m/s^2 and suppressing the units, we obtain the quadratic equation

$$4.9t^2 + 120t - 2400 = 0$$

Using the quadratic formula, we obtain $t = 13$ s. Therefore, we find that

$$x = (v_0 \cos\ 30.0°)t = (240\ \text{m/s})(\cos 30.0°)(13\ \text{s}) = 2700\ \text{m}$$

Equation (1) then gives

$$\theta = \tan^{-1}\left(\frac{2400\ \text{m}}{2700\ \text{m}}\right) = \boxed{42°}$$

---

41. ***REASONING AND SOLUTION*** In the absence of air resistance, the bullet exhibits projectile motion. The x component of the motion has zero acceleration while the y component of the motion is subject to the acceleration due to gravity. The horizontal distance traveled by the bullet is given by Equation 3.5a (with $a_x = 0$):

$$x = v_{0x}t = (v_0 \cos\theta)t$$

with t equal to the time required for the bullet to reach the target. The time t can be found by considering the vertical motion. From Equation 3.3b,

$$v_y = v_{0y} + a_y t$$

When the bullet reaches the target, $v_y = -v_{0y}$. Assuming that up and to the right are the positive directions, we have

$$t = \frac{-2v_{0y}}{a_y} = \frac{-2v_0 \sin\theta}{a_y} \qquad \text{and} \qquad x = (v_0 \cos\theta)\left(\frac{-2v_0 \sin\theta}{a_y}\right)$$

Using the fact that $2\sin\theta\ \cos\theta = \sin\ 2\theta$, we have

$$x = -\frac{2v_0^2 \cos\theta\ \sin\theta}{a_y} = -\frac{v_0^2 \sin 2\theta}{a_y}$$

Thus, we find that

$$\sin 2\theta = -\frac{x\,a_y}{v_0^2} = -\frac{(91.4\text{ m})\,(-9.80\text{ m/s}^2)}{(427\text{ m/s})^2} = 4.91\times10^{-3}$$

and

$$2\theta = 0.281° \quad \text{or} \quad 2\theta = 180.000° - 0.281° = 179.719°$$

Therefore,

$$\theta = \boxed{0.141° \text{ and } 89.860°}$$

---

43. ***REASONING*** Since the horizontal motion is not accelerated, we know that the x component of the velocity remains constant at 340 m/s. Thus, we can use Equation 3.5a (with $a_x = 0$) to determine the time that the bullet spends in the building before it is embedded in the wall. Since we know the vertical displacement of the bullet after it enters the building, we can use the flight time in the building and Equation 3.5b to find the y component of the velocity of the bullet as it enters the window. Then, Equation 3.6b can be used (with $v_{0y} = 0$) to determine the vertical displacement y of the bullet as it passes between the buildings. We can determine the distance H by adding the magnitude of y to the vertical distance of 0.50 m within the building.

Once we know the vertical displacement of the bullet as it passes between the buildings, we can determine the time t_1 required for the bullet to reach the window using Equation 3.4b. Since the motion in the x direction is not accelerated, the distance D can then be found from $D = v_{0x}t_1$.

SOLUTION Assuming that the direction to the right is positive, we find that the time that the bullet spends in the building is (according to Equation 3.5a)

$$t = \frac{x}{v_{0x}} = \frac{6.9\text{ m}}{340\text{ m/s}} = 0.0203\text{ s}$$

The vertical displacement of the bullet after it enters the building is, taking down as the negative direction, equal to –0.50 m. Therefore, the vertical component of the velocity of the bullet as it passes through the window is, from Equation 3.5b,

$$v_{0y(\text{window})} = \frac{y - \frac{1}{2}a_y t^2}{t} = \frac{y}{t} - \frac{1}{2}a_y t = \frac{-0.50\text{ m}}{0.0203\text{ s}} - \frac{1}{2}(-9.80\text{ m/s}^2)(0.0203\text{ s}) = -24.5\text{ m/s}$$

The vertical displacement of the bullet as it travels between the buildings is (according to Equation 3.6b with $v_{0y} = 0$)

$$y = \frac{v_y^2}{2a_y} = \frac{(-24.5\text{ m/s})^2}{2(-9.80\text{ m/s}^2)} = -30.6\text{ m}$$

Therefore, the distance H is

$$H = 30.6\text{ m} + 0.50\text{ m} = \boxed{31\text{ m}}$$

The time for the bullet to reach the window, according to Equation 3.4b, is

$$t_1 = \frac{2y}{v_{0y} + v_y} = \frac{2y}{v_y} = \frac{2(-30.6\text{ m})}{(-24.5\text{ m/s})} = 2.50\text{ s}$$

Hence, the distance D is given by

$$D = v_{0x}t_1 = (340\text{ m/s})(2.50\text{ s}) = \boxed{850\text{ m}}$$

---

47. ***REASONING*** The velocity $\mathbf{v}_{SG}$ of the swimmer relative to the ground is the vector sum of the velocity $\mathbf{v}_{SW}$ of the swimmer relative to the water and the velocity $\mathbf{v}_{WG}$ of the water relative to the ground as shown at the right: $\mathbf{v}_{SG} = \mathbf{v}_{SW} + \mathbf{v}_{WG}$.

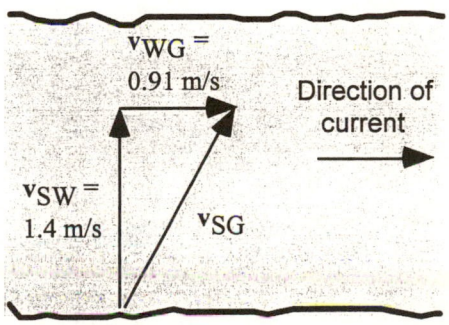

The component of $\mathbf{v}_{SG}$ that is parallel to the width of the river determines how fast the swimmer is moving across the river; this parallel component is v_{SW}. The time for the swimmer to cross the river is equal to the width of the river divided by the magnitude of this velocity component.

The component of $\mathbf{v}_{SG}$ that is parallel to the direction of the current determines how far the swimmer is carried down stream; this component is v_{WG}. Since the motion occurs with constant velocity, the distance that the swimmer is carried downstream while crossing the river is equal to the magnitude of v_{WG} multiplied by the time it takes for the swimmer to cross the river.

SOLUTION

a. The time t for the swimmer to cross the river is

$$t = \frac{\text{width}}{v_{SW}} = \frac{2.8\times10^3\text{ m}}{1.4\text{ m/s}} = \boxed{2.0\times10^3\text{ s}}$$

b. The distance x that the swimmer is carried downstream while crossing the river is

$$x = v_{WG}t = (0.91 \text{ m/s})(2.0\times10^3 \text{ s}) = \boxed{1.8\times10^3 \text{ m}}$$

---

51. ***REASONING*** The velocity $\mathbf{v}_{AB}$ of train **A** relative to train **B** is the vector sum of the velocity $\mathbf{v}_{AG}$ of train **A** relative to the ground and the velocity $\mathbf{v}_{GB}$ of the ground relative to train **B**, as indicated by Equation 3.7: $\mathbf{v}_{AB} = \mathbf{v}_{AG} + \mathbf{v}_{GB}$. The values of $\mathbf{v}_{AG}$ and $\mathbf{v}_{BG}$ are given in the statement of the problem. We must also make use of the fact that $\mathbf{v}_{GB} = -\mathbf{v}_{BG}$.

SOLUTION

a. Taking east as the positive direction, the velocity of **A** relative to **B** is, according to Equation 3.7,

$$\mathbf{v}_{AB} = \mathbf{v}_{AG} + \mathbf{v}_{GB} = \mathbf{v}_{AG} - \mathbf{v}_{BG} = (+13 \text{ m/s}) - (-28 \text{ m/s}) = \boxed{+41 \text{ m/s}}$$

The positive sign indicates that the direction of $\mathbf{v}_{AB}$ is $\boxed{\text{due east}}$.

b. Similarly, the velocity of **B** relative to **A** is

$$\mathbf{v}_{BA} = \mathbf{v}_{BG} + \mathbf{v}_{GA} = \mathbf{v}_{BG} - \mathbf{v}_{AG} = (-28 \text{ m/s}) - (+13 \text{ m/s}) = \boxed{-41 \text{ m/s}}$$

The negative sign indicates that the direction of $\mathbf{v}_{BA}$ is $\boxed{\text{due west}}$.

---

57. ***REASONING*** The velocity $\mathbf{v}_{PM}$ of the puck relative to Mario is the vector sum of the velocity $\mathbf{v}_{PI}$ of the puck relative to the ice and the velocity $\mathbf{v}_{IM}$ of the ice relative to Mario as indicated by Equation 3.7: $\mathbf{v}_{PM} = \mathbf{v}_{PI} + \mathbf{v}_{IM}$. The values of $\mathbf{v}_{MI}$ and $\mathbf{v}_{PI}$ are given in the statement of the problem. In order to use the data, we must make use of the fact that $\mathbf{v}_{IM} = -\mathbf{v}_{MI}$, with the result that $\mathbf{v}_{PM} = \mathbf{v}_{PI} - \mathbf{v}_{MI}$.

SOLUTION The first two rows of the following table give the east/west and north/south components of the vectors $\mathbf{v}_{PI}$ and $-\mathbf{v}_{MI}$. The third row gives the components of their resultant $\mathbf{v}_{PM} = \mathbf{v}_{PI} - \mathbf{v}_{MI}$. Due east and due north have been taken as positive.

Vector	*East/West Component*	*North/South Component*
$\mathbf{v}_{PI}$	$-(11.0\text{ m/s})\sin 22° = -4.1\text{ m/s}$	$-(11.0\text{ m/s})\cos 22° = -10.2\text{ m/s}$
$-\mathbf{v}_{MI}$	0	+7.0 m/s
$\mathbf{v}_{PM} = \mathbf{v}_{PI} - \mathbf{v}_{MI}$	–4.1 m/s	–3.2 m/s

Now that the components of $\mathbf{v}_{PM}$ are known, the Pythagorean theorem can be used to find the magnitude.

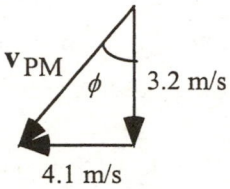

$$v_{PM} = \sqrt{(-4.1\text{ m/s})^2 + (-3.2\text{ m/s})^2} = \boxed{5.2\text{ m/s}}$$

The direction of $\mathbf{v}_{PM}$ is found from

$$\phi = \tan^{-1}\left(\frac{4.1\text{ m/s}}{3.2\text{ m/s}}\right) = \boxed{52°\text{ west of south}}$$

---

59. ***REASONING*** The velocity $\mathbf{v}_{OW}$ of the object relative to the water is the vector sum of the velocity $\mathbf{v}_{OS}$ of the object relative to the ship and the velocity $\mathbf{v}_{SW}$ of the ship relative to the water, as indicated by Equation 3.7: $\mathbf{v}_{OW} = \mathbf{v}_{OS} + \mathbf{v}_{SW}$. The value of $\mathbf{v}_{SW}$ is given in the statement of the problem. We can find the value of $\mathbf{v}_{OS}$ from the fact that we know the position of the object relative to the ship at two different times. The initial position is $\mathbf{r}_{OS1}$, and the final position is $\mathbf{r}_{OS2}$. Since the object moves with constant velocity,

$$\mathbf{v}_{OS} = \frac{\Delta\mathbf{r}_{OS}}{\Delta t} = \frac{\mathbf{r}_{OS2} - \mathbf{r}_{OS1}}{\Delta t} \qquad (1)$$

SOLUTION The first two rows of the following table give the east/west and north/south components of the vectors $\mathbf{r}_{OS2}$ and $-\mathbf{r}_{OS1}$. The third row of the table gives the components of $\Delta\mathbf{r}_{OS} = \mathbf{r}_{OS2} - \mathbf{r}_{OS1}$. Due east and due north have been taken as positive.

Vector	*East/West Component*	*North/South Component*
$\mathbf{r}_{OS2}$	$-(1120\text{ m})\cos 57.0°$ $= -6.10\times10^2\text{ m}$	$-(1120\text{ m})\sin 57.0°$ $= -9.39\times10^2\text{ m}$
$-\mathbf{r}_{OS1}$	$-(2310\text{ m})\cos 32.0°$ $= -1.96\times10^3\text{ m}$	$+(2310\text{ m})\sin 32.0°$ $= 1.22\times10^3\text{ m}$
$\Delta\mathbf{r}_{OS} = \mathbf{r}_{OS2} - \mathbf{r}_{OS1}$	$-2.57\times10^3\text{ m}$	$2.81\times10^2\text{ m}$

Now that the components of $\Delta\mathbf{r}_{OS}$ are known, the Pythagorean theorem can be used to find the magnitude.

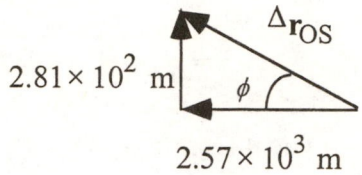

$$\Delta r_{OS} = \sqrt{(-2.57\times10^3\text{ m})^2 + (2.81\times10^2\text{ m})^2} = 2.59\times10^3\text{ m}$$

The direction of $\Delta\mathbf{r}_{OS}$ is found from

$$\phi = \tan^{-1}\left(\frac{2.81\times10^2\text{ m}}{2.57\times10^3\text{ m}}\right) = 6.24°$$

Therefore, from Equation (1),

$$\mathbf{v}_{OS} = \frac{\Delta\mathbf{r}_{OS}}{\Delta t} = \frac{\mathbf{r}_{OS2} - \mathbf{r}_{OS1}}{\Delta t} = \frac{2.59\times10^3\text{ m}}{360\text{ s}} = 7.19\text{ m/s, } 6.24°\text{ north of west}$$

Now that $\mathbf{v}_{OS}$ is known, we can find $\mathbf{v}_{OW}$, as indicated by Equation 3.7: $\mathbf{v}_{OW} = \mathbf{v}_{OS} + \mathbf{v}_{SW}$. The following table summarizes the vector addition:

Vector	*East/West Component*	*North/South Component*
$\mathbf{v}_{OS}$	$-(7.19\text{ m/s})\cos 6.24° = -7.15\text{ m/s}$	$(7.19\text{ m/s})\sin 6.24° = 0.782\text{ m/s}$
$\mathbf{v}_{SW}$	$+4.20\text{ m/s}$	0 m/s
$\mathbf{v}_{OW} = \mathbf{v}_{OS} + \mathbf{v}_{SW}$	-2.95 m/s	0.782 m/s

Now that the components of $\mathbf{v}_{OW}$ are known, the Pythagorean theorem can be used to find the magnitude.

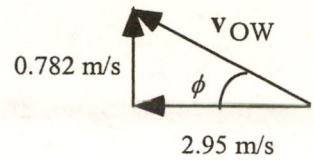

$$v_{OW} = \sqrt{(-2.95\text{ m/s})^2 + (0.782\text{ m/s})^2} = \boxed{3.05\text{ m/s}}$$

The direction of $\mathbf{v}_{OW}$ is found from

$$\phi = \tan^{-1}\left(\frac{0.782\text{ m/s}}{2.95\text{ m/s}}\right) = \boxed{14.8°\text{ north of west}}$$

---

63. ***REASONING*** The time that the ball spends in the air is determined by its vertical motion. The time required for the ball to reach the lake can be found by solving Equation 3.5b for t. The motion of the golf ball is characterized by constant velocity in the x direction and accelerated motion (due to gravity) in the y direction. Thus, the x component of the velocity of the golf ball is constant, and the y component of the velocity at any time t can be found from Equation 3.3b. Once the x and y components of the velocity are known for a particular time t, the speed can be obtained from $v = \sqrt{v_x^2 + v_y^2}$.

SOLUTION

a. Since the ball rolls off the cliff horizontally, $v_{0y} = 0$. If the origin is chosen at top of the cliff and upward is assumed to be the positive direction, then the vertical component of the ball's displacement is $y = -15.5$ m. Thus, Equation 3.5b gives

$$t = \sqrt{\frac{2y}{a_y}} = \sqrt{\frac{2(-15.5\text{ m})}{(-9.80\text{ m/s}^2)}} = \boxed{1.78\text{ s}}$$

b. Since there is no acceleration in the x direction, $v_x = v_{0x} = 11.4$ m/s. The y component of the velocity of the ball just before it strikes the water is, according to Equation 3.3b,

$$v_y = v_{0y} + a_y t = \left[0 + (-9.80\text{ m/s}^2)(1.78\text{ s})\right] = -17.4\text{ m/s}$$

The speed of the ball just before it strikes the water is, therefore,

$$v = \sqrt{v_x^2 + v_y^2} = \sqrt{(11.4\text{ m/s})^2 + (-17.4\text{ m/s})^2} = \boxed{20.8\text{ m/s}}$$

---

65. ***REASONING AND SOLUTION*** The water exhibits projectile motion. The x component of the motion has zero acceleration while the y component is subject to the acceleration due to gravity. In order to reach the highest possible fire, the displacement of the hose from the building is x, where, according to Equation 3.5a (with $a_x = 0$),

$$x = v_{0x}t = (v_0 \cos\theta)t$$

with t equal to the time required for the water the reach its maximum vertical displacement. The time t can be found by considering the vertical motion. From Equation 3.3b,

$$v_y = v_{0y} + a_y t$$

When the water has reached its maximum vertical displacement, $v_y = 0$. Taking up and to the right as the positive directions, we find that

$$t = \frac{-v_{0y}}{a_y} = \frac{-v_0 \sin\theta}{a_y}$$

and

$$x = (v_0 \cos\theta)\left(\frac{-v_0 \sin\theta}{a_y}\right)$$

Therefore, we have

$$x = -\frac{v_0^2 \cos\theta \ \sin\theta}{a_y} = -\frac{(25.0 \text{ m/s})^2 \cos 35.0° \ \sin 35.0°}{-9.80 \text{ m/s}^2} = \boxed{30.0 \text{ m}}$$

---

69. ***REASONING*** The velocity $\mathbf{v}_{PG}$ of the plane relative to the ground is the vector sum of the velocity $\mathbf{v}_{PA}$ of the plane relative to the air and the velocity $\mathbf{v}_{AG}$ of the air relative to the ground, as indicated by Equation 3.7: $\mathbf{v}_{PG} = \mathbf{v}_{PA} + \mathbf{v}_{AG}$, or $\mathbf{v}_{AG} = \mathbf{v}_{PG} - \mathbf{v}_{PA}$. We are given $\mathbf{v}_{PA}$ and can find $\mathbf{v}_{PG}$ from the given displacement and travel time. Thus,

$$\mathbf{v}_{PG} = \frac{81.0 \times 10^3 \text{ m}}{9.00 \times 10^2 \text{ s}} = 90.0 \text{ m/s, } 45° \text{ west of south}$$

SOLUTION The first two rows of the following table give the east/west and north/south components of the vectors $\mathbf{v}_{PG}$ and $-\mathbf{v}_{PA}$. The third row gives the components of $\mathbf{v}_{AG} = \mathbf{v}_{PG} - \mathbf{v}_{PA}$. Due east and due north have been chosen as the positive directions.

Vector	*East/West Component*	*North/South Component*
$\mathbf{v}_{PG}$	–(90.0 m/s) sin 45° = –63.6 m/s	–(90.0 m/s) cos 45° = –63.6 m/s
$-\mathbf{v}_{PA}$	0 m/s	+57.8 m/s
$\mathbf{v}_{AG} = \mathbf{v}_{PG} - \mathbf{v}_{PA}$	–63.6 m/s	–5.8 m/s

The magnitude of the velocity of the wind with respect to the ground can be obtained using the Pythagorean theorem:

$$v_{AG} = \sqrt{(-63.6\ \text{m/s})^2 + (-5.8\ \text{m/s})^2} = \boxed{63.9\ \text{m/s}}$$

The direction of $\mathbf{v}_{AG}$ is found from

$$\phi = \tan^{-1}\left(\frac{63.6\ \text{m/s}}{5.8\ \text{m/s}}\right) = \boxed{85°\ \text{west of south}}$$

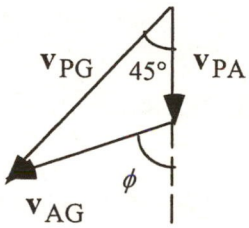

71. ***REASONING*** The horizontal distance covered by stone **1** is equal to the distance covered by stone **2** after it passes point **P** in the following diagram. Thus, the distance Δx between the points where the stones strike the ground is equal to x_2, the horizontal distance covered by stone **2** when it reaches **P**. In the diagram, we assume up and to the right are positive.

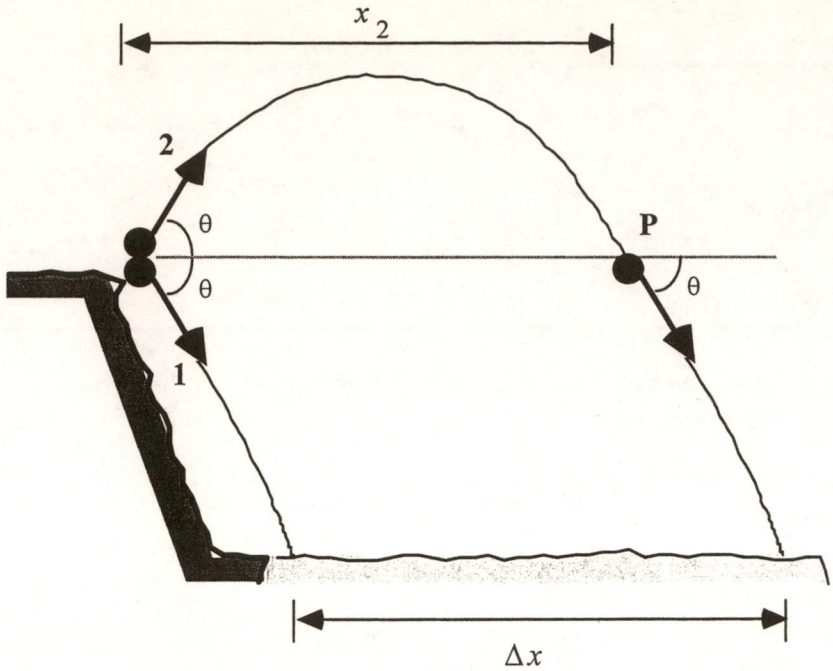

SOLUTION If t_P is the time required for stone **2** to reach **P**, then

$$x_2 = v_{0x}t_P = (v_0 \cos\theta)t_P$$

For the vertical motion of stone **2**, $v_y = v_0 \sin\theta + a_y t$. Solving for t gives

$$t = \frac{v_y - v_0 \sin\theta}{a_y}$$

When stone **2** reaches **P**, $v_y = -v_0 \sin\theta$, so the time required to reach **P** is

$$t_P = \frac{-2v_0 \sin\theta}{a_y}$$

Then,

$$x_2 = v_{0x}t_P = (v_0 \cos\theta)\left(\frac{-2v_0 \sin\theta}{a_y}\right)$$

$$x_2 = \frac{-2v_0^2 \sin\theta \ \cos\theta}{a_y} = \frac{-2(13.0 \text{ m/s})^2 \sin\ 30.0° \cos\ 30.0°}{-9.80 \text{ m/s}^2} = \boxed{14.9 \text{ m}}$$

CHAPTER 4 | *FORCES AND NEWTON'S LAWS OF MOTION*

CONCEPTUAL QUESTIONS

---

3. ***REASONING AND SOLUTION*** If the net external force acting on an object is zero, it is possible for the object to be traveling with a nonzero velocity. According to Newton's second law, $\Sigma\mathbf{F} = m\mathbf{a}$, if the net external force $\Sigma\mathbf{F}$ is zero, the acceleration $\mathbf{a}$ is also zero. If the acceleration is zero, the velocity must be constant, both in magnitude and in direction. Thus, an object can move with a constant nonzero velocity when the net external force is zero.

---

16. ***REASONING AND SOLUTION*** The apparent weight will differ from the true weight only in an accelerating elevator. When the scale in an elevator reads the true weight, the only conclusion that can be made is that the elevator has zero acceleration. Therefore, one cannot conclude whether the elevator is moving with any constant velocity upward, any constant velocity downward, or whether the elevator is at rest, since each of these conditions involves zero acceleration.

---

23. ***REASONING AND SOLUTION*** An object is in equilibrium when its acceleration is zero.

a. If a single nonzero force acts on an object, the object will accelerate according to Newton's second law. The object is not in equilibrium.

b. If two forces that point in mutually perpendicular directions act on an object, the object will experience a net force. By Newton's second law, the object will, therefore, have a nonzero acceleration. The object is not in equilibrium.

c. If two forces that point in directions that are not perpendicular act on the object, the object may or may not be in equilibrium, depending on how the forces are oriented. In general, the resultant of two such forces is nonzero, the object will accelerate, and it is not in equilibrium. In the special case where the two forces point in opposite directions and have the same magnitude, the net force is zero, the object has zero acceleration, and the object is, therefore, in equilibrium.

CHAPTER 4 | *FORCES AND NEWTON'S LAWS OF MOTION*

PROBLEMS

---

5. ***REASONING*** We can use the appropriate equation of kinematics to find the acceleration of the bullet. Then Newton's second law can be used to find the average net force on the bullet.

SOLUTION According to Equation 2.4, the acceleration of the bullet is

$$a = \frac{v - v_0}{t} = \frac{715 \text{ m/s} - 0 \text{ m/s}}{2.50 \times 10^{-3} \text{ s}} = 2.86 \times 10^5 \text{ m/s}^2$$

Therefore, the net average force on the bullet is

$$\sum F = ma = (15 \times 10^{-3} \text{ kg})(2.86 \times 10^5 \text{ m/s}^2) = \boxed{4290 \text{ N}}$$

---

7. ***REASONING*** According to Newton's second law of motion, the net force applied to the fist is equal to the mass of the fist multiplied by its acceleration. The data in the problem gives the final velocity of the fist and the time it takes to acquire that velocity. The average acceleration can be obtained directly from these data using the definition of average acceleration given in Equation 2.4.

SOLUTION The magnitude of the average net force applied to the fist is, therefore,

$$\sum \overline{F} = ma = m\left(\frac{\Delta v}{\Delta t}\right) = (0.70 \text{ kg})\left(\frac{8.0 \text{ m/s} - 0 \text{ m/s}}{0.15 \text{ s}}\right) = \boxed{37 \text{ N}}$$

---

9. ***REASONING*** Let due east be chosen as the positive direction. Then, when both forces point due east, Newton's second law gives

$$\underbrace{F_A + F_B}_{\Sigma F} = ma_1 \qquad (1)$$

where $a_1 = 0.50 \text{ m/s}^2$. When F_A points due east and F_B points due west, Newton's second law gives

$$\underbrace{F_A - F_B}_{\Sigma F} = ma_2 \qquad (2)$$

where $a_2 = 0.40\ \text{m/s}^2$. These two equations can be used to find the magnitude of each force.

SOLUTION

a. Adding Equations 1 and 2 gives

$$F_A = \frac{m(a_1 + a_2)}{2} = \frac{(8.0\ \text{kg})(0.50\ \text{m/s}^2 + 0.40\ \text{m/s}^2)}{2} = \boxed{3.6\ \text{N}}$$

b. Subtracting Equation 2 from Equation 1 gives

$$F_B = \frac{m(a_1 - a_2)}{2} = \frac{(8.0\ \text{kg})(0.50\ \text{m/s}^2 - 0.40\ \text{m/s}^2)}{2} = \boxed{0.40\ \text{N}}$$

---

13. ***REASONING*** According to Newton's second law ($\sum \mathbf{F} = m\mathbf{a}$), the acceleration of the object is given by $\mathbf{a} = \sum \mathbf{F}/m$, where $\sum \mathbf{F}$ is the net force that acts on the object. We must first find the net force that acts on the object, and then determine the acceleration using Newton's second law.

SOLUTION The following table gives the x and y components of the two forces that act on the object. The third row of that table gives the components of the net force.

Force	*x-Component*	*y-Component*
$\mathbf{F}_1$	40.0 N	0 N
$\mathbf{F}_2$	(60.0 N) cos 45.0° = 42.4 N	(60.0 N) sin 45.0° = 42.4 N
$\sum \mathbf{F} = \mathbf{F}_1 + \mathbf{F}_2$	82.4 N	42.4 N

The magnitude of $\sum \mathbf{F}$ is given by the Pythagorean theorem as

$$\Sigma F = \sqrt{(82.4\ \text{N})^2 + (42.4)^2} = 92.7\ \text{N}$$

The angle θ that $\sum\mathbf{F}$ makes with the $+x$ axis is

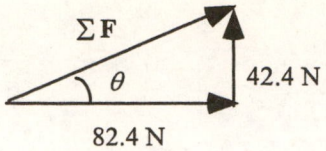

$$\theta = \tan^{-1}\left(\frac{42.4\text{ N}}{82.4\text{ N}}\right) = 27.2°$$

According to Newton's second law, the magnitude of the acceleration of the object is

$$a = \frac{\sum F}{m} = \frac{92.7\text{ N}}{3.00\text{ kg}} = \boxed{30.9\text{ m/s}^2}$$

Since Newton's second law is a vector equation, we know that the direction of the right hand side must be equal to the direction of the left hand side. In other words, the direction of the acceleration a is the same as the direction of the net force $\sum\mathbf{F}$. Therefore, the direction of the acceleration of the object is $\boxed{27.2° \text{ above the } +x \text{ axis}}$.

---

17. ***REASONING*** We first determine the acceleration of the boat. Then, using Newton's second law, we can find the net force $\sum\mathbf{F}$ that acts on the boat. Since two of the three forces are known, we can solve for the unknown force $\mathbf{F}_W$ once the net force $\sum\mathbf{F}$ is known.

SOLUTION Let the direction due east be the positive x direction and the direction due north be the positive y direction. The x and y components of the initial velocity of the boat are then

$$v_{0x} = (2.00\text{ m/s})\cos 15.0° = 1.93\text{ m/s}$$

$$v_{0y} = (2.00\text{ m/s})\sin 15.0° = 0.518\text{ m/s}$$

Thirty seconds later, the x and y velocity components of the boat are

$$v_x = (4.00\text{ m/s})\cos 35.0° = 3.28\text{ m/s}$$

$$v_y = (4.00\text{ m/s})\sin 35.0° = 2.29\text{ m/s}$$

Therefore, according to Equations 3.3a and 3.3b, the x and y components of the acceleration of the boat are

$$a_x = \frac{v_x - v_{0x}}{t} = \frac{3.28\text{ m/s} - 1.93\text{ m/s}}{30.0\text{ s}} = 4.50\times10^{-2}\text{ m/s}^2$$

$$a_y = \frac{v_y - v_{0y}}{t} = \frac{2.29\text{ m/s} - 0.518\text{ m/s}}{30.0\text{ s}} = 5.91\times10^{-2}\text{ m/s}^2$$

Thus, the x and y components of the net force that act on the boat are

$$\sum F_x = ma_x = (325\text{ kg})\,(4.50\times10^{-2}\text{ m/s}^2) = 14.6\text{ N}$$

$$\sum F_y = ma_y = (325\text{ kg})\,(5.91\times10^{-2}\text{ m/s}^2) = 19.2\text{ N}$$

The following table gives the x and y components of the net force $\sum\mathbf{F}$ and the two known forces that act on the boat. The fourth row of that table gives the components of the unknown force $\mathbf{F}_W$.

Force	*x-Component*	*y-Component*
$\sum\mathbf{F}$	14.6 N	19.2 N
$\mathbf{F}_1$	(31.0 N) cos 15.0° = 29.9 N	(31.0 N) sin 15.0° = 8.02 N
$\mathbf{F}_2$	–(23.0 N) cos 15.0° = –22.2 N	–(23.0 N) sin 15.0° = –5.95 N
$\mathbf{F}_W = \sum\mathbf{F} - \mathbf{F}_1 - \mathbf{F}_2$	14.6 N – 29.9 N + 22.2 N = 6.9 N	19.2 N – 8.02 N + 5.95 N = 17.1 N

The magnitude of $\mathbf{F}_W$ is given by the Pythagorean theorem as

$$F_W = \sqrt{(6.9\text{ N})^2 + (17.1\text{ N})^2} = \boxed{18.4\text{ N}}$$

The angle θ that $\mathbf{F}_W$ makes with the x axis is

$$\theta = \tan^{-1}\left(\frac{17.1\text{ N}}{6.9\text{ N}}\right) = 68°$$

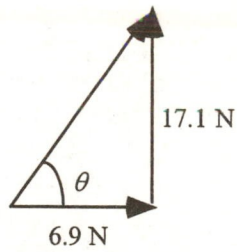

Therefore, the direction of $\mathbf{F}_W$ is $\boxed{68°\text{, north of east}}$.

---

21. ***REASONING AND SOLUTION***

a. Combining Equations 4.4 and 4.5, we see that the acceleration due to gravity on the surface of Saturn can be calculated as follows:

$$g_{\text{Saturn}} = G\frac{M_{\text{Saturn}}}{r_{\text{Saturn}}^2} = \left(6.67\times10^{-11}\text{ N}\cdot\text{m}^2/\text{kg}^2\right)\frac{\left(5.67\times10^{26}\text{ kg}\right)}{(6.00\times10^7\text{ m})^2} = \boxed{10.5\text{ m/s}^2}$$

b. The ratio of the person's weight on Saturn to that on earth is

$$\frac{W_{\text{Saturn}}}{W_{\text{earth}}} = \frac{mg_{\text{Saturn}}}{mg_{\text{earth}}} = \frac{g_{\text{Saturn}}}{g_{\text{earth}}} = \frac{10.5\ \text{m/s}^2}{9.80\ \text{m/s}^2} = \boxed{1.07}$$

---

23. ***REASONING AND SOLUTION*** According to Equations 4.4 and 4.5, the weight of an object of mass m at a distance r from the *center* of the earth is

$$mg = \frac{GM_{\text{E}}m}{r^2}$$

In a circular orbit that is 3.59×10^7 m above the surface of the earth (radius $= 6.38\times10^6$ m, mass $= 5.98\times10^{24}$ kg), the total distance from the center of the earth is $r = 3.59\times10^7\ \text{m} + 6.38\times10^6\ \text{m}$. Thus the acceleration g due to gravity is

$$g = \frac{GM_{\text{E}}}{r^2} = \frac{(6.67\times10^{-11}\text{N}\cdot\text{m}^2/\text{kg}^2)(5.98\times10^{24}\text{kg})}{(3.59\times10^7\text{m} + 6.38\times10^6\text{m})^2} = \boxed{0.223\ \text{m/s}^2}$$

---

27. ***REASONING AND SOLUTION***

a. According to Equation 4.4, the weight of an object of mass m on the surface of Mars would be given by

$$W = \frac{GM_{\text{M}}m}{R_{\text{M}}^2}$$

where M_{M} is the mass of Mars and R_{M} is the radius of Mars. On the surface of Mars, the weight of the object can be given as $W = mg$ (see Equation 4.5), so

$$mg = \frac{GM_{\text{M}}m}{R_{\text{M}}^2} \quad \text{or} \quad g = \frac{GM_{\text{M}}}{R_{\text{M}}^2}$$

Substituting values, we have

$$g = \frac{(6.67\times10^{-11}\text{N}\cdot\text{m}^2/\text{kg}^2)(6.46\times10^{23}\ \text{kg})}{(3.39\times10^6\ \text{m})^2} = \boxed{3.75\ \text{m/s}^2}$$

b. According to Equation 4.5,

$$W = mg = (65\ \text{kg})(3.75\ \text{m/s}^2) = \boxed{2.4\times10^2\ \text{N}}$$

---

29. ***REASONING AND SOLUTION*** There are two forces that act on the balloon; they are, the combined weight of the balloon and its load, Mg, and the upward buoyant force F_B. If we take upward as the positive direction, then, initially when the balloon is motionless, Newton's second law gives $F_B - Mg = 0$. If an amount of mass m is dropped overboard so that the balloon has an upward acceleration, Newton's second law for this situation is

$$F_B - (M - m)g = (M - m)a$$

But $F_B = mg$, so that

$$Mg - (M - m)g = mg = (M - m)a$$

Solving for the mass m that should be dropped overboard, we obtain

$$m = \frac{Ma}{g + a} = \frac{(310\text{ kg})(0.15\text{ m/s}^2)}{9.80\text{ m/s}^2 + 0.15\text{ m/s}^2} = \boxed{4.7\text{ kg}}$$

---

35. ***REASONING*** The book is kept from falling as long as the total static frictional force balances the weight of the book. The forces that act on the book are shown in the following free-body diagram, where P is the pressing force applied by each hand.

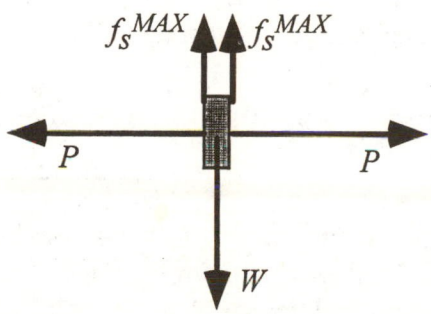

In this diagram, note that there are two pressing forces, one from each hand. Each hand also applies a static frictional force, and, therefore, two static frictional forces are shown. The maximum static frictional force is related in the usual way to a normal force F_N, but in this problem the normal force is provided by the pressing force, so that $F_N = P$.

SOLUTION Since the frictional forces balance the weight, we have

$$2f_s^{MAX} = 2(\mu_s F_N) = 2(\mu_s P) = W$$

Solving for P, we find that

$$P = \frac{W}{2\mu_s} = \frac{31\text{ N}}{2(0.40)} \doteq \boxed{39\text{ N}}$$

---

39. ***REASONING*** In order to start the crate moving, an external agent must supply a force that is at least as large as the maximum value $f_s^{MAX} = \mu_s F_N$, where μ_s is the coefficient of static friction (see Equation 4.7). Once the crate is moving, the magnitude of the frictional force is very nearly constant at the value $f_k = \mu_k F_N$, where μ_k is the coefficient of kinetic friction (see Equation 4.8). In both cases described in the problem statement, there are only two vertical forces that act on the crate; they are the upward normal force F_N, and the downward pull of gravity (the weight) mg. Furthermore, the crate has no vertical acceleration in either case. Therefore, if we take upward as the positive direction, Newton's second law in the vertical direction gives $F_N - mg = 0$, and we see that, in both cases, the magnitude of the normal force is $F_N = mg$.

SOLUTION

a. Therefore, the applied force needed to start the crate moving is

$$f_s^{MAX} = \mu_s mg = (0.760)(60.0\text{ kg})(9.80\text{ m/s}^2) = \boxed{447\text{ N}}$$

b. When the crate moves in a straight line at constant speed, its velocity does not change, and it has zero acceleration. Thus, Newton's second law in the horizontal direction becomes $P - f_k = 0$, where P is the required pushing force. Thus, the applied force required to keep the crate sliding across the dock at a constant speed is

$$P = f_k = \mu_k mg = (0.410)(60.0\text{ kg})(9.80\text{ m/s}^2) = \boxed{241\text{ N}}$$

---

43. ***REASONING*** The free-body diagrams for the large cube (mass = M) and the small cube (mass = m) are shown in the following drawings. In the case of the large cube, we have omitted the weight and the normal force from the surface, since the play no role in the solution (although they do balance).

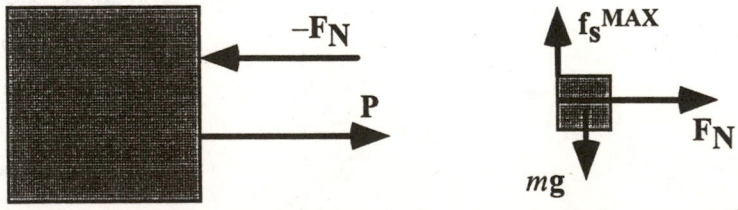

In these diagrams, note that the two blocks exert a normal force on each other; the large block exerts the force F_N on the smaller block, while the smaller block exerts the force $-F_N$ on the larger block. In accord with Newton's third law these forces have opposite directions and equal magnitudes F_N. Under the influence of the forces shown, the two blocks have the same acceleration a. We begin our solution by applying Newton's second law to each one.

SOLUTION According to Newton's second law, we have

$$\underbrace{\Sigma F = P - F_N = Ma}_{\text{Large block}} \qquad \underbrace{F_N = ma}_{\text{Small block}}$$

Substituting $F_N = ma$ into the large-block expression and solving for P gives

$$P = (M + m)\,a$$

For the smaller block to remain in place against the larger block, the static frictional force must balance the weight of the smaller block, so that $f_s^{MAX} = mg$. But f_s^{MAX} is given by $f_s^{MAX} = \mu_s F_N$, where, from the Newton's second law, we know that $F_N = ma$. Thus, we have $\mu_s ma = mg$ or $a = g/\mu_s$. Using this result in the expression for P gives

$$P = (M+m)a = \frac{(M+m)g}{\mu_s} = \frac{(25\text{ kg} + 4.0\text{ kg})(9.80\text{ m/s}^2)}{0.71} = \boxed{4.0\times10^2\text{ N}}$$

---

49. ***REASONING*** In order for the object to move with constant velocity, the net force on the object must be zero. Therefore, the north/south component of the third force must be equal in magnitude and opposite in direction to the 80.0 N force, while the east/west component of the third force must be equal in magnitude and opposite in direction to the 60.0 N force. Therefore, the third force has components: 80.0 N due south and 60.0 N due east. We can use the Pythagorean theorem and trigonometry to find the magnitude and direction of this third force.

SOLUTION The magnitude of the third force is

$$F_3 = \sqrt{(80.0\text{ N})^2 + (60.0\text{ N})^2} = \boxed{1.00\times10^2\text{ N}}$$

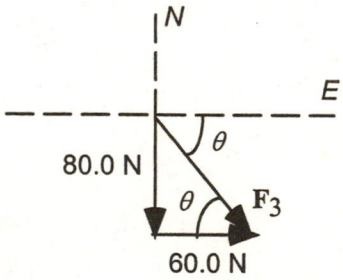

The direction of F_3 is specified by the angle θ where

$$\theta = \tan^{-1}\left(\frac{80.0\text{ N}}{60.0\text{ N}}\right) = \boxed{53.1°\text{, south of east}}$$

---

53. ***REASONING AND SOLUTION*** The figure at the right shows the forces that act on the wine bottle. Newton's second law applied in the horizontal and vertical directions gives

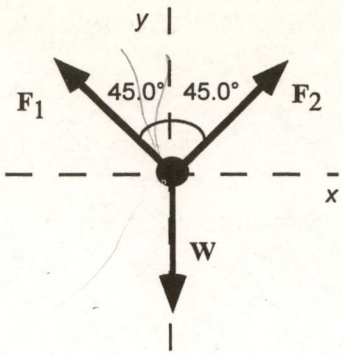

$$\sum F_y = F_1 \cos\ 45.0° + F_2 \cos\ 45.0° - W = 0 \qquad (1)$$

$$\sum F_x = F_2 \sin\ 45.0° - F_1 \sin\ 45.0° = 0 \qquad (2)$$

From Equation (2), we see that $F_1 = F_2$. According to Equation (1), we have

$$F_1 = \frac{W}{2 \cos 45.0°} = \frac{mg}{2 \cos 45.0°}$$

Therefore,

$$F_1 = F_2 = \frac{(1.40 \text{ kg})\ (9.80 \text{ m/s}^2)}{2 \cos 45.0°} = \boxed{9.70 \text{ N}}$$

---

57 ***REASONING*** There are four forces that act on the chandelier; they are the forces of tension T in each of the three wires, and the downward force of gravity mg. Under the influence of these forces, the chandelier is at rest and, therefore, in equilibrium. Consequently, the sum of the x components as well as the sum of the y components of the forces must each be zero. The figure below shows the free-body diagram for the chandelier and the force components for a suitable system of x, y axes. Note that the free-body diagram only shows one of the forces of tension; the second and third tension forces are not shown. The triangle at the right shows the geometry of one of the cords, where ℓ is the length of the cord, and d is the distance from the ceiling.

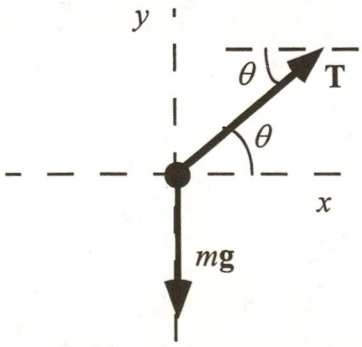

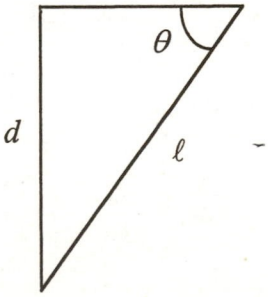

We can use the forces in the y direction to find the magnitude T of the tension in any one wire.

SOLUTION Remembering that there are three tension forces, we see from the free-body diagram that

$$3T\sin\theta = mg \qquad \text{or} \qquad T = \frac{mg}{3\sin\theta} = \frac{mg}{3(d/\ell)} = \frac{mg\ell}{3d}$$

Therefore, the magnitude of the tension in any one of the cords is

$$T = \frac{(44\text{ kg})(9.80\text{ m/s}^2)(2.0\text{ m})}{3(1.5\text{ m})} = \boxed{1.9\times10^2\text{N}}$$

---

61. ***REASONING*** When the bicycle is coasting straight down the hill, the forces that act on it are the normal force F_N exerted by the surface of the hill, the force of gravity mg, and the force of air resistance R. When the bicycle climbs the hill, there is one additional force; it is the applied force that is required for the bicyclist to climb the hill at constant speed. We can use our knowledge of the motion of the bicycle down the hill to find R. Once R is known, we can analyze the motion of the bicycle as it climbs the hill.

SOLUTION The figure to the left below shows the free-body diagram for the forces during the downhill motion. The hill is inclined at an angle θ above the horizontal. The figure to the right shows these forces resolved into components parallel to and perpendicular to the line of motion.

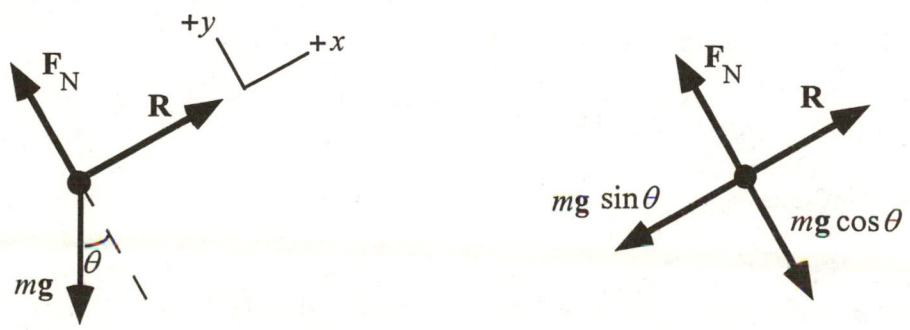

Since the bicyclist is traveling at a constant velocity, his acceleration is zero. Therefore, according to Newton's second law, we have $\sum F_x = 0$ and $\sum F_y = 0$. Taking the direction up the hill as positive, we have $\sum F_x = R - mg\sin\theta = 0$, or

$$R = mg\sin\theta = (80.0\text{ kg})(9.80\text{ m/s}^2)\sin 15.0° = 203\text{ N}$$

When the bicyclist climbs the same hill at constant speed, an applied force P must push the system up the hill, and the force of air resistance will oppose the motion by pointing down the hill.

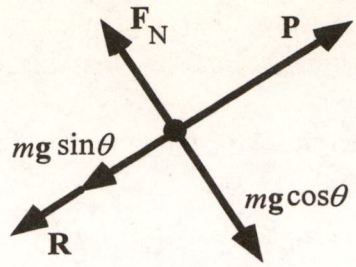

The figure at the right shows the resolved forces that act on the system during the uphill motion.

Using the same sign convention as above, we have $\sum F_x = P - mg\sin\theta - R = 0$, or

$$P = R + mg\sin\theta = 203\text{ N} + 203\text{ N} = \boxed{406\text{ N}}$$

---

65. ***REASONING*** If we assume that the acceleration is constant, we can use Equation 2.4 ($v = v_0 + at$) to find the acceleration of the car. Once the acceleration is known, Newton's second law ($\sum \mathbf{F} = m\mathbf{a}$) can be used to find the magnitude and direction of the net force that produces the deceleration of the car.

SOLUTION The average acceleration of the car is, according to Equation 2.4,

$$a = \frac{v - v_0}{t} = \frac{17.0\text{ m/s} - 27.0\text{ m/s}}{8.00\text{ s}} = -1.25\text{ m/s}^2$$

where the minus sign indicates that the direction of the acceleration is opposite to the direction of motion; therefore, the acceleration points due west.

According to Newton's Second law, the net force on the car is

$$\sum F = ma = (1380\text{ kg})(-1.25\text{ m/s}^2) = -1730\text{ N}$$

The magnitude of the net force is $\boxed{1730\text{ N}}$. From Newton's second law, we know that the direction of the force is the same as the direction of the acceleration, so the force also points $\boxed{\text{due west}}$.

---

69. ***REASONING*** The speed of the skateboarder at the bottom of the ramp can be found by solving Equation 2.9 ($v^2 = v_0^2 + 2ax$, where x is the distance that the skater moves down the ramp) for v. The figure at the right shows the free-body diagram for the skateboarder. The net force ΣF, which accelerates the skateboarder down the ramp, is the component of

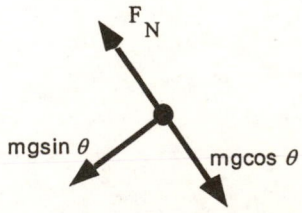

the weight that is parallel to the incline: $\sum F = mg\sin\theta$. Therefore, we know from Newton's second law that the acceleration of the skateboarder down the ramp is

$$a = \frac{\Sigma F}{m} = \frac{mg\sin\theta}{m} = g\sin\theta$$

SOLUTION Thus, the speed of the skateboarder at the bottom of the ramp is

$$v = \sqrt{v_0^2 + 2ax} = \sqrt{v_0^2 + 2gx\ \sin\theta} = \sqrt{(2.6\text{ m/s})^2 + 2(9.80\text{ m/s}^2)(6.0\text{ m})\sin 18°} = \boxed{6.6\text{ m/s}}$$

---

71. ***REASONING AND SOLUTION***

a. Each cart has the same mass and acceleration; therefore, the net force acting on any one of the carts is, according to Newton's second law

$$\sum F = ma = (26\text{ kg})(0.050\text{ m/s}^2) = \boxed{1.3\text{ N}}$$

b. The fifth cart must essentially push the sixth, seventh, eight, ninth and tenth cart. In other words, it must exert on the sixth cart a total force of

$$\sum F = ma = 5(26\text{ kg})(0.050\text{ m/s}^2) = \boxed{6.5\text{ N}}$$

---

75. ***REASONING*** The shortest time to pull the person from the cave corresponds to the maximum acceleration, a_y, that the rope can withstand. We first determine this acceleration and then use kinematic Equation 3.5b ($y = v_{0y}t + \frac{1}{2}a_y t^2$) to find the time t.

SOLUTION As the person is being pulled from the cave, there are two forces that act on him; they are the tension T in the rope that points vertically upward, and the weight of the person mg that points vertically downward. Thus, if we take upward as the positive direction, Newton's second law gives $\sum F_y = T - mg = ma_y$. Solving for a_y, we have

$$a_y = \frac{T}{m} - g = \frac{T}{W/g} - g = \frac{569\text{ N}}{(5.20\times10^2\text{ N})/(9.80\text{ m/s}^2)} - 9.80\text{ m/s}^2 = 0.92\text{ m/s}^2$$

Therefore, from Equation 3.5b with $v_{0y} = 0$ m/s, we have $y = \frac{1}{2}a_y t^2$. Solving for t, we find

$$t = \sqrt{\frac{2y}{a_y}} = \sqrt{\frac{2(35.1\text{ m})}{0.92\text{ m/s}^2}} = \boxed{8.7\text{ s}}$$

---

81. ***REASONING*** The tension in each coupling bar is responsible for accelerating the objects behind it. The masses of the cars are m_1, m_2, and m_3. We can use Newton's second law to express the tension in each coupling bar, since friction is negligible:

$$\underbrace{T_A = (m_1 + m_2 + m_3)a}_{\text{Coupling bar A}} \qquad \underbrace{T_B = (m_2 + m_3)a}_{\text{Coupling bar B}} \qquad \underbrace{T_C = m_3 a}_{\text{Coupling bar C}}$$

In these expressions $a = 0.12\ \text{m/s}^2$ remains constant. Consequently, the tension in a given bar will change only if the total mass of the objects accelerated by that bar changes as a result of the luggage transfer. Using Δ (Greek capital delta) to denote a change in the usual fashion, we can express the changes in the above tensions as follows:

$$\underbrace{\Delta T_A = \left[\Delta(m_1 + m_2 + m_3)\right]a}_{\text{Coupling bar A}} \qquad \underbrace{\Delta T_B = \left[\Delta(m_2 + m_3)\right]a}_{\text{Coupling bar B}} \qquad \underbrace{\Delta T_C = (\Delta m_3)a}_{\text{Coupling bar C}}$$

SOLUTION

a. Moving luggage from car 2 to car 1 does not change the total mass $m_1 + m_2 + m_3$, so $\Delta(m_1 + m_2 + m_3) = 0$ kg and $\boxed{\Delta T_A = 0\ \text{N}}$.

The transfer from car 2 to car 1 causes the total mass $m_2 + m_3$ to decrease by 39 kg, so $\Delta(m_2 + m_3) = -39$ kg and

$$\Delta T_B = \left[\Delta(m_2 + m_3)\right]a = (-39\ \text{kg})(0.12\ \text{m/s}^2) = \boxed{-4.7\ \text{N}}$$

The transfer from car 2 to car 1 does not change the mass m_3, so $\Delta m_3 = 0$ kg and $\boxed{\Delta T_C = 0\ \text{N}}$.

b. Moving luggage from car 2 to car 3 does not change the total mass $m_1 + m_2 + m_3$, so $\Delta(m_1 + m_2 + m_3) = 0$ kg and $\boxed{\Delta T_A = 0\ \text{N}}$.

The transfer from car 2 to car 3 does not change the total mass $m_2 + m_3$, so $\Delta(m_2 + m_3) = 0$ kg and $\boxed{\Delta T_B = 0\ \text{N}}$.

The transfer from car 2 to car 3 causes the mass m_3 to increase by 39 kg, so $\Delta m_3 = +39$ kg and

$$\Delta T_C = (\Delta m_3)a = (+39\ \text{kg})(0.12\ \text{m/s}^2) = \boxed{+4.7\ \text{N}}$$

85. ***REASONING AND SOLUTION*** The penguin comes to a halt on the horizontal surface because the kinetic frictional force opposes the motion and causes it to slow down. The time required for the penguin to slide to a halt (v = 0 m/s) after entering the horizontal patch of ice is, according to Equation 2.4,

$$t = \frac{v - v_0}{a_x} = \frac{-v_0}{a_x}$$

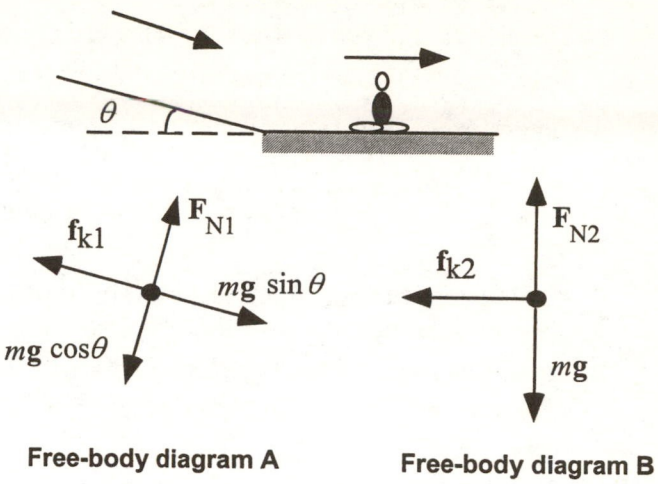

We must, therefore, determine the acceleration of the penguin as it slides along the horizontal patch.

For the penguin sliding on the horizontal patch of ice, we find from free-body diagram B and Newton's second law in the x direction (motion to the right is taken as positive) that

$$\sum F_x = -f_{k2} = ma_x \qquad \text{or} \qquad a_x = \frac{-f_{k2}}{m} = \frac{-\mu_k F_{N2}}{m}$$

In the y direction in free-body diagram B, we have $\sum F_y = F_{N2} - mg = 0$, or $F_{N2} = mg$. Therefore, the acceleration of the penguin is

$$a_x = \frac{-\mu_k mg}{m} = -\mu_k g \qquad (1)$$

Equation (1) indicates that, in order to find the acceleration a_x, we must find the coefficient of kinetic friction.

We are told in the problem statement that the coefficient of kinetic friction between the penguin and the ice is the same for the incline as for the horizontal patch. Therefore, we can use the motion of the penguin on the incline to determine the coefficient of friction and use it in Equation (1).

For the penguin sliding down the incline, we find from free-body diagram A and Newton's second law (taking the direction of motion as positive) that

$$\sum F_x = mg\sin\theta - f_{k1} = ma_x = 0 \qquad \text{or} \qquad f_{k1} = mg\sin\theta \qquad (2)$$

Here, we have used the fact that the penguin slides down the incline with a constant velocity, so that it has zero acceleration. From Equation 4.8, we know that $f_{k1} = \mu_k F_{N1}$. Applying Newton's second law in the direction perpendicular to the incline, we have

$$\Sigma F_y = F_{N1} - mg\cos\theta = 0 \qquad \text{or} \qquad F_{N1} = mg\cos\theta$$

Therefore, $f_{k1} = \mu_k mg\ \cos\theta$, so that according to Equation (2), we find

$$f_{k1} = \mu_k mg\cos\theta = mg\sin\theta$$

Solving for the coefficient of kinetic friction, we have

$$\mu_k = \frac{\sin\theta}{\cos\theta} = \tan\theta$$

Finally, the time required for the penguin to slide to a halt after entering the horizontal patch of ice is

$$t = \frac{-v_0}{a_x} = \frac{-v_0}{-\mu_k g} = \frac{v_0}{g\tan\theta} = \frac{1.4\text{ m/s}}{(9.80\text{ m/s}^2)\tan\ 6.9°} = \boxed{1.2\text{ s}}$$

---

87. ***REASONING AND SOLUTION***

a. According to Equation 4.5, the weight of the space traveler of mass $\boxed{m = 115\text{ kg}}$ on earth is

$$W = mg = (115\text{ kg})\ (9.80\text{ m/s}^2) = \boxed{1.13\times10^3\text{ N}}$$

b. In interplanetary space where there are no nearby planetary objects, the gravitational force exerted on the space traveler is zero and $g = 0\text{ m/s}^2$. Therefore, the weight is $\boxed{W = 0\text{ N}}$. Since the mass of an object is an intrinsic property of the object and is independent of its location in the universe, the mass of the space traveler is still $\boxed{m = 115\text{ kg}}$.

---

89. ***REASONING AND SOLUTION*** According to Equation 3.3b, the acceleration of the astronaut is $a_y = (v_y - v_{0y})/t = v_y/t$. The apparent weight and the true weight of the astronaut are related according to Equation 4.6. Direct substitution gives

$$\underbrace{F_N}_{\text{Apparent weight}} = \underbrace{mg}_{\text{True weight}} + ma_y = m\ (g + a_y) = m\left(g + \frac{v_y}{t}\right)$$

$$= (57\text{ kg})\left(9.80\text{ m/s}^2 + \frac{45\text{ m/s}}{15\text{ s}}\right) = \boxed{7.3\times10^2\text{ N}}$$

---

91. ***REASONING AND SOLUTION*** Four forces act on the sled. They are the pulling force **P**, the force of kinetic friction $\mathbf{f}_k$, the weight $m\mathbf{g}$ of the sled, and the normal force $\mathbf{F}_N$ exerted on the sled by the surface on which it slides. The following figures show free-body diagrams for the sled. In the diagram on the right, the forces have been resolved into their x and y components.

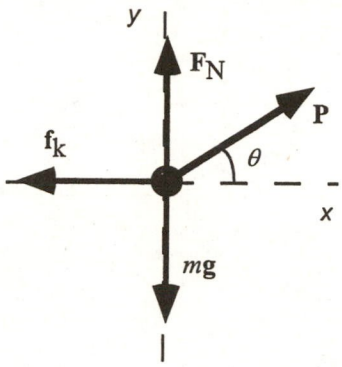

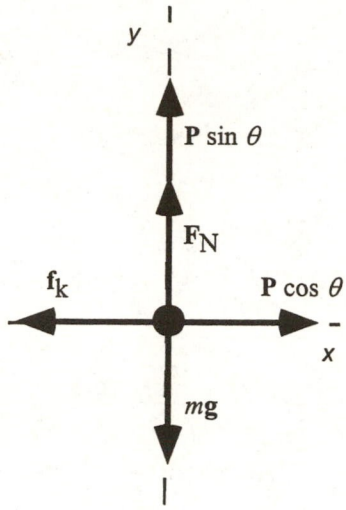

Since the sled is pulled at constant velocity, its acceleration is zero, and Newton's second law in the direction of motion is (with right chosen as the positive direction)

$$\sum F_x = P\cos\theta - f_k = ma_x = 0$$

From Equation 4.8, we know that $f_k = \mu_k F_N$, so that the above expression becomes

$$P\cos\theta - \mu_k F_N = 0 \qquad (1)$$

In the vertical direction,

$$\sum F_y = P\sin\theta + F_N - mg = ma_y = 0 \qquad (2)$$

Solving Equation (2) for the normal force, and substituting into Equation (1), we obtain

$$P\cos\theta - \mu_k\left(mg - P\sin\theta\right) = 0$$

Solving for μ_k, the coefficient of kinetic friction, we find

$$\mu_k = \frac{P\cos\theta}{mg - P\sin\theta} = \frac{(80.0\text{ N})\cos 30.0°}{(20.0\text{ kg})(9.80\text{ m/s}^2) - (80.0\text{ N})\sin 30.0°} = \boxed{0.444}$$

95. ***REASONING*** The net force acting on the ball can be calculated using Newton's second law. Before we can use Newton's second law, however, we must use Equation 2.9 from the equations of kinematics to determine the acceleration of the ball.

SOLUTION According to Equation 2.9, the acceleration of the ball is given by

$$a = \frac{v^2 - v_0^2}{2x}$$

Thus, the magnitude of the net force on the ball is given by

$$\sum F = ma = m\left(\frac{v^2 - v_0^2}{2x}\right) = (0.058\text{ kg})\left[\frac{(45\text{ m/s})^2 - (0\text{ m/s})^2}{2(0.44\text{ m})}\right] = \boxed{130\text{ N}}$$

---

99. ***REASONING AND SOLUTION*** The system is shown in the drawing. We will let $m_1 = 21.0\text{ kg}$, and $m_2 = 45.0\text{ kg}$. Then, m_1 will move upward, and m_2 will move downward. There are two forces that act on each object; they are the tension T in the cord and the weight mg of the object. The forces are shown in the free-body diagrams at the far right.

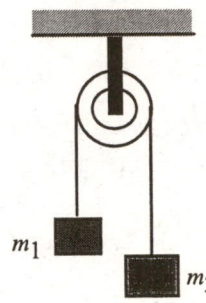

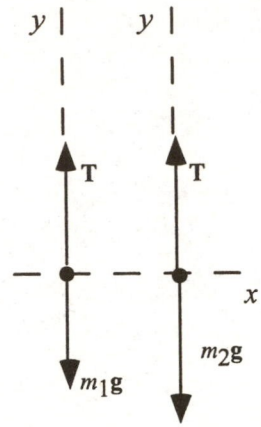

We will take up as the positive direction. If the acceleration of m_1 is a, then the acceleration of m_2 must be $-a$.

From Newton's second law, we have for m_1

$$\sum F_y = T - m_1 g = m_1 a \qquad (1)$$

and for m_2

$$\sum F_y = T - m_2 g = -m_2 a \qquad (2)$$

a. Eliminating T between these two equations, we obtain

$$a = \frac{m_2 - m_1}{m_2 + m_1} g = \left(\frac{45.0\text{ kg} - 21.0\text{ kg}}{45.0\text{ kg} + 21.0\text{ kg}}\right)(9.80\text{ m/s}^2) = \boxed{3.56\text{ m/s}^2}$$

b. Eliminating a between Equations (1) and (2), we find

$$T = \frac{2m_1m_2}{m_1+m_2}g = \left[\frac{2(21.0\text{ kg})(45.0\text{ kg})}{21.0\text{ kg}+45.0\text{ kg}}\right](9.80\text{ m/s}^2) = \boxed{281\text{ N}}$$

---

103. ***REASONING AND SOLUTION*** Three forces act on the man. They are two upward forces of tension of equal magnitude T, and the force of gravity $m\mathbf{g}$. Therefore, if we take up as the positive direction, Newton's second law gives

$$\sum F = 2T - mg = ma$$

Solving for the acceleration a, we find

$$a = \frac{2T-mg}{m} = \frac{2T}{m} - g = \frac{2(358\text{ N})}{72.0\text{ kg}} - 9.80\text{ m/s}^2 = \boxed{0.14\text{ m/s}^2}$$

---

107. ***REASONING AND SOLUTION*** The free-body diagram is shown at the right. The forces that act on the picture are the pressing force P, the normal force $\mathbf{F}_N$ exerted on the picture by the wall, the weight mg of the picture, and the force of static friction $\mathbf{f}_s^{MAX}$. The maximum magnitude for the frictional force is given by Equation 4.7: $f_s^{MAX} = \mu_s F_N$. The picture is in equilibrium, and, if we take the directions to the right and up as positive, we have in the x direction

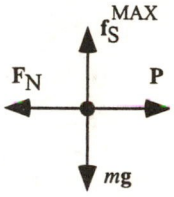

$$\sum F_x = P - F_N = 0 \qquad \text{or} \qquad P = F_N$$

and in the y direction

$$\sum F_y = f_s^{MAX} - mg = 0 \qquad \text{or} \qquad f_s^{MAX} = mg$$

Therefore,

$$f_s^{MAX} = \mu_s F_N = mg$$

But since $F_N = P$, we have

$$\mu_s P = mg$$

Solving for P, we have

$$P = \frac{mg}{\mu_s} = \frac{(1.10\text{ kg})(9.80\text{ m/s}^2)}{0.660} = \boxed{16.3\text{ N}}$$

---

109. ***REASONING AND SOLUTION***

a. The left mass (mass 1) has a tension T_1 pulling it up. Newton's second law gives

$$T_1 - m_1 g = m_1 a \qquad (1)$$

The right mass (mass 3) has a different tension, T_3, trying to pull it up. Newton's second for it is

$$T_3 - m_3 g = -m_3 a \qquad (2)$$

The middle mass (mass 2) has both tensions acting on it along with friction. Newton's second law for its horizontal motion is

$$T_3 - T_1 - \mu_k m_2 g = m_2 a \qquad (3)$$

Solving Equation (1) and Equation (2) for T_1 and T_3, respectively, and substituting into Equation (3) gives

$$a = \frac{(m_3 - m_1 - \mu_k m_2) g}{m_1 + m_2 + m_3}$$

Hence,

$$a = \frac{\left[25.0\text{ kg} - 10.0\text{ kg} - (0.100)(80.0\text{ kg})\right](9.80\text{ m/s}^2)}{10.0\text{ kg} + 80.0\text{ kg} + 25.0\text{ kg}} = \boxed{0.60\text{ m/s}^2}$$

b. From part a:

$$T_1 = m_1(g + a) = (10.0\text{ kg})(9.80\text{ m/s}^2 + 0.60\text{ m/s}^2) = \boxed{104\text{ N}}$$

$$T_3 = m_3(g - a) = (25.0\text{ kg})(9.80\text{ m/s}^2 - 0.60\text{ m/s}^2) = \boxed{230\text{ N}}$$

CHAPTER 5 | *DYNAMICS OF UNIFORM CIRCULAR MOTION*

CONCEPTUAL QUESTIONS

---

7. ***REASONING AND SOLUTION*** A bug lands on a windshield wiper. The wipers are turned on. Since the wipers move along the arc of a circle, the bug will experience a centripetal acceleration, and hence, a centripetal force must be present. The magnitude of the centripetal force is given by $F_c = mv^2 / r$. In order for the bug to remain at rest on the wiper blade, the force of static friction between the bug and the wiper blade must contribute in a major way to the centripetal force. Without the centripetal force, the bug will be dislodged. When the wipers are turned on at a higher setting, v is larger, and the centripetal force required to keep the bug moving along the arc of the circle is larger than if the wipers are turned on the low setting. Since the high setting requires a larger centripetal force to keep the bug on the wiper, the bug is more likely to be dislodged at that setting than at the low setting.

---

9. ***REASONING AND SOLUTION*** Since the speed and radius of the circle are constant, the centripetal acceleration is constant. As the water leaks out, however, the mass of the object undergoing the uniform circular motion decreases. Centripetal force is mass times the centripetal acceleration, so that the centripetal force applied to the container must be decreasing. It is the tension in the rope that provides the centripetal force. You are holding the free end of the rope and pulling on it in order to create the tension. Therefore, you must be reducing your pull as the water leaks out. In turn, according to Newton's third law, the rope must be pulling back on your hand with a force of decreasing magnitude, and you feel this pull weakening as time passes.

---

13. ***REASONING AND SOLUTION***

a. Referring to Figure 5.10 in the text, we can see that the centripetal force on the plane is $L \sin\theta = mv^2 / r$, where L is the magnitude of the lifting force. In addition, the vertical component of the lifting force must balance the weight of the plane, so that $L \cos\theta = mg$. Dividing these two equations reveals that $\tan\theta = v^2 / (rg)$.

b. The banking condition for a car traveling at speed v around a curve of radius r, banked at angle θ is $\tan\theta = v^2 / (rg)$, according to Equation 5.4 in the text.

c. The speed v of a satellite in a circular orbit of radius r about the earth is given by $v = \sqrt{GM_E / r}$, according to Equation 5.5 in the text.

d. The minimum speed required for a loop-the-loop trick around a loop of radius r is $v = \sqrt{rg}$, according to the discussion in Section 5.7 of the text. According to Equations 4.4 and 4.5, $g = GM_E / r^2$. Thus, any expression that depends on g also depends on M_E and would be affected by a change in the earth's mass. Such is the case for each of the four situations discussed above.

CHAPTER 5 DYNAMICS OF UNIFORM CIRCULAR MOTION

PROBLEMS

1. ***REASONING*** The speed of the plane is given by Equation 5.1: $v = 2\pi r / T$, where T is the period or the time required for the plane to complete one revolution.

SOLUTION Solving Equation 5.1 for T we have

$$T = \frac{2\pi r}{v} = \frac{2\pi(2850\text{ m})}{110\text{ m/s}} = \boxed{160\text{ s}}$$

5. ***REASONING AND SOLUTION*** In each case, the magnitude of the centripetal acceleration is given by Equation 5.2, $a_c = v^2 / r$. Therefore,

$$\frac{a_{cA}}{a_{cB}} = \frac{v_A^2 / r_A}{v_B^2 / r_B}$$

Since each boat experiences the same centripetal acceleration, $a_{cA} / a_{cB} = 1$. Solving for the ratio of the speeds gives

$$\frac{v_A}{v_B} = \sqrt{\frac{r_A}{r_B}} = \sqrt{\frac{120\text{ m}}{240\text{ m}}} = \boxed{0.71}$$

9. ***REASONING*** The magnitude of the centripetal acceleration of any point on the helicopter blade is given by Equation 5.2, $a_C = v^2 / r$, where r is the radius of the circle on which that point moves. From Equation 5.1: $v = 2\pi r / T$. Combining these two expressions, we obtain

$$a_C = \frac{4\pi^2 r}{T^2}$$

All points on the blade move with the same period T.

SOLUTION The ratio of the centripetal acceleration at the end of the blade (point 1) to that which exists at a point located 3.0 m from the center of the circle (point 2) is

$$\frac{a_{C1}}{a_{C2}} = \frac{4\pi^2 r_1 / T^2}{4\pi^2 r_2 / T^2} = \frac{r_1}{r_2} = \frac{6.7\text{ m}}{3.0\text{ m}} = \boxed{2.2}$$

11. ***REASONING*** In Example 3, it was shown that the magnitudes of the centripetal acceleration for the two cases are

$$[\text{Radius} = 33\text{ m}] \qquad a_C = 35\text{ m/s}^2$$
$$[\text{Radius} = 24\text{ m}] \qquad a_C = 48\text{ m/s}^2$$

According to Newton's second law, the centripetal force is $F_C = ma_C$ (see Equation 5.3).

SOLUTION a. Therefore, when the sled undergoes the turn of radius 33 m,

$$F_C = ma_C = (350\text{ kg})(35\text{ m/s}^2) = \boxed{1.2 \times 10^4\text{ N}}$$

b. Similarly, when the radius of the turn is 24 m,

$$F_C = ma_C = (350\text{ kg})(48\text{ m/s}^2) = \boxed{1.7 \times 10^4\text{ N}}$$

---

17. ***REASONING*** Let v_0 be the initial speed of the ball as it begins its projectile motion. Then, the centripetal force is given by Equation 5.3: $F_C = mv_0^2/r$. We are given the values for m and r; however, we must determine the value of v_0 from the details of the projectile motion after the ball is released.

In the absence of air resistance, the x component of the projectile motion has zero acceleration, while the y component of the motion is subject to the acceleration due to gravity. The horizontal distance traveled by the ball is given by Equation 3.5a (with $a_x = 0$):

$$x = v_{0x}t = (v_0 \cos\theta)t$$

with t equal to the flight time of the ball while it exhibits projectile motion. The time t can be found by considering the vertical motion. From Equation 3.3b,

$$v_y = v_{0y} + a_y t$$

After a time t, $v_y = -v_{0y}$. Assuming that up and to the right are the positive directions, we have

$$t = \frac{-2v_{0y}}{a_y} = \frac{-2v_0 \sin\theta}{a_y}$$

and

$$x = (v_0 \cos\theta)\left(\frac{-2v_0 \sin\theta}{a_y}\right)$$

Using the fact that $2\sin\theta\cos\theta = \sin 2\theta$, we have

$$x = -\frac{2v_0^2 \cos\theta \sin\theta}{a_y} = -\frac{v_0^2 \sin 2\theta}{a_y} \qquad (1)$$

Equation (1) (with upward and to the right chosen as the positive directions) can be used to determine the speed v_0 with which the ball begins its projectile motion. Then Equation 5.3 can be used to find the centripetal force.

SOLUTION Solving equation (1) for v_0, we have

$$v_0 = \sqrt{\frac{-x\,a_y}{\sin 2\theta}} = \sqrt{\frac{-(86.75\text{ m})(-9.80\text{ m/s}^2)}{\sin 2(41°)}} = 29.3\text{ m/s}$$

Then, from Equation 5.3,

$$F_C = \frac{mv_0^2}{r} = \frac{(7.3\text{ kg})(29.3\text{ m/s})^2}{1.8\text{ m}} = \boxed{3500\text{ N}}$$

---

21. ***REASONING AND SOLUTION*** Equation 5.4 gives the relationship between the speed v the angle of banking, and the radius of curvature. Solving for v, we obtain

$$v = \sqrt{r\,g\,\tan\theta} = \sqrt{(120\text{ m})(9.80\text{ m/s}^2)\tan 18°} = \boxed{2.0\times 10^1\text{ m/s}}$$

---

25. ***REASONING*** Refer to Figure 5.10 in the text. The horizontal component of the lift **L** is the centripetal force that holds the plane in the circle. Thus,

$$L\sin\theta = \frac{mv^2}{r} \qquad (1)$$

The vertical component of the lift supports the weight of the plane; therefore,

$$L\cos\theta = mg \qquad (2)$$

Dividing the first equation by the second gives

$$\tan\theta = \frac{v^2}{rg} \qquad (3)$$

Equation (3) can be used to determine the angle θ of banking. Once θ is known, then the magnitude of L can be found from either equation (1) or equation (2).

SOLUTION Solving equation (3) for θ gives

$$\theta = \tan^{-1}\left[\frac{(123\text{ m/s})^2}{(3810\text{ m})(9.80\text{ m/s}^2)}\right] = 22.1°$$

The lifting force is, from equation (2),

$$L = \frac{mg}{\cos\theta} = \frac{(2.00\times10^5\text{ kg})(9.80\text{ m/s}^2)}{\cos 22.1°} = \boxed{2.12\times10^6\text{ N}}$$

---

27. ***REASONING*** Equation 5.5 gives the orbital speed for a satellite in a circular orbit around the earth. It can be modified to determine the orbital speed around any planet **P** by replacing the mass of the earth M_E by the mass of the planet M_P: $v = \sqrt{GM_P/r}$.

SOLUTION The ratio of the orbital speeds is, therefore,

$$\frac{v_2}{v_1} = \frac{\sqrt{GM_P/r_2}}{\sqrt{GM_P/r_1}} = \sqrt{\frac{r_1}{r_2}}$$

Solving for v_2 gives

$$v_2 = v_1\sqrt{\frac{r_1}{r_2}} = (1.70\times10^4\text{ m/s})\sqrt{\frac{5.25\times10^6\text{ m}}{8.60\times10^6\text{ m}}} = \boxed{1.33\times10^4\text{ m/s}}$$

---

33. ***REASONING*** The true weight of the satellite when it is at rest on the planet's surface can be found from Equation 4.4: $W = (GM_P m)/r^2$ where M_P and m are the masses of the planet and the satellite, respectively, and r is the radius of the planet. However, before we can use Equation 4.4, we must determine the mass M_P of the planet.

The mass of the planet can be found by replacing M_E by M_P in Equation 5.6 and solving for M_P. When using Equation 5.6, we note that r corresponds to the radius of the circular orbit *relative to the center of the planet.*

SOLUTION The period of the satellite is $T = 2.00\text{ h} = 7.20\times10^3\text{ s}$. From Equation 5.6,

$$M_P = \frac{4\pi^2 r^3}{GT^2} = \frac{4\pi^2\left[(4.15\times10^6\text{ m}) + (4.1\times10^5\text{ m})\right]^3}{(6.67\times10^{-11}\text{ N}\cdot\text{m}^2/\text{kg}^2)(7.20\times10^3\text{ s})^2} = 1.08\times10^{24}\text{ kg}$$

Using Equation 4.4, we have

$$W = \frac{GM_P m}{r^2} = \frac{(6.67\times10^{-11}\ \text{N}\cdot\text{m}^2/\text{kg}^2)(1.08\times10^{24}\ \text{kg})(5850\ \text{kg})}{(4.15\times10^6\ \text{m})^2} = \boxed{2.45\times10^4\ \text{N}}$$

---

37. ***REASONING AND SOLUTION*** Since the tension serves the same purpose as the normal force at point 1 in Figure 5.21, we have, using the equation for the situation at point 1 with F_{N1} replaced by T,

$$\frac{mv^2}{r} = T - mg$$

Solving for T gives

$$T = \frac{mv^2}{r} + mg = m\left(\frac{v^2}{r} + g\right) = (2100\ \text{kg})\left[\frac{(7.6\ \text{m/s})^2}{15\ \text{m}} + (9.80\ \text{m/s}^2)\right] = \boxed{2.9\times10^4\ \text{N}}$$

---

39. ***REASONING*** The centripetal force is the name given to the net force pointing toward the center of the circular path. At point 3 at the top the net force pointing toward the center of the circle consists of the normal force and the weight, both pointing toward the center. At point 1 at the bottom the net force consists of the normal force pointing upward toward the center and the weight pointing downward or away from the center. In either case the centripetal force is given by Equation 5.3 as $F_c = mv^2/r$.

SOLUTION At point 3 we have

$$F_c = F_N + mg = \frac{mv_3^2}{r}$$

At point 1 we have

$$F_c = F_N - mg = \frac{mv_1^2}{r}$$

Subtracting the second equation from the first gives

$$2mg = \frac{mv_3^2}{r} - \frac{mv_1^2}{r}$$

Rearranging gives

$$v_3^2 = 2gr + v_1^2$$

Thus, we find that

$$v_3 = \sqrt{2(9.80\ \text{m/s}^2)(3.0\ \text{m}) + (15\ \text{m/s})^2} = \boxed{17\ \text{m/s}}$$

---

43. ***REASONING AND SOLUTION*** The magnitude of the centripetal force on the ball is given by Equation 5.3: $F_C = mv^2/r$. Solving for v, we have

$$v = \sqrt{\frac{F_C r}{m}} = \sqrt{\frac{(0.028\text{ N})(0.25\text{ m})}{0.015\text{ kg}}} = \boxed{0.68\text{ m/s}}$$

---

47. ***REASONING AND SOLUTION*** Since the magnitude of the centripetal acceleration is given by Equation 5.2, $a_C = v^2/r$, we can solve for r and find that

$$r = \frac{v^2}{a_C} = \frac{(98.8\text{ m/s})^2}{3.00(9.80\text{ m/s}^2)} = \boxed{332\text{ m}}$$

---

49. ***REASONING*** As the motorcycle passes over the top of the hill, it will experience a centripetal force, the magnitude of which is given by Equation 5.3: $F_C = mv^2/r$. The centripetal force is provided by the net force on the cycle + driver system. At that instant, the net force on the system is composed of the normal force, which points upward, and the weight, which points downward. Taking the direction toward the center of the circle (downward) as the positive direction, we have $F_C = mg - F_N$. This expression can be solved for F_N, the normal force.

SOLUTION

a. The magnitude of the centripetal force is

$$F_C = \frac{mv^2}{r} = \frac{(342\text{ kg})(25.0\text{ m/s})^2}{126\text{ m}} = \boxed{1.70\times10^3\text{ N}}$$

b. The magnitude of the normal force is

$$F_N = mg - F_C = (342\text{ kg})(9.80\text{ m/s}^2) - 1.70\times10^3\text{ N} = \boxed{1.66\times10^3\text{ N}}$$

---

55. ***REASONING*** If the effects of gravity are not ignored in Example 5, the plane will make an angle θ with the vertical as shown in figure **A** below. The figure **B** shows the forces that act on the plane, and figure **C** shows the horizontal and vertical components of these forces.

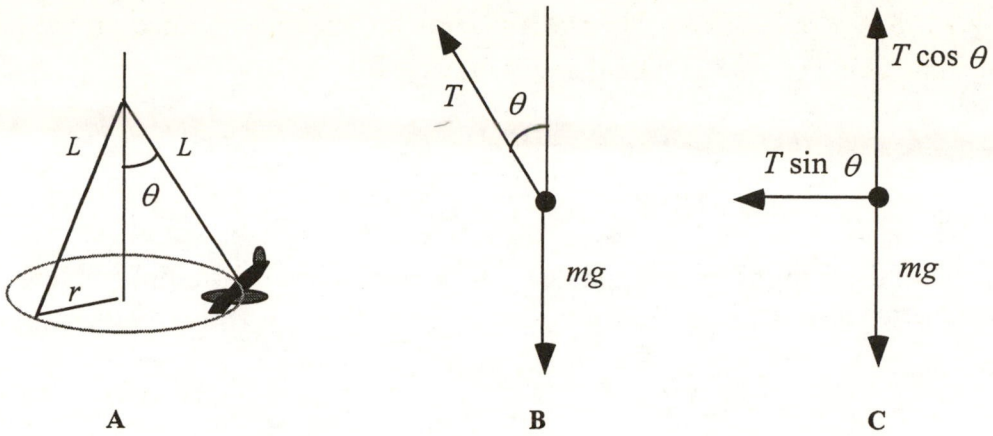

From figure **C** we see that the resultant force in the horizontal direction is the horizontal component of the tension in the guideline and provides the centripetal force. Therefore,

$$T \sin \theta = \frac{mv^2}{r}$$

From figure **A**, the radius r is related to the length L of the guideline by $r = L \sin\theta$; therefore,

$$T \sin \theta = \frac{mv^2}{L \sin \theta} \tag{1}$$

The resultant force in the vertical direction is zero: $T\cos\theta - mg = 0$, so that

$$T\cos\theta = mg \tag{2}$$

From equation (2) we have

$$T = \frac{mg}{\cos \theta} \tag{3}$$

Equation (3) contains two unknown, T and θ. First we will solve equations (1) and (3) simultaneously to determine the value(s) of the angle θ. Once θ is known, we can calculate the tension using equation (3).

SOLUTION Substituting equation (3) into equation (1):

$$\left(\frac{mg}{\cos \theta}\right) \sin \theta = \frac{mv^2}{L \sin \theta}$$

Thus,

$$\frac{\sin^2 \theta}{\cos \theta} = \frac{v^2}{gL} \tag{4}$$

Using the fact that $\cos^2 \theta + \sin^2 \theta = 1$, equation (4) can be written

$$\frac{1 - \cos^2 \theta}{\cos \theta} = \frac{v^2}{gL}$$

or

$$\frac{1}{\cos \theta} - \cos \theta = \frac{v^2}{gL}$$

This can be put in the form of an equation that is quadratic in cos θ. Multiplying both sides by cos θ and rearranging yields:

$$\cos^2 \theta + \frac{v^2}{gL} \cos \theta - 1 = 0 \tag{5}$$

Equation (5) is of the form

$$ax^2 + bx + c = 0 \tag{6}$$

with $x = \cos \theta$, $a = 1$, $b = v^2/(gL)$, and $c = -1$. The solution to equation (6) is found from the quadratic formula:

$$x = \frac{-b \pm \sqrt{b^2 - 4ac}}{2a}$$

When $v = 19.0$ m/s, $b = 2.17$. The positive root from the quadratic formula gives $x = \cos \theta = 0.391$. Substitution into equation (3) yields

$$T = \frac{mg}{\cos \theta} = \frac{(0.900 \text{ kg})(9.80 \text{ m/s}^2)}{0.391} = \boxed{23 \text{ N}}$$

When $v = 38.0$ m/s, $b = 8.67$. The positive root from the quadratic formula gives $x = \cos \theta = 0.114$. Substitution into equation (3) yields

$$T = \frac{mg}{\cos \theta} = \frac{(0.900 \text{ kg})(9.80 \text{ m/s}^2)}{0.114} = \boxed{77 \text{ N}}$$

CHAPTER 6 | *WORK AND ENERGY*

CONCEPTUAL QUESTIONS

---

4. ***REASONING AND SOLUTION*** The sailboat moves at constant velocity, and, therefore, has zero acceleration. From Newton's second law, we know that the net external force on the sailboat must be zero.

   a. There is no work done on the sailboat by a zero net external force.

   b. Work is done by the individual forces that act on the boat; namely the wind that propels the boat forward and the water that resists the motion of the boat. Since the wind propels the boat forward, it does positive work on the boat. Since the force of the water is a resistive force, it acts opposite to the displacement of the boat, and, therefore, it does negative work. Since the total work done on the boat is zero, each force must do an equal amount of work with one quantity being positive, and the other being negative.

   Note: The answer to part (a) could have been deduced from the work-energy theorem as well. Since the velocity of the boat is constant, the kinetic energy of the boat does not change and the total work done on the boat is zero.

---

9. ***REASONING AND SOLUTION*** A shopping bag hangs straight down from the hand of a person who walks across a horizontal floor at constant velocity.

   a. The force exerted on the shopping bag by the hand is perpendicular to the direction of motion; hence, the work done by the hand is zero.

   b. If the person then rides up an escalator at constant velocity, a component of the force exerted by the hand will point in the direction of the displacement and do a positive amount of work.

---

13. ***REASONING AND SOLUTION*** Yes, the speed of the Steel Dragon would be 3.0 m/s at the top of the second hill. As the roller coaster descends the first hill, its total mechanical energy remains constant at all times. Its gravitational potential energy at the top of the hill is converted entirely into kinetic energy at the bottom. When the roller coaster ascends the second hill, its total mechanical energy also remains constant, and is equal to what it had during the descent of the first hill. As the roller coaster moves up the hill, kinetic energy is converted into gravitational potential energy. By the time it reaches the top of the second (identical) hill, the kinetic energy gained in descending the hill has been converted back into gravitational potential energy. Therefore, its kinetic energy at the top of the second hill is the same as it was at the top of the first hill, and its speed is 3.0 m/s at the top.

CHAPTER 6 | *WORK AND ENERGY*

PROBLEMS

---

1. ***REASONING AND SOLUTION*** The work done by the retarding force is given by Equation 6.1: $W = (F\cos\theta)s$. Since the force is a retarding force, it must point opposite to the direction of the displacement, so that $\theta = 180°$. Thus, we have

$$W = (F\cos\theta)s = (3.0\times10^3\ \text{N})(\cos 180°)(850\ \text{m}) = \boxed{-2.6\times10^6\ \text{J}}$$

The work done by this force is $\boxed{\text{negative}}$, because the retarding force is directed opposite to the direction of the displacement of the truck.

---

5. ***REASONING AND SOLUTION*** We will assume that the tug-of-war rope remains parallel to the ground, so that the force that moves team B is in the same direction as the displacement. According to Equation 6.1, the work done by team A is

$$W = (F\cos\theta)s = (1100\ \text{N})(\cos 0°)(2.0\ \text{m}) = \boxed{2.2\times10^3\ \text{J}}$$

---

9. ***REASONING AND SOLUTION*** According to Equation 6.1, the work done by the husband and wife are, respectively,

$$[\text{Husband}] \qquad W_\text{H} = (F_\text{H}\cos\theta_\text{H})s$$

$$[\text{Wife}] \qquad W_\text{W} = (F_\text{W}\cos\theta_\text{W})s$$

Since both the husband and the wife do the same amount of work,

$$(F_\text{H}\cos\theta_\text{H})s = (F_\text{W}\cos\theta_\text{W})s$$

Since the displacement has the same magnitude s in both cases, the magnitude of the force exerted by the wife is

$$F_\text{W} = F_\text{H}\frac{\cos\theta_\text{H}}{\cos\theta_\text{W}} = (67\ \text{N})\frac{\cos 58°}{\cos 38°} = \boxed{45\ \text{N}}$$

---

13. ***REASONING AND SOLUTION*** The work required to bring each car up to speed is, from the work-energy theorem, $W = \text{KE}_\text{f} - \text{KE}_0 = \frac{1}{2}mv_\text{f}^2 - \frac{1}{2}mv_0^2$. Therefore,

$$W_\text{B} = \tfrac{1}{2}m\left(v_\text{f}^2 - v_0^2\right) = \tfrac{1}{2}(1.20\times10^3\ \text{kg})\left[(40.0\ \text{m/s})^2 - (0\ \text{m/s})^2\right] = 9.60\times10^5\ \text{J}$$

$$W_\text{B} = \tfrac{1}{2}m\left(v_\text{f}^2 - v_0^2\right) = \tfrac{1}{2}(2.00\times10^3\ \text{kg})\left[(40.0\ \text{m/s})^2 - (0\ \text{m/s})^2\right] = 1.60\times10^6\ \text{J}$$

The *additional* work required to bring car B up to speed is, therefore,

$$W_\text{B} - W_\text{A} = (1.6\times10^6\ \text{J}) - (9.60\times10^5\ \text{J}) = \boxed{6.4\times10^5\ \text{J}}$$

---

17. ***REASONING*** The work done to launch either object can be found from Equation 6.3, the work-energy theorem, $W = \text{KE}_\text{f} - \text{KE}_0 = \frac{1}{2}mv_\text{f}^2 - \frac{1}{2}mv_0^2$.

SOLUTION

a. The work required to launch the hammer is

$$W = \tfrac{1}{2}mv_\text{f}^2 - \tfrac{1}{2}mv_0^2 = \tfrac{1}{2}m\left(v_\text{f}^2 - v_0^2\right) = \tfrac{1}{2}(7.3\ \text{kg})\left[(29\ \text{m/s})^2 - (0\ \text{m/s})^2\right] = \boxed{3.1\times10^3\ \text{J}}$$

b. Similarly, the work required to launch the bullet is

$$W = \tfrac{1}{2}m\left(v_\text{f}^2 - v_0^2\right) = \tfrac{1}{2}(0.0026\ \text{kg})\left[(410\ \text{m/s})^2 - (0\ \text{m/s})^2\right] = \boxed{2.2\times10^2\ \text{J}}$$

---

19. ***REASONING*** According to the work-energy theorem, the kinetic energy of the sled increases in each case because work is done on the sled. The work-energy theorem is given by Equation 6.3: $W = \text{KE}_\text{f} - \text{KE}_0 = \frac{1}{2}mv_\text{f}^2 - \frac{1}{2}mv_0^2$. The work done on the sled is given by Equation 6.1: $W = (F\cos\theta)s$. The work done in each case can, therefore, be expressed as

$$W_1 = (F\cos 0°)s = \frac{1}{2}mv_\text{f}^2 - \frac{1}{2}mv_0^2 = \Delta\text{KE}_1$$

and

$$W_2 = (F\cos 62°)s = \frac{1}{2}mv_\text{f}^2 - \frac{1}{2}mv_0^2 = \Delta\text{KE}_2$$

The fractional increase in the kinetic energy of the sled when $\theta = 0°$ is

$$\frac{\Delta\text{KE}_1}{\text{KE}_0} = \frac{(F\cos 0°)s}{\text{KE}_0} = 0.38$$

Therefore,

$$Fs = (0.38)\ \mathrm{KE}_0 \tag{1}$$

The fractional increase in the kinetic energy of the sled when $\theta = 62°$ is

$$\frac{\Delta \mathrm{KE}_2}{\mathrm{KE}_0} = \frac{(F\cos 62°)s}{\mathrm{KE}_0} = \frac{Fs}{\mathrm{KE}_0}(\cos 62°) \tag{2}$$

Equation (1) can be used to substitute in Equation (2) for Fs.

SOLUTION Combining Equations (1) and (2), we have

$$\frac{\Delta \mathrm{KE}_2}{\mathrm{KE}_0} = \frac{Fs}{\mathrm{KE}_0}(\cos 62°) = \frac{(0.38)\ \mathrm{KE}_0}{\mathrm{KE}_0}(\cos 62°) = (0.38)(\cos 62°) = 0.18$$

Thus, the sled's kinetic energy would increase by $\boxed{18\ \%}$.

---

21. ***REASONING*** When the satellite goes from the first to the second orbit, its kinetic energy changes. The net work that the external force must do to change the orbit can be found from the work-energy theorem: $W = \mathrm{KE_f} - \mathrm{KE}_0 = \frac{1}{2}mv_\mathrm{f}^2 - \frac{1}{2}mv_0^2$. The speeds v_f and v_0 can be obtained from Equation 5.5 for the speed of a satellite in a circular orbit of radius r. Given the speeds, the work energy theorem can be used to obtain the work.

SOLUTION According to Equation 5.5, $v = \sqrt{GM_\mathrm{E}/\mathrm{r}}$. Substituting into the work-energy theorem, we have

$$W = \tfrac{1}{2}mv_\mathrm{f}^2 - \tfrac{1}{2}mv_0^2 = \tfrac{1}{2}m\left(v_\mathrm{f}^2 - v_0^2\right) = \tfrac{1}{2}m\left[\left(\sqrt{\frac{GM_\mathrm{E}}{r_\mathrm{f}}}\right)^2 - \left(\sqrt{\frac{GM_\mathrm{E}}{r_0}}\right)^2\right] = \frac{GM_\mathrm{E}m}{2}\left(\frac{1}{r_\mathrm{f}} - \frac{1}{r_0}\right)$$

Therefore,

$$\mathrm{W} = \frac{(6.67\times10^{-11}\ \mathrm{N\cdot m^2/kg^2})(5.98\times10^{24}\ \mathrm{kg})(6200\ \mathrm{kg})}{2}$$

$$\times\left(\frac{1}{7.0\times10^6\ \mathrm{m}} - \frac{1}{3.3\times10^7\ \mathrm{m}}\right) = \boxed{1.4\times10^{11}\ \mathrm{J}}$$

---

27. ***REASONING*** During each portion of the trip, the work done by the resistive force is given by Equation 6.1, $W = (F\cos\theta)s$. Since the resistive force always points opposite to the displacement of the bicyclist, $\theta = 180°$; hence, on each part of the trip,

$W = (F\cos 180°)s = -Fs$. The work done by the resistive force during the round trip is the algebraic sum of the work done during each portion of the trip.

SOLUTION

a. The work done during the round trip is, therefore,

$$W_{\text{total}} = W_1 + W_2 = -F_1 s_1 - F_2 s_2$$

$$= -(3.0\text{ N})(5.0\times 10^3\text{ m}) - (3.0\text{ N})(5.0\times 10^3\text{ m}) = \boxed{-3.0\times 10^4\text{ J}}$$

b. Since the work done by the resistive force over the closed path is *not* zero, we can conclude that $\boxed{\text{the resistive force is } \textit{not} \text{ a conservative force}}$.

---

31. ***REASONING*** The only nonconservative force that acts on rocket man is the force generated by the propulsion unit. Thus, Equation 6.7b can be used to determine the net work done by this force.

SOLUTION From Equation 6.7b, we have

$$W_{\text{nc}} = \Delta\text{KE} + \Delta\text{PE} = \left(\tfrac{1}{2}mv_{\text{f}}^2 - \tfrac{1}{2}mv_0^2\right) + \left(mgh_{\text{f}} - mgh_0\right)$$

Since rocket man starts from rest, $v_0 = 0$ m/s. If we take $h_0 = 0$ m on the ground, we have

$$W_{\text{nc}} = \tfrac{1}{2}mv_{\text{f}}^2 + mgh_{\text{f}} = m\left(\tfrac{1}{2}v_{\text{f}}^2 + gh_{\text{f}}\right)$$

$$= (136\text{ kg})\left[\tfrac{1}{2}(5.0\text{ m/s})^2 + (9.80\text{ m/s}^2)(16\text{ m})\right] = \boxed{2.3\times 10^4\text{ J}}$$

---

35. ***REASONING*** No forces other than gravity act on the rock since air resistance is being ignored. Thus, the net work done by nonconservative forces is zero, $W_{\text{nc}} = 0$ J. Consequently, the principle of conservation of mechanical energy holds, so the total mechanical energy remains constant as the rock falls.

If we take $h = 0$ m at ground level, the gravitational potential energy at any height h is, according to Equation 6.5, $\text{PE} = mgh$. The kinetic energy of the rock is given by Equation 6.2: $\text{KE} = \frac{1}{2}mv^2$. In order to use Equation 6.2, we must have a value for v^2 at each desired height h. The quantity v^2 can be found from Equation 2.9 with $v_0 = 0$ m/s, since the rock is released from rest. At a height h, the rock has fallen through a distance $(20.0\text{ m}) - h$, and according to Equation 2.9, $v^2 = 2ay = 2a[(20.0\text{ m}) - h]$. Therefore, the kinetic energy at any height h is given by $\text{KE} = ma[(20.0\text{ m}) - h]$. The total energy of the rock at any height h is the sum of the kinetic energy and potential energy at the particular height.

SOLUTION The calculations are performed below for $h = 10.0$ m. The table that follows also shows the results for $h = 20.0$ m and $h = 0$ m.

$$\text{PE} = mgh = (2.00\text{ kg})(9.80\text{ m/s}^2)(10.0\text{ m}) = \boxed{196\text{ J}}$$

$$\text{KE} = ma[(20.0\text{ m}) - h] = (2.00\text{ kg})(9.80\text{ m/s}^2)(20.0\text{ m} - 10.0\text{ m}) = \boxed{196\text{ J}}$$

$$E = \text{KE} + \text{PE} = 196\text{ J} + 196\text{ J} = \boxed{392\text{ J}}$$

h (m)	KE (J)	PE (J)	E (J)
20.0	0	392	392
10.0	196	196	392
0	392	0	392

Note that in each case, the value of E is the same, because mechanical energy is conserved.

---

41. ***REASONING*** Gravity is the only force acting on the vaulters, since friction and air resistance are being ignored. Therefore, the net work done by the nonconservative forces is zero, and the principle of conservation of mechanical energy holds.

SOLUTION Let E_{2f} represent the total mechanical energy of the second vaulter at the ground, and E_{20} represent the total mechanical energy of the second vaulter at the bar. Then, the principle of mechanical energy is written for the second vaulter as

$$\underbrace{\frac{1}{2}mv_{2f}^2 + mgh_{2f}}_{E_{2f}} = \underbrace{\frac{1}{2}mv_{20}^2 + mgh_{20}}_{E_{20}}$$

Since the mass m of the vaulter appears in every term of the equation, m can be eliminated algebraically. The quantity $h_{20} = h$, where h is the height of the bar. Furthermore, when the vaulter is at ground level, $h_{2f} = 0$ m. Solving for v_{20} we have

$$v_{20} = \sqrt{v_{2f}^2 - 2gh} \qquad (1)$$

In order to use Equation (1), we must first determine the height h of the bar. The height h can be determined by applying the principle of conservation of mechanical energy to the first vaulter on the ground and at the bar. Using notation similar to that above, we have

$$\underbrace{\frac{1}{2}mv_{1f}^2 + mgh_{1f}}_{E_{1f}} = \underbrace{\frac{1}{2}mv_{10}^2 + mgh_{10}}_{E_{10}}$$

where E_{10} corresponds to the total mechanical energy of the first vaulter at the bar. The height of the bar is, therefore,

$$h = h_{10} = \frac{v_{1f}^2 - v_{10}^2}{2g} = \frac{(8.90\text{ m/s})^2 - (1.00\text{ m/s})^2}{2(9.80\text{ m/s}^2)} = 3.99\text{ m}$$

The speed at which the second vaulter clears the bar is, from Equation (1),

$$v_{20} = \sqrt{v_{2f}^2 - 2gh} = \sqrt{(9.00\text{ m/s})^2 - 2(9.80\text{ m/s}^2)(3.99\text{ m})} = \boxed{1.7\text{ m/s}}$$

---

47. ***REASONING AND SOLUTION*** The work-energy theorem can be used to determine the change in kinetic energy of the car according to Equation 6.8:

$$W_{nc} = \underbrace{\left(\tfrac{1}{2}mv_f^2 + mgh_f\right)}_{E_f} - \underbrace{\left(\tfrac{1}{2}mv_0^2 + mgh_0\right)}_{E_0}$$

The nonconservative forces are the forces of friction and the force due to the chain mechanism. Thus, we can rewrite Equation 6.8 as

$$W_{\text{friction}} + W_{\text{chain}} = (\text{KE}_f + mgh_f) - (\text{KE}_0 + mgh_0)$$

We will measure the heights from ground level. Solving for the change in kinetic energy of the car, we have

$$\text{KE}_f - \text{KE}_0 = W_{\text{friction}} + W_{\text{chain}} - mgh_f + mgh_0$$

$$= -2.00\times10^4\text{ J} + 3.00\times10^4\text{ J} - (375\text{ kg})(9.80\text{ m/s}^2)(20.0\text{ m})$$
$$+ (375\text{ kg})(9.80\text{ m/s}^2)(5.00\text{ m}) = \boxed{-4.51\times10^4\text{ J}}$$

---

49. ***REASONING*** The work-energy theorem can be used to determine the net work done on the car by the nonconservative forces of friction and air resistance. The work-energy theorem is, according to Equation 6.8,

$$W_{nc} = \left(\tfrac{1}{2}mv_f^2 + mgh_f\right) - \left(\tfrac{1}{2}mv_0^2 + mgh_0\right)$$

The nonconservative forces are the force of friction, the force of air resistance, and the force provided by the engine. Thus, we can rewrite Equation 6.8 as

$$W_{\text{friction}} + W_{\text{air}} + W_{\text{engine}} = \left(\tfrac{1}{2}mv_{\text{f}}^2 + mgh_{\text{f}}\right) - \left(\tfrac{1}{2}mv_0^2 + mgh_0\right)$$

This expression can be solved for $W_{\text{friction}} + W_{\text{air}}$.

SOLUTION We will measure the heights from sea level, where $h_0 = 0$ m. Since the car starts from rest, $v_0 = 0$ m/s. Thus, we have

$$W_{\text{friction}} + W_{\text{air}} = m\left(\tfrac{1}{2}v_{\text{f}}^2 + gh_{\text{f}}\right) - W_{\text{engine}}$$

$$= (1.50\times10^3\text{ kg})\left[\tfrac{1}{2}(27.0\text{ m/s})^2 + (9.80\text{ m/s}^2)(2.00\times10^2\text{ m})\right] - (4.70\times10^6\text{ J})$$

$$= \boxed{-1.21\times10^6\text{ J}}$$

---

53. ***REASONING AND SOLUTION*** According to the work-energy theorem as given in Equation 6.8, we have

$$W_{\text{nc}} = \left(\tfrac{1}{2}mv_{\text{f}}^2 + mgh_{\text{f}}\right) - \left(\tfrac{1}{2}mv_0^2 + mgh_0\right)$$

The metal piece starts at rest and is at rest just as it barely strikes the bell, so that $v_{\text{f}} = v_0 = 0$ m/s. In addition, $h_{\text{f}} = h$ and $h_0 = 0$ m, while $W_{\text{nc}} = 0.25\left(\tfrac{1}{2}Mv^2\right)$, where M and v are the mass and speed of the hammer. Thus, the work-energy theorem becomes

$$0.25\left(\tfrac{1}{2}Mv^2\right) = mgh$$

Solving for the speed of the hammer, we find

$$v = \sqrt{\frac{2mgh}{0.25M}} = \sqrt{\frac{2(0.400\text{ kg})(9.80\text{ m/s}^2)(5.00\text{ m})}{0.25\ (9.00\text{ kg})}} = \boxed{4.17\text{ m/s}}$$

---

55. ***REASONING AND SOLUTION*** One kilowatt·hour is the amount of work or energy generated when one kilowatt of power is supplied for a time of one hour. From Equation 6.10a, we know that $W = \bar{P}t$. Using the fact that $1\text{ kW} = 1.0\times10^3$ J/s and that $1\text{h} = 3600$ s, we have

$$1.0\text{ kWh} = (1.0\times10^3\text{ J/s})(1\text{ h}) = (1.0\times10^3\text{ J/s})(3600\text{ s}) = \boxed{3.6\times10^6\text{ J}}$$

---

61. ***REASONING AND SOLUTION*** In the drawings below, the positive direction is to the right. When the boat is not pulling a skier, the engine must provide a force F_1 to overcome the resistive force of the water F_R. Since the boat travels with constant speed, these forces must be equal in magnitude and opposite in direction.

$$-F_R + F_1 = ma = 0$$

or

$$F_R = F_1 \qquad (1)$$

When the boat is pulling a skier, the engine must provide a force F_2 to balance the resistive force of the water F_R and the tension in the tow rope T.

$$-F_R + F_2 - T = ma = 0$$

or

$$F_2 = F_R + T \qquad (2)$$

The average power is given by $\overline{P} = F\overline{v}$, according to Equation 6.11 in the text. Since the boat moves with the same speed in both cases, $v_1 = v_2$, and we have

$$\frac{\overline{P}_1}{F_1} = \frac{\overline{P}_2}{F_2} \qquad \text{or} \qquad F_2 = F_1\frac{\overline{P}_2}{\overline{P}_1}$$

Using Equations (1) and (2), this becomes

$$F_1 + T = F_1\frac{\overline{P}_2}{\overline{P}_1}.$$

Solving for T gives

$$T = F_1\left(\frac{\overline{P}_2}{\overline{P}_1} - 1\right)$$

The force F_1 can be determined from $\overline{P}_1 = F_1 v_1$, thereby giving

$$T = \frac{\overline{P}_1}{v_1}\left(\frac{\overline{P}_2}{\overline{P}_1} - 1\right) = \frac{7.50\times10^4\ \text{W}}{12\ \text{m/s}}\left(\frac{8.30\times10^4\ \text{W}}{7.50\times10^4\ \text{W}} - 1\right) = \boxed{6.7\times10^2\ \text{N}}$$

---

63. ***REASONING*** The area under the force-versus-displacement graph over any displacement interval is equal to the work done over that displacement interval.

SOLUTION

a. Since the area under the force-versus-displacement graph from $s = 0$ to 0.50 m is greater for bow 1, it follows that $\boxed{\text{bow 1 requires more work to draw the bow fully}}$.

b. The work required to draw bow 1 is equal to the area of the triangular region under the force-versus-displacement curve for bow 1. Since the area of a triangle is equal one-half times the base of the triangle times the height of the triangle, we have

$$W_1 = \frac{1}{2}(0.50\text{m})(350\text{ N}) = 88\text{ J}$$

For bow 2, we note that each small square under the force-versus-displacement graph has an area of

$$(0.050\text{ m})(40.0\text{ N}) = 2.0\text{ J}$$

We estimate that there are approximately 31.3 squares under the force-versus-displacement graph for bow 2; therefore, the total work done is

$$(31.3\text{ squares})\left(\frac{2.0\text{ J}}{\text{square}}\right) = 63\text{ J}$$

Therefore, the additional work required to stretch bow 1 as compared to bow 2 is

$$88\text{ J} - 63\text{ J} = \boxed{25\text{ J}}$$

---

69. ***REASONING AND SOLUTION*** Since we are ignoring friction and air resistance, the net work done by the nonconservative forces is zero, and the principle of conservation of mechanical energy holds. Let E_0 be the total mechanical energy of the vaulter at take-off and E_f be the total mechanical energy as he clears the bar. The principle of conservation of mechanical energy gives

$$\underbrace{\frac{1}{2}mv_f^2 + mgh_f}_{E_f} = \underbrace{\frac{1}{2}mv_0^2 + mgh_0}_{E_0}$$

For maximum height h_f, the vaulter just clears the bar with zero speed, $v_f = 0$ m/s. If we let $h_0 = 0$ m at ground level, we have

$$mgh_f = \frac{1}{2}mv_0^2$$

Solving for h_f, we find

$$h_f = \frac{v_0^2}{2g} = \frac{(9.00\text{ m/s})^2}{2(9.80\text{ m/s}^2)} = \boxed{4.13\text{ m}}$$

---

71. ***REASONING AND SOLUTION*** Solving Equation 6.1 for the angle θ, we obtain

$$\theta = \cos^{-1}\left(\frac{W}{Fs}\right) = \cos^{-1}\left[\frac{1.10\times10^3\ \text{J}}{(30.0\ \text{N})(50.0\ \text{m})}\right] = \boxed{42.8°}$$

---

77. ***REASONING*** Friction and air resistance are being ignored. The normal force from the slide is perpendicular to the motion, so it does no work. Thus, no net work is done by nonconservative forces, and the principle of conservation of mechanical energy applies.

SOLUTION Applying the principle of conservation of mechanical energy to the swimmer at the top and the bottom of the slide, we have

$$\underbrace{\tfrac{1}{2}mv_f^2 + mgh_f}_{E_f} = \underbrace{\tfrac{1}{2}mv_0^2 + mgh_0}_{E_0}$$

If we let h be the height of the bottom of the slide above the water, $h_f = h$, and $h_0 = H$. Since the swimmer starts from rest, $v_0 = 0$ m/s, and the above expression becomes

$$\tfrac{1}{2}v_f^2 + gh = gH$$

Solving for H, we obtain

$$H = h + \frac{v_f^2}{2g}$$

Before we can calculate H, we must find v_f and h. Since the velocity in the horizontal direction is constant,

$$v_f = \frac{\Delta x}{\Delta t} = \frac{5.00\ \text{m}}{0.500\ \text{s}} = 10.0\ \text{m/s}$$

The vertical displacement of the swimmer after leaving the slide is, from Equation 3.5b (with down being negative),

$$y = \tfrac{1}{2}a_y t^2 = \tfrac{1}{2}(-9.80\ \text{m/s}^2)(0.500\ \text{s})^2 = -1.23\ \text{m}$$

Therefore, h = 1.23 m. Using these values of v_f and h in the above expression for H, we find

$$H = h + \frac{v_f^2}{2g} = 1.23\ \text{m} + \frac{(10.0\ \text{m/s})^2}{2(9.80\ \text{m/s}^2)} = \boxed{6.33\ \text{m}}$$

---

79. ***REASONING*** After the wheels lock, the only nonconservative force acting on the truck is friction. The work done by this conservative force can be determined from the work-energy theorem. According to Equation 6.8,

$$W_{\text{nc}} = E_{\text{f}} - E_0 = \left(\tfrac{1}{2}mv_{\text{f}}^2 + mgh_{\text{f}}\right) - \left(\tfrac{1}{2}mv_0^2 + mgh_0\right) \tag{1}$$

where W_{nc} is the work done by friction. According to Equation 6.1, $W = (F\cos\theta)s$; since the force of kinetic friction points opposite to the displacement, $\theta = 180°$. According to Equation 4.8, the kinetic frictional force has a magnitude of $f_{\text{k}} = \mu_{\text{k}}F_{\text{N}}$, where μ_{k} is the coefficient of kinetic friction and F_{N} is the magnitude of the normal force. Thus,

$$W_{\text{nc}} = (f_{\text{k}}\cos\theta)s = \mu_{\text{k}}F_{\text{N}}(\cos 180°)s = -\mu_{\text{k}}F_{\text{N}}s \tag{2}$$

Since the truck is sliding down an incline, we refer to the free-body diagram in order to determine the magnitude of the normal force F_{N}. The free-body diagram at the right shows the three forces that act on the truck. They are the normal force $\mathbf{F}_{\text{N}}$, the force of kinetic friction $\mathbf{f}_{\text{k}}$, and the weight $m\mathbf{g}$. The weight has been resolved into its components, and these vectors are shown as dashed arrows. From the free body diagram and the fact that the truck does not accelerate perpendicular to the incline, we can see that the magnitude of the normal force is given by

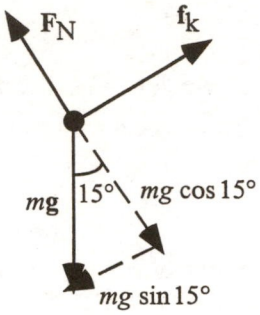

$$F_{\text{N}} = mg\cos 15.0°$$

Therefore, Equation (2) becomes

$$W_{\text{nc}} = -\mu_{\text{k}}\left(mg\cos 15.0°\right)s \tag{3}$$

Combining Equations (1), (2), and (3), we have

$$-\mu_{\text{k}}\left(mg\cos 15.0°\right)s = \left(\tfrac{1}{2}mv_{\text{f}}^2 + mgh_{\text{f}}\right) - \left(\tfrac{1}{2}mv_0^2 + mgh_0\right) \tag{4}$$

This expression may be solved for s.

SOLUTION When the car comes to a stop, $v_{\text{f}} = 0$ m/s. If we take $h_{\text{f}} = 0$ m at ground level, then, from the drawing at the right, we see that $h_0 = s\sin 15.0°$. Equation (4) then gives

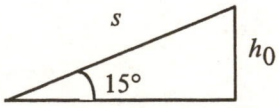

$$s = \frac{v_0^2}{2g\left(\mu_{\text{k}}\cos 15.0° - \sin 15.0°\right)} = \frac{(11.1\text{ m/s})^2}{2\left(9.80\text{ m/s}^2\right)\left[(0.750)\cos 15.0° - \sin 15.0°\right]} = \boxed{13.5\text{ m}}$$

CHAPTER 7 | *IMPULSE AND MOMENTUM*

CONCEPTUAL QUESTIONS

---

2. ***REASONING AND SOLUTION*** Since linear momentum is a vector quantity, the total linear momentum of any system is the resultant of the linear momenta of the constituents. The people who are standing around have zero momentum. Those who move randomly carry momentum randomly in all directions. Since there is such a large number of people, there is, on average, just as much linear momentum in any one direction as in any other. On average, the resultant of this random distribution is zero. Therefore, the approximate linear momentum of the Times Square system is zero.

---

13. ***REASONING AND SOLUTION***

a. No. The person overhead jumps straight down, and only the force of gravity acts on him, which is a vertical (not a horizontal) force. Since friction and air resistance are negligible, no horizontal force is applied to the boat. Therefore, no net horizontal force is being applied to the system composed of the boat plus the jumper. According to the impulse-momentum theorem, this means that the horizontal momentum of the boat plus the jumper cannot change. Note that when the jumper lands in the boat, he applies a force to the boat, and the boat applies a force to him. However, these forces are internal forces of the boat/jumper system and cannot change the total momentum of the system.

b. The speed of the boat (with the jumper in it) decreases relative to the speed of the boat before the jumper lands in it. Before the jumper lands in the boat, the total horizontal momentum initially is that due to the empty boat. After the jumper lands, the total horizontal moment finally is that due to the boat with the jumper in it. As explained in part (a), these initial and final values are the same. Since the mass of the boat plus the jumper is greater than the mass of the empty boat, the magnitude of the velocity (that is, the speed) of the boat with the jumper in it must decrease, because momentum is mass times velocity.

---

20. ***REASONING AND SOLUTION*** A sunbather is lying on a floating, stationary raft. She then gets up and walks to one end of the raft. The sunbather and the raft are considered as an isolated system.

a. As the sunbather walks to one end of the raft, she exerts a force on the raft; however, the force is internal to the isolated system. Since there are no external forces acting on the system, the linear momentum of the system cannot change. Since the linear momentum of the system is initially zero, it must remain zero. Therefore, the velocity of the center of mass of the system must be zero.

b. The sunbather has linear momentum as she walks to one end of the raft. Since the linear momentum of the isolated system must remain zero, the raft must acquire a linear momentum that is equal in magnitude and opposite in direction to that of the sunbather. From the definition of linear momentum, $\mathbf{p} = m\mathbf{v}$, we know that the direction of the linear momentum of an object is the same as the direction of the velocity of the object. Thus, the raft acquires a velocity that is opposite to the direction of motion of the sunbather.

CHAPTER 7 | *IMPULSE AND MOMENTUM*

PROBLEMS

---

1. ***REASONING*** According to Equation 7.1, the impulse **J** produced by an average force $\overline{\mathbf{F}}$ is $\mathbf{J} = \overline{\mathbf{F}}\Delta t$, where Δt is the time interval during which the force acts. We will apply this definition for each of the forces and then set the two impulses equal to one another. The fact that one average force has a magnitude that is three times as large as that of the other average force will then be used to obtain the desired time interval.

SOLUTION Applying Equation 7.1, we write the impulse of each average force as follows:

$$\mathbf{J}_1 = \overline{\mathbf{F}}_1 \Delta t_1 \quad \text{and} \quad \mathbf{J}_2 = \overline{\mathbf{F}}_2 \Delta t_2$$

But the impulses $\mathbf{J}_1$ and $\mathbf{J}_2$ are the same, so we have that $\overline{\mathbf{F}}_1 \Delta t_1 = \overline{\mathbf{F}}_2 \Delta t_2$. Writing this result in terms of the magnitudes of the forces gives

$$\overline{F}_1 \Delta t_1 = \overline{F}_2 \Delta t_2 \quad \text{or} \quad \Delta t_2 = \left(\frac{\overline{F}_1}{\overline{F}_2}\right)\Delta t_1$$

The ratio of the force magnitudes is given as $\overline{F}_1 / \overline{F}_2 = 3$, so we find that

$$\Delta t_2 = \left(\frac{\overline{F}_1}{\overline{F}_2}\right)\Delta t_1 = 3(3.2\text{ ms}) = \boxed{9.6\text{ ms}}$$

---

5. ***REASONING*** The impulse that the volleyball player applies to the ball can be found from the impulse-momentum theorem, Equation 7.4. Two forces act on the volleyball while it's being spiked: an average force $\overline{\mathbf{F}}$ exerted by the player, and the weight of the ball. As in Example 1, we will assume that $\overline{\mathbf{F}}$ is much greater than the weight of the ball, so the weight can be neglected. Thus, the net average force $(\Sigma\overline{\mathbf{F}})$ is equal to $\overline{\mathbf{F}}$.

SOLUTION From Equation 7.4, the impulse that the player applies to the volleyball is

$$\underbrace{\overline{\mathbf{F}}\,\Delta t}_{\text{Impulse}} = \underbrace{m\mathbf{v}_f}_{\substack{\text{Final}\\\text{momentum}}} - \underbrace{m\mathbf{v}_0}_{\substack{\text{Initial}\\\text{momentum}}}$$

$$= m(\mathbf{v}_f - \mathbf{v}_0) = (0.35\text{ kg})[(-21\text{ m/s}) - (+4.0\text{ m/s})] = \boxed{-8.7\text{ kg}\cdot\text{m/s}}$$

The minus sign indicates that the direction of the impulse is the same as that of the final velocity of the ball.

---

13. ***REASONING*** The impulse applied to the golf ball by the floor can be found from Equation 7.4, the impulse-momentum theorem: $(\Sigma\overline{\mathbf{F}})\Delta t = m\mathbf{v}_f - m\mathbf{v}_0$. Two forces act on the golf ball, the average force $\overline{\mathbf{F}}$ exerted by the floor, and the weight of the golf ball. Since $\overline{\mathbf{F}}$ is much greater than the weight of the golf ball, the net average force $(\Sigma\overline{\mathbf{F}})$ is equal to $\overline{\mathbf{F}}$.

Only the vertical component of the ball's momentum changes during impact with the floor. In order to use Equation 7.4 directly, we must first find the vertical components of the initial and final velocities. We begin, then, by finding these velocity components.

SOLUTION The figures below show the initial and final velocities of the golf ball.

If we take up as the positive direction, then the vertical components of the initial and final velocities are, respectively, $\mathbf{v}_{0y} = -v_0 \cos 30.0°$ and $\mathbf{v}_{fy} = +v_f \cos 30.0°$. Then, from Equation 7.4 the impulse is

$$\overline{\mathbf{F}}\,\Delta t = m(\mathbf{v}_{fy} - \mathbf{v}_{0y}) = m\left[(+v_f \cos\ 30.0°) - (-v_0 \cos\ 30.0°)\right]$$

Since $v_0 = v_f = 45\ \text{m/s}$, the impulse applied to the golf ball by the floor is

$$\overline{\mathbf{F}}\Delta t = 2mv_0 \cos\ 30.0° = 2(0.047\ \text{kg})(45\ \text{m/s})(\cos 30.0°) = \boxed{3.7\ \text{N}\cdot\text{s}}$$

---

17. ***REASONING*** The system consists of the lumberjack and the log. For this system, the sum of the external forces is zero. This is because the weight of the system is balanced by the corresponding normal force (provided by the buoyant force of the water) and the water is assumed to be frictionless. The lumberjack and the log, then, constitute an isolated system, and the principle of conservation of linear momentum holds.

SOLUTION

a. The total linear momentum of the system before the lumberjack begins to move is zero, since all parts of the system are at rest. Momentum conservation requires that the total momentum remains zero during the motion of the lumberjack.

$$\underbrace{m_1 v_{f1} + m_2 v_{f2}}_{\substack{\text{Total momentum} \\ \text{just before the jump}}} = \underbrace{0}_{\text{Initial momentum}}$$

Here the subscripts "1" and "2" refer to the first log and lumberjack, respectively. Let the direction of motion of the lumberjack be the positive direction. Then, solving for v_{1f} gives

$$v_{f1} = -\frac{m_2 v_{f2}}{m_1} = -\frac{(98\text{ kg})(+3.6\text{ m/s})}{230\text{ kg}} = \boxed{-1.5\text{ m/s}}$$

The minus sign indicates that the first log recoils as the lumberjack jumps off.

b. Now the system is composed of the lumberjack, just before he lands on the second log, and the second log. Gravity acts on the system, but for the short time under consideration while the lumberjack lands, the effects of gravity in changing the linear momentum of the system are negligible. Therefore, to a very good approximation, we can say that the linear momentum of the system is very nearly conserved. In this case, the initial momentum is not zero as it was in part (a); rather the initial momentum of the system is the momentum of the lumberjack just before he lands on the second log. Therefore,

$$\underbrace{m_1 v_{f1} + m_2 v_{f2}}_{\substack{\text{Total momentum} \\ \text{just after lumberjack lands}}} = \underbrace{m_1 v_{01} + m_2 v_{02}}_{\text{Initial momentum}}$$

In this expression, the subscripts "1" and "2" now represent the second log and lumberjack, respectively. Since the second log is initially at rest, $v_{01} = 0$. Furthermore, since the lumberjack and the second log move with a common velocity, $v_{f1} = v_{f2} = v_f$. The statement of momentum conservation then becomes

$$m_1 v_f + m_2 v_f = m_2 v_{02}$$

Solving for v_f, we have

$$v_f = \frac{m_2 v_{02}}{m_1 + m_2} = \frac{(98\text{ kg})(+3.6\text{ m/s})}{230\text{ kg} + 98\text{ kg}} = \boxed{+1.1\text{ m/s}}$$

The positive sign indicates that the system moves in the same direction as the original direction of the lumberjack's motion.

---

21. ***REASONING AND SOLUTION*** The comet piece and Jupiter constitute an isolated system, since no external forces act on them. Therefore, the head-on collision obeys the conservation of linear momentum:

$$\underbrace{(m_{\text{comet}} + m_{\text{Jupiter}})v_{\text{f}}}_{\substack{\text{Total momentum}\\ \text{after collision}}} = \underbrace{m_{\text{comet}}v_{\text{comet}} + m_{\text{Jupiter}}v_{\text{Jupiter}}}_{\substack{\text{Total momentum}\\ \text{before collision}}}$$

where v_{f} is the common velocity of the comet piece and Jupiter after the collision. We assume initially that Jupiter is moving in the $+x$ direction so $v_{\text{Jupiter}} = +1.3\times10^{4}\ \text{m/s}$. The comet piece must be moving in the opposite direction so $v_{\text{comet}} = -6.0\times10^{4}\ \text{m/s}$. The final velocity v_{f} of Jupiter and the comet piece can be written as $v_{\text{f}} = v_{\text{Jupiter}} + \Delta v$ where Δv is the change in velocity of Jupiter due to the collision. Substituting this expression into the conservation of momentum equation gives

$$(m_{\text{comet}} + m_{\text{Jupiter}})(v_{\text{Jupiter}} + \Delta v) = m_{\text{comet}}v_{\text{comet}} + m_{\text{Jupiter}}v_{\text{Jupiter}}$$

Multiplying out the left side of this equation, algebraically canceling the term $m_{\text{Jupiter}}v_{\text{Jupiter}}$ from both sides of the equation, and solving for Δv yields

$$\Delta v = \frac{m_{\text{comet}}(v_{\text{comet}} - v_{\text{Jupiter}})}{m_{\text{comet}} + m_{\text{Jupiter}}}$$

$$= \frac{(4.0\times10^{12}\ \text{kg})(-6.0\times10^{4}\ \text{m/s} - 1.3\times10^{4}\ \text{m/s})}{4.0\times10^{12}\ \text{kg} + 1.9\times10^{27}\ \text{kg}} = -1.5\times10^{-10}\ \text{m/s}$$

The change in Jupiter's speed is $\boxed{1.5\times10^{-10}\ \text{m/s}}$.

---

23. ***REASONING*** We will define the system to be the platform, the two people and the ball. Since the ball travels nearly horizontally, the effects of gravity are negligible. Momentum is conserved. Since the initial momentum of the system is zero, it must remain zero as the ball is thrown and caught. While the ball is in motion, the platform will recoil in such a way that the total momentum of the system remains zero. As the ball is caught, the system must come to rest so that the total momentum remains zero. The distance that the platform moves before coming to rest again can be determined by using the expressions for momentum conservation and the kinematic description for this situation.

SOLUTION While the ball is in motion, we have

$$MV + mv = 0 \qquad (1)$$

where M is the combined mass of the platform and the two people, V is the recoil velocity of the platform, m is the mass of the ball, and v is the velocity of the ball.

The distance that the platform moves is given by

$$x = Vt \qquad (2)$$

where t is the time that the ball is in the air. The time that the ball is in the air is given by

$$t = \frac{L}{v - V} \tag{3}$$

where L is the length of the platform, and the quantity $(v - V)$ is the velocity of the ball relative to the platform. Remember, both the ball and the platform are moving while the ball is in the air. Combining equations (2) and (3) gives

$$x = \left(\frac{V}{v - V}\right)L \tag{4}$$

From equation (1) the ratio of the velocities is $V/v = -m/M$. Equation (4) then gives

$$x = \frac{(V/v)L}{1-(V/v)} = \frac{(-m/M)L}{1+(m/M)} = -\frac{mL}{M+m} = -\frac{(6.0\text{ kg})(2.0\text{ m})}{118\text{ kg}+6.0\text{ kg}} = -0.097\text{ m}$$

The minus sign indicates that displacement of the platform is in the opposite direction to the displacement of the ball. The distance moved by the platform is the magnitude of this displacement, or $\boxed{0.097\text{ m}}$.

---

27. ***REASONING*** Since all of the collisions are elastic, the total mechanical energy of the ball is conserved. However, since gravity affects its vertical motion, its linear momentum is *not* conserved. If h_f is the maximum height of the ball on its final bounce, conservation of energy gives

$$\underbrace{\tfrac{1}{2}mv_f^2 + mgh_f}_{E_f} = \underbrace{\tfrac{1}{2}mv_0^2 + mgh_0}_{E_0}$$

Solving this equation for h_f gives

$$h_f = \frac{v_0^2 - v_f^2}{2g} + h_0$$

SOLUTION In order to use this expression, we must obtain the values for the velocities v_0 and v_f. The initial velocity has only a horizontal component, $v_0 = v_{0x}$. The final velocity also has only a horizontal component since the ball is at the top of its trajectory, $v_f = v_{fx}$. No forces act in the horizontal direction so the momentum of the ball in this direction is conserved, hence $v_0 = v_f$. Therefore,

$$h_f = h_0 = \boxed{3.00\text{ m}}$$

---

31. ***REASONING*** Batman and the boat with the criminal constitute the system. Gravity acts on this system as an external force; however, gravity acts vertically, and we are concerned only with the horizontal motion of the system. If we neglect air resistance and friction, there are no external forces that act horizontally; therefore, the total linear momentum in the horizontal direction is conserved. When Batman collides with the boat, the horizontal component of his velocity is zero, so the statement of conservation of linear momentum in the horizontal direction can be written as

$$\underbrace{(m_1 + m_2)v_f}_{\substack{\text{Total horizontal momentum}\\ \text{after collision}}} = \underbrace{m_1 v_{01} + 0}_{\substack{\text{Total horizontal momentum}\\ \text{before collision}}}$$

Here, m_1 is the mass of the boat, and m_2 is the mass of Batman. This expression can be solved for v_f, the velocity of the boat after Batman lands in it.

SOLUTION Solving for v_f gives

$$v_f = \frac{m_1 v_{01}}{m_1 + m_2} = \frac{(510\text{ kg})(+11\text{ m/s})}{510\text{ kg} + 91\text{ kg}} = \boxed{+9.3\text{ m/s}}$$

The plus sign indicates that the boat continues to move in its initial direction of motion.

---

35. ***REASONING*** The two skaters constitute the system. Since the net external force acting on the system is zero, the total linear momentum of the system is conserved. In the x direction (the east/west direction), conservation of linear momentum gives $P_{fx} = P_{0x}$, or

$$(m_1 + m_2)v_f \cos\theta = m_1 v_{01}$$

Note that since the skaters hold onto each other, they move away with a common velocity v_f. In the y direction, $P_{fy} = P_{0y}$, or

$$(m_1 + m_2)v_f \sin\theta = m_2 v_{02}$$

These equations can be solved simultaneously to obtain both the angle θ and the velocity v_f.

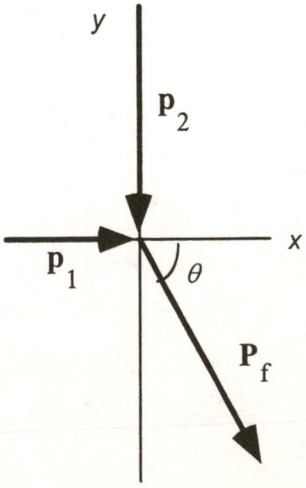

SOLUTION

a. Division of the equations above gives

$$\theta = \tan^{-1}\left(\frac{m_2 v_{02}}{m_1 v_{01}}\right) = \tan^{-1}\left[\frac{(70.0\text{ kg})(7.00\text{ m/s})}{(50.0\text{ kg})(3.00\text{ m/s})}\right] = \boxed{73.0°}$$

b. Solution of the first of the momentum equations gives

$$v_f = \frac{m_1 v_{01}}{(m_1 + m_2)\cos\theta} = \frac{(50.0\text{ kg})(3.00\text{ m/s})}{(50.0\text{ kg} + 70.0\text{ kg})(\cos 73.0°)} = \boxed{4.28\text{ m/s}}$$

---

39. ***REASONING*** The two balls constitute the system. The tension in the wire is the only nonconservative force that acts on the ball. The tension does no work since it is perpendicular to the displacement of the ball. Since $W_{nc} = 0$ J, the principle of conservation of mechanical energy holds and can be used to find the speed of the 1.50-kg ball just before the collision. Momentum is conserved during the collision, so the principle of conservation of momentum can be used to find the velocities of both balls just after the collision. Once the collision has occurred, energy conservation can be used to determine how high each ball rises.

SOLUTION

a. Applying the principle of energy conservation to the 1.50-kg ball, we have

$$\underbrace{\tfrac{1}{2}mv_f^2 + mgh_f}_{E_f} = \underbrace{\tfrac{1}{2}mv_0^2 + mgh_0}_{E_0}$$

If we measure the heights from the lowest point in the swing, $h_f = 0$ m, and the expression above simplifies to

$$\tfrac{1}{2}mv_f^2 = \tfrac{1}{2}mv_0^2 + mgh_0$$

Solving for v_f, we have

$$v_f = \sqrt{v_0^2 + 2gh_0} = \sqrt{(5.00\text{ m/s})^2 + 2(9.80\text{ m/s}^2)(0.300\text{ m})} = \boxed{5.56\text{ m/s}}$$

b. If we assume that the collision is elastic, then the velocities of both balls just after the collision can be obtained from Equations 7.8a and 7.8b:

$$v_{f1} = \left(\frac{m_1 - m_2}{m_1 + m_2}\right)v_{01} \qquad \text{and} \qquad v_{f2} = \left(\frac{2m_1}{m_1 + m_2}\right)v_{01}$$

Since v_{01} corresponds to the speed of the 1.50-kg ball just before the collision, it is equal to the quantity v_f calculated in part (a). With the given values of $m_1 = 1.50$ kg and $m_2 = 4.60$ kg, and the value of v_{01} = 5.56 m/s obtained in part (a), Equations 7.8a and 7.8b yield the following values:

$$\boxed{v_{f1} = -2.83\text{ m/s}} \qquad \text{and} \qquad \boxed{v_{f2} = +2.73\text{ m/s}}$$

The minus sign in v_{f1} indicates that the first ball reverses its direction as a result of the collision.

c. If we apply the conservation of mechanical energy to either ball after the collision we have

$$\underbrace{\tfrac{1}{2}mv_f^2 + mgh_f}_{E_f} = \underbrace{\tfrac{1}{2}mv_0^2 + mgh_0}_{E_0}$$

where v_0 is the speed of the ball just after the collision, and h_f is the final height to which the ball rises. For either ball, $h_0 = 0$ m, and when either ball has reached its maximum height, $v_f = 0$ m/s. Therefore, the expression of energy conservation reduces to

$$gh_f = \tfrac{1}{2}v_0^2 \qquad \text{or} \qquad h_f = \frac{v_0^2}{2g}$$

Thus, the heights to which each ball rises after the collision are

[1.50-kg ball] $$h_f = \frac{v_0^2}{2g} = \frac{(2.83\text{ m/s})^2}{2(9.80\text{ m/s}^2)} = \boxed{0.409\text{ m}}$$

[4.60-kg ball] $$h_f = \frac{v_0^2}{2g} = \frac{(2.73\text{ m/s})^2}{2(9.80\text{ m/s}^2)} = \boxed{0.380\text{ m}}$$

---

41. ***REASONING AND SOLUTION*** The location of the center of mass for a two-body system is given by Equation 7.10:

$$x_{cm} = \frac{m_1x_1 + m_2x_2}{m_1 + m_2}$$

where the subscripts "1" and "2" refer to the earth and the moon, respectively. For convenience, we will let the center of the earth be coincident with the origin so that $x_1 = 0$ and $x_2 = d$, the center-to-center distance between the earth and the moon. Direct calculation then gives

$$x_{cm} = \frac{m_2d}{m_1 + m_2} = \frac{(7.35\times10^{22}\text{ kg})(3.85\times10^{8}\text{ m})}{5.98\times10^{24}\text{ kg} + 7.35\times10^{22}\text{ kg}} = \boxed{4.67\times10^{6}\text{ m}}$$

---

45. ***REASONING*** Let m be Al's mass, which means that Jo's mass is 168 kg – m. Since friction is negligible and since the downward-acting weight of each person is balanced by the upward-acting normal force from the ice, the net external force acting on the two-person

system is zero. Therefore, the system is isolated, and the conservation of linear momentum applies. The initial total momentum must be equal to the final total momentum.

SOLUTION Applying the principle of conservation of linear momentum and assuming that the direction in which Al moves is the positive direction, we find

$$\underbrace{m(0\text{ m/s})+(168\text{ kg}-m)(0\text{ m/s})}_{\text{Initial total momentum}}=\underbrace{m(0.90\text{ m/s})+(168\text{ kg}-m)(-1.2\text{ m/s})}_{\text{Final total momentum}}$$

Solving this equation for m and suppressing the units in the interests of clarity, we find

$$0=m(0.90)-(168)(1.2)+m(1.2)$$

$$m=\frac{(168)(1.2)}{0.90+1.2}=\boxed{96\text{ kg}}$$

---

47. ***REASONING*** The system consists of the two balls. The total linear momentum of the two-ball system is conserved because the net external force acting on it is zero. The principle of conservation of linear momentum applies whether or not the collision is elastic.

$$\underbrace{m_1v_{f1}+m_2v_{f2}}_{\substack{\text{Total momentum}\\ \text{after collision}}}=\underbrace{m_1v_{01}+0}_{\substack{\text{Total momentum}\\ \text{before collision}}}$$

When the collision is elastic, the kinetic energy is also conserved during the collision

$$\underbrace{\tfrac{1}{2}m_1v_{f1}^2+\tfrac{1}{2}m_2v_{f2}^2}_{\substack{\text{Total kinetic energy}\\ \text{after collision}}}=\underbrace{\tfrac{1}{2}m_1v_{01}^2+0}_{\substack{\text{Total kinetic energy}\\ \text{before collision}}}$$

SOLUTION

a. The final velocities for an elastic collision are determined by simultaneously solving the above equations for the final velocities. The procedure is discussed in Example 7 in the text, and leads to Equations 7.8a and 7.8b. According to Equation 7.8:

$$v_{f1}=\left(\frac{m_1-m_2}{m_1+m_2}\right)v_{01}\qquad\text{and}\qquad v_{f2}=\left(\frac{2m_1}{m_1+m_2}\right)v_{01}$$

Let the initial direction of motion of the 5.00-kg ball define the positive direction. Substituting the values given in the text, these equations give

$$[5.00\text{-kg ball}]\qquad v_{f1}=\left(\frac{5.00\text{ kg}-7.50\text{ kg}}{5.00\text{ kg}+7.50\text{ kg}}\right)(2.00\text{ m/s})=\boxed{-0.400\text{ m/s}}$$

[7.50-kg ball] $$v_{f2} = \left(\frac{2(5.00\text{ kg})}{5.00\text{ kg} + 7.50\text{ kg}}\right)(2.00\text{ m/s}) = \boxed{+1.60\text{ m/s}}$$

The signs indicate that, after the collision, the 5.00-kg ball reverses its direction of motion, while the 7.50-kg ball moves in the direction in which the 5.00-kg ball was initially moving.

b. When the collision is completely inelastic, the balls stick together, giving a composite body of mass $m_1 + m_2$ that moves with a velocity v_f. The statement of conservation of linear momentum then becomes

$$\underbrace{(m_1 + m_2)v_f}_{\substack{\text{Total momentum}\\ \text{after collision}}} = \underbrace{m_1 v_{01} + 0}_{\substack{\text{Total momentum}\\ \text{before collision}}}$$

The final velocity of the two balls after the collision is, therefore,

$$v_f = \frac{m_1 v_{01}}{m_1 + m_2} = \frac{(5.00\text{ kg})(2.00\text{ m/s})}{5.00\text{ kg} + 7.50\text{ kg}} = \boxed{+0.800\text{ m/s}}$$

---

51. ***REASONING*** The two-stage rocket constitutes the system. The forces that act to cause the separation during the explosion are, therefore, forces that are internal to the system. Since no external forces act on this system, it is isolated and the principle of conservation of linear momentum applies:

$$\underbrace{m_1 v_{f1} + m_2 v_{f2}}_{\substack{\text{Total momentum}\\ \text{after separation}}} = \underbrace{(m_1 + m_2)v_0}_{\substack{\text{Total momentum}\\ \text{before separation}}}$$

where the subscripts "1" and "2" refer to the lower and upper stages, respectively. This expression can be solved for v_{f1}.

SOLUTION Solving for v_{f1} gives

$$v_{f1} = \frac{(m_1 + m_2)v_0 - m_2 v_{f2}}{m_1}$$

$$= \frac{[2400\text{ kg} + 1200\text{ kg}](4900\text{ m/s}) - (1200\text{ kg})(5700\text{ m/s})}{2400\text{ kg}} = \boxed{+4500\text{ m/s}}$$

Since v_{f1} is positive, $\boxed{\text{its direction is the same as the rocket before the explosion}}$.

---

53. ***REASONING*** No net external force acts on the plate parallel to the floor; therefore, the component of the momentum of the plate that is parallel to the floor is conserved as the plate breaks and flies apart. Initially, the total momentum parallel to the floor is zero. After the collision with the floor, the component of the total momentum parallel to the floor must remain zero. The drawing in the text shows the pieces in the plane parallel to the floor just after the collision. Clearly, the linear momentum in the plane parallel to the floor has two components; therefore the linear momentum of the plate must be conserved in each of these two mutually perpendicular directions. Using the drawing in the text, with the positive directions taken to be up and to the right, we have

[x direction] $$-m_1 v_1 (\sin 25.0°) + m_2 v_2 (\cos 45.0°) = 0$$

[y direction] $$m_1 v_1 (\cos 25.0°) + m_2 v_2 (\sin 45.0°) - m_3 v_3 = 0$$

These equations can be solved simultaneously for the masses m_1 and m_2.

SOLUTION Using the values given in the drawing for the velocities after the plate breaks, we have, suppressing units,

$$-1.27m_1 + 1.27m_2 = 0 \qquad (1)$$

$$2.72m_1 + 1.27m_2 - 3.99 = 0 \qquad (2)$$

Subtracting (1) from (2) gives $\boxed{m_1 = 1.00 \text{ kg}}$. Substituting this value into either (1) or (2) then yields $\boxed{m_2 = 1.00 \text{ kg}}$.

---

57. ***REASONING*** The cannon and the shell constitute the system. Since no external force hinders the motion of the system after the cannon is unbolted, conservation of linear momentum applies in that case. If we assume that the burning gun powder imparts the same kinetic energy to the system in each case, we have sufficient information to develop a mathematical description of this situation, and solve it for the velocity of the shell fired by the loose cannon.

SOLUTION For the case where the cannon is unbolted, momentum conservation gives

$$\underbrace{m_1 v_{f1} + m_2 v_{f2}}_{\substack{\text{Total momentum}\\ \text{after shell is fired}}} = \underbrace{0}_{\substack{\text{Initial momentum}\\ \text{of system}}} \qquad (1)$$

where the subscripts "1" and "2" refer to the cannon and shell, respectively. In both cases, the burning gun power imparts the same kinetic energy to the system. When the cannon is bolted to the ground, only the shell moves and the kinetic energy imparted to the system is

$$\text{KE} = \tfrac{1}{2} m_{\text{shell}} v_{\text{shell}}^2 = \tfrac{1}{2}(85.0 \text{ kg})(551 \text{ m/s})^2 = 1.29 \times 10^7 \text{ J}$$

The kinetic energy imparted to the system when the cannon is unbolted has the same value and can be written using the same notation as in equation (1):

$$\mathrm{KE} = \tfrac{1}{2} m_1 v_{f1}^2 + \tfrac{1}{2} m_2 v_{f2}^2 \tag{2}$$

Solving equation (1) for v_{f1}, the velocity of the cannon after the shell is fired, and substituting the resulting expression into Equation (2) gives

$$\mathrm{KE} = \frac{m_2^2 v_{f2}^2}{2m_1} + \tfrac{1}{2} m_2 v_{f2}^2 \tag{3}$$

Solving equation (3) for v_{f2} gives

$$v_{f2} = \sqrt{\frac{2\mathrm{KE}}{m_2\left(\dfrac{m_2}{m_1}+1\right)}} = \sqrt{\frac{2(1.29\times10^7\ \mathrm{J})}{(85\ \mathrm{kg})\left(\dfrac{85\ \mathrm{kg}}{5.80\times10^3\ \mathrm{kg}}+1\right)}} = \boxed{+547\ \mathrm{m/s}}$$

CHAPTER 8 | *ROTATIONAL KINEMATICS*

CONCEPTUAL QUESTIONS

---

5. ***REASONING AND SOLUTION***
a. A thin rod rotates at a constant angular speed about an axis of rotation that is perpendicular to the rod at its center. As the rod rotates, each point at a distance r from the center on one half of the rod has the same tangential speed as the point at a distance r from the center on the other half of the rod. This is true for all values of r for $0 < r \leq (L/2)$ where L is the length of the rod.

b. If the rod rotates about an axis that is perpendicular to the rod at one end, no two points are the same distance from the axis of rotation. Therefore, no two points on the rod have the same tangential speed.

---

9. ***REASONING AND SOLUTION*** The centripetal acceleration of any point in the space station is given by $a_c = r\omega^2$ (Equation 8.11),where r is the distance from the point to the axis of rotation of the space station. If the centripetal acceleration is adjusted to be g at a given value of r (such as the astronaut's feet), the centripetal acceleration will be different at other values of r. Therefore, if the adjustment is made so that the centripetal acceleration at the astronaut's feet equals g, then the centripetal acceleration will not equal g at the astronaut's head.

---

13. ***REASONING AND SOLUTION*** The speedometer of a truck uses a device that measures the angular speed of the tires. The angular speed is related to the linear speed of the truck by $v = r\omega$.

Suppose two trucks are traveling side-by-side along a highway at the same linear speed v, and one truck has larger wheels than the other. The angular speed of the larger-diameter wheels is less than that of the smaller-diameter wheels. Since the speedometer uses a device that measures the angular speed of the tires, the speedometer reading on the truck with the larger wheels will indicate a smaller linear speed than that on the truck with the smaller wheels.

CHAPTER 8 | *ROTATIONAL KINEMATICS*

PROBLEMS

---

1. ***REASONING AND SOLUTION*** Since there are 2π radians per revolution, and it is stated in the problem that there are 100 grads in one-quarter of a circle, we find that the number of grads in one radian is

$$(1.00\text{ rad})\left(\frac{1\text{ rev}}{2\pi\text{ rad}}\right)\left(\frac{100\text{ grad}}{0.250\text{ rev}}\right) = \boxed{63.7\text{ grad}}$$

---

3. ***REASONING*** The average angular velocity is equal to the angular displacement divided by the elapsed time (Equation 8.2). Thus, the angular displacement of the baseball is equal to the product of the average angular velocity and the elapsed time. However, the problem gives the travel time in seconds and asks for the displacement in radians, while the angular velocity is given in revolutions per minute. Thus, we will begin by converting the angular velocity into radians per second.

SOLUTION Since 2π rad = 1 rev and 1 min = 60 s, the average angular velocity $\bar{\omega}$ (in rad/s) of the baseball is

$$\bar{\omega} = \left(\frac{330\text{ rev}}{\text{min}}\right)\left(\frac{2\pi\text{ rad}}{1\text{ rev}}\right)\left(\frac{1\text{ min}}{60\text{ s}}\right) = 35\text{ rad/s}$$

Since the average angular velocity of the baseball is equal to the angular displacement $\Delta\theta$ divided by the elapsed time Δt, the angular displacement is

$$\Delta\theta = \bar{\omega}\,\Delta t = (35\text{ rad/s})(0.60\text{ s}) = \boxed{21\text{ rad}} \qquad (8.2)$$

---

5. ***REASONING AND SOLUTION*** Equation 8.4 gives the desired result. Assuming $t_0 = 0$ s, the final angular velocity is

$$\omega = \omega_0 + \alpha t = 0\text{ rad/s} + (328\text{ rad/s}^2)(1.50\text{ s}) = \boxed{492\text{ rad/s}}$$

---

13. ***REASONING*** The time required for the bullet to travel the distance d is equal to the time required for the discs to undergo an angular displacement of 0.240 rad. The time can be found from Equation 8.2; once the time is known, the speed of the bullet can be found using Equation 2.2.

SOLUTION From the definition of average angular velocity:

$$\bar{\omega} = \frac{\Delta\theta}{\Delta t}$$

the required time is

$$\Delta t = \frac{\Delta\theta}{\bar{\omega}} = \frac{0.240\text{ rad}}{95.0\text{ rad/s}} = 2.53\times10^{-3}\text{ s}$$

Note that $\bar{\omega} = \omega$ because the angular speed is constant. The (constant) speed of the bullet can then be determined from the definition of average speed:

$$\bar{v} = \frac{\Delta x}{\Delta t} = \frac{d}{\Delta t} = \frac{0.850\text{ m}}{2.53\times10^{-3}\text{ s}} = \boxed{336\text{ m/s}}$$

---

17. ***REASONING AND SOLUTION***

a. From Equation 8.7 we obtain

$$\theta = \omega_0 t + \tfrac{1}{2}\alpha t^2 = (5.00\text{ rad/s})(4.00\text{ s}) + \tfrac{1}{2}(2.50\text{ rad/s}^2)(4.00\text{ s})^2 = \boxed{4.00\times10^1\text{ rad}}$$

b. From Equation 8.4, we obtain

$$\omega = \omega_0 + \alpha t = 5.00\text{ rad/s} + (2.50\text{ rad/s}^2)(4.00\text{ s}) = \boxed{15.0\text{ rad/s}}$$

---

19. ***REASONING AND SOLUTION***

a. Since the flywheel comes to rest, its final angular velocity is zero. Furthermore, if the initial angular velocity ω_0 is assumed to be a positive number and the flywheel decelerates, the angular acceleration must be a negative number. Solving Equation 8.8 for θ, we obtain

$$\theta = \frac{\omega^2 - \omega_0^2}{2\alpha} = \frac{(0\text{ rad/s})^2 - (220\text{ rad/s})^2}{2(-2.0\text{ rad/s}^2)} = \boxed{1.2\times10^4\text{ rad}}$$

b. The time required for the flywheel to come to rest can be found from Equation 8.4. Solving for t, we obtain

$$t = \frac{\omega - \omega_0}{\alpha} = \frac{0\text{ rad/s} - 220\text{ rad/s}}{-2.0\text{ rad/s}^2} = \boxed{1.1\times10^2\text{ s}}$$

---

25. ***REASONING*** The time required for the change in the angular velocity to occur can be found by solving Equation 8.4 for t. In order to use Equation 8.4, however, we must know the initial angular velocity ω_0. Equation 8.6 can be used to find the initial angular velocity.

SOLUTION From Equation 8.6 we have

$$\theta = \tfrac{1}{2}(\omega_0 + \omega)t$$

Solving for ω_0 gives

$$\omega_0 = \frac{2\theta}{t} - \omega$$

Since the angular displacement θ is zero, $\omega_0 = -\omega$. Solving Equation 8.4 for t, and using the fact that $\omega_0 = -\omega$, gives

$$t = \frac{2\omega}{\alpha} = \frac{2(-25.0 \text{ rad/s})}{-4.00 \text{ rad/s}^2} = \boxed{12.5 \text{ s}}$$

---

27. ***REASONING*** The angular displacement of the child when he catches the horse is, from Equation 8.2, $\theta_c = \omega_c t$. In the same time, the angular displacement of the horse is, from Equation 8.7 with $\omega_0 = 0$ rad/s, $\theta_h = \tfrac{1}{2}\alpha t^2$. If the child is to catch the horse $\theta_c = \theta_h + (\pi/2)$.

SOLUTION Using the above conditions yields

$$\tfrac{1}{2}\alpha t^2 - \omega_c t + \tfrac{1}{2}\pi = 0$$

or

$$\tfrac{1}{2}(0.0100 \text{ rad/s}^2)t^2 - (0.250 \text{ rad/s})t + \tfrac{1}{2}(\pi \text{ rad}) = 0$$

The quadratic formula yields t = 7.37 s and 42.6 s; therefore, the shortest time needed to catch the horse is $\boxed{t = 7.37 \text{ s}}$.

---

29. ***REASONING*** The angular speed ω and tangential speed v_T are related by Equation 8.9 ($v_T = r\omega$), and this equation can be used to determine the radius r. However, we must remember that this relationship is only valid if we use radian measure. Therefore, it will be necessary to convert the given angular speed in rev/s into rad/s.

SOLUTION Solving Equation 8.9 for the radius gives

$$r = \frac{v_T}{\omega} = \frac{54 \text{ m/s}}{\underbrace{(47 \text{ rev/s})\left(\frac{2\pi \text{ rad}}{1 \text{ rev}}\right)}_{\text{Conversion from rev/s into rad/s}}} = \boxed{0.18 \text{ m}}$$

where we have used the fact that 1 rev corresponds to 2π rad to convert the given angular speed from rev/s into rad/s.

---

35. ***REASONING AND SOLUTION*** According to Equation 8.9, the crank handle has an angular speed of

$$\omega = \frac{v_T}{r} = \frac{1.20 \text{ m/s}}{\frac{1}{2}(0.400 \text{ m})} = 6.00 \text{ rad/s}$$

The crank barrel must have the same angular speed as the handle; therefore, the tangential speed of a point on the barrel is, again using Equation 8.9,

$$v_T = r\omega = (5.00 \times 10^{-2} \text{ m})(6.00 \text{ rad/s}) = 0.300 \text{ m/s}$$

If we assume that the rope does not slip, then all points on the rope must move with the same speed as the tangential speed of any point on the barrel. Therefore the linear speed with which the bucket moves down the well is $\boxed{v = 0.300 \text{ m/s}}$.

---

39. ***REASONING*** Since the car is traveling with a constant speed, its tangential acceleration must be zero. The radial or centripetal acceleration of the car can be found from Equation 5.2. Since the tangential acceleration is zero, the total acceleration of the car is equal to its radial acceleration.

SOLUTION

a. Using Equation 5.2, we find that the car's radial acceleration, and therefore its total acceleration, is

$$a = a_R = \frac{v_T^2}{r} = \frac{(75.0 \text{ m/s})^2}{625 \text{ m}} = \boxed{9.00 \text{ m/s}^2}$$

b. The direction of the car's total acceleration is the same as the direction of its radial acceleration. That is, the direction is $\boxed{\text{radially inward}}$.

---

41. ***REASONING*** The top of the racket has both tangential and centripetal acceleration components given by Equations 8.10 and 8.11, respectively: $a_T = r\alpha$ and $a_c = r\omega^2$. The total acceleration of the top of the racket is the resultant of these two components. Since

these acceleration components are mutually perpendicular, their resultant can be found by using the Pythagorean theorem.

SOLUTION Employing the Pythagorean theorem, we obtain

$$a = \sqrt{a_T^2 + a_c^2} = \sqrt{(r\alpha)^2 + (r\omega^2)^2} = r\sqrt{\alpha^2 + \omega^4}$$

Therefore,

$$a = (1.5\text{ m})\sqrt{(160\text{ rad/s}^2)^2 + (14\text{ rad/s})^4} = \boxed{380\text{ m/s}^2}$$

---

45. ***REASONING*** The tangential acceleration and the centripetal acceleration of a point at a distance r from the rotation axis are given by Equations 8.10 and 8.11, respectively: $a_T = r\alpha$ and $a_c = r\omega^2$. After the drill has rotated through the angle in question, $a_c = 2a_T$, or

$$r\omega^2 = 2r\alpha$$

This expression can be used to find the angular acceleration α. Once the angular acceleration is known, Equation 8.8 can be used to find the desired angle.

SOLUTION Solving the expression obtained above for α gives

$$\alpha = \frac{\omega^2}{2}$$

Solving Equation 8.8 for θ (with $\omega_0 = 0$ rad/s since the drill starts from rest), and using the expression above for the angular acceleration α gives

$$\theta = \frac{\omega^2}{2\alpha} = \frac{\omega^2}{2(\omega^2/2)} = \left(\frac{\omega^2}{2}\right)\left(\frac{2}{\omega^2}\right) = \boxed{1.00\text{ rad}}$$

Note that since both Equations 8.10 and 8.11 require that the angles be expressed in radians, the final result for θ is in radians.

---

47. ***REASONING AND SOLUTION*** From Equation 2.4, the linear acceleration of the motorcycle is

$$a = \frac{v - v_0}{t} = \frac{22.0\text{ m/s} - 0\text{ m/s}}{9.00\text{ s}} = 2.44\text{ m/s}^2$$

Since the tire rolls without slipping, the linear acceleration equals the tangential acceleration of a point on the outer edge of the tire: $a = a_T$. Solving Equation 8.13 for α gives

$$\alpha = \frac{a_T}{r} = \frac{2.44 \text{ m/s}^2}{0.280 \text{ m}} = \boxed{8.71 \text{ rad/s}^2}$$

---

51. ***REASONING*** The angle through which the tire rotates is equal to its average angular velocity $\bar{\omega}$ multiplied by the elapsed time t, $\theta = \bar{\omega} t$. According to Equation 8.6, this angle is related to the initial and final angular velocities of the tire by

$$\theta = \bar{\omega} t = \tfrac{1}{2}(\omega_0 + \omega)t$$

The tire is assumed to roll at a constant angular velocity, so that $\omega_0 = \omega$ and $\theta = \omega t$. Since the tire is rolling, its angular speed is related to its linear speed v by Equation 8.12, $v = r\omega$, where r is the radius of the tire. The angle of rotation then becomes

$$\theta = \omega t = \left(\frac{v}{r}\right)t$$

The time t that it takes for the tire to travel a distance x is equal to $t = x/v$, according to Equation 2.1. Thus, the angle that the tire rotates through is

$$\theta = \left(\frac{v}{r}\right)t = \left(\frac{v}{r}\right)\left(\frac{x}{v}\right) = \frac{x}{r}$$

SOLUTION Since 1 rev = 2π rad, the angle (in revolutions) is

$$\theta = \frac{x}{r} = \frac{96\,000 \times 10^3 \text{ m}}{0.31 \text{ m}} = 3.1 \times 10^8 \text{ rad}$$

$$\theta = \left(3.1 \times 10^8 \text{ rad}\right)\left(\frac{1 \text{ rev}}{2\pi \text{ rad}}\right) = \boxed{4.9 \times 10^7 \text{ rev}}$$

---

55. ***REASONING AND SOLUTION*** By inspection, the distance traveled by the "axle" or the center of the moving quarter is

$$d = 2\pi(2r) = 4\pi r$$

where r is the radius of the quarter. The distance d traveled by the "axle" of the moving quarter must be equal to the circular arc length s along the outer edge of the quarter. This arc length is $s = r\theta$, where θ is the angle through which the quarter rotates. Thus,

$$4\pi r = r\theta$$

so that $\theta = 4\pi \text{ rad}$. This is equivalent to

$$(4\pi \text{ rad})\left(\frac{1 \text{ rev}}{2\pi \text{ rad}}\right) = \boxed{2 \text{ revolutions}}$$

---

57. ***REASONING*** The tangential acceleration $\mathbf{a}_T$ of the speedboat can be found by using Newton's second law, $\mathbf{F}_T = m\mathbf{a}_T$, where $\mathbf{F}_T$ is the net tangential force. Once the tangential acceleration of the boat is known, Equation 2.4 can be used to find the tangential speed of the boat 2.0 s into the turn. With the tangential speed and the radius of the turn known, Equation 5.2 can then be used to find the centripetal acceleration of the boat.

SOLUTION

a. From Newton's second law, we obtain

$$a_T = \frac{F_T}{m} = \frac{550 \text{ N}}{220 \text{ kg}} = \boxed{2.5 \text{ m/s}^2}$$

b. The tangential speed of the boat 2.0 s into the turn is, according to Equation 2.4,

$$v_T = v_{0T} + a_T t = 5.0 \text{ m/s} + (2.5 \text{ m/s}^2)(2.0 \text{ s}) = 1.0 \times 10^1 \text{ m/s}$$

The centripetal acceleration of the boat is then

$$a_c = \frac{v_T^2}{r} = \frac{(1.0 \times 10^1 \text{ m/s})^2}{32 \text{ m}} = \boxed{3.1 \text{ m/s}^2}$$

---

59. ***REASONING AND SOLUTION*** Since the angular speed of the fan decreases, the sign of the angular acceleration must be opposite to the sign for the angular velocity. Taking the angular velocity to be positive, the angular acceleration, therefore, must be a negative quantity. Using Equation 8.4 we obtain

$$\omega_0 = \omega - \alpha t = 83.8 \text{ rad/s} - (-42.0 \text{ rad/s}^2)(1.75 \text{ s}) = \boxed{157.3 \text{ rad/s}}$$

---

63. ***REASONING AND SOLUTION***

a. From Equation 8.9, and the fact that 1 revolution = 2π radians, we obtain

$$v_T = r\omega = (0.0568 \text{ m})\left(3.50 \frac{\text{rev}}{\text{s}}\right)\left(\frac{2\pi \text{ rad}}{1 \text{ rev}}\right) = \boxed{1.25 \text{ m/s}}$$

b. Since the disk rotates at *constant* tangential speed,

$$v_{T1} = v_{T2} \qquad \text{or} \qquad \omega_1 r_1 = \omega_2 r_2$$

Solving for ω_2, we obtain

$$\omega_2 = \frac{\omega_1 r_1}{r_2} = \frac{(3.50 \text{ rev/s})(0.0568 \text{ m})}{0.0249 \text{ m}} = \boxed{7.98 \text{ rev/s}}$$

CHAPTER 9 | *ROTATIONAL DYNAMICS*

CONCEPTUAL QUESTIONS

---

8. ***REASONING AND SOLUTION*** For a rigid body in equilibrium, there can be no net torque acting on the object *with respect to any axis*. Consider an axis perpendicular to the plane of the paper at the right end of the rod. The lines of action of both $\mathbf{F}_1$ and $\mathbf{F}_3$ pass through this axis. Therefore, no torque is created by either of these forces. However, the force $\mathbf{F}_2$ does create a torque with respect to this axis. As a result, there is a net torque acting on the object with respect to at least one axis, and the three forces shown in the drawing cannot keep the object in equilibrium.

---

14. ***REASONING AND SOLUTION*** When you are lying down with your hands behind your head, your moment of inertia I about your stomach is greater than it is when your hands are on your stomach. Sit-ups are then more difficult because, with a greater I, you must exert a greater torque about your stomach to give your upper torso the same angular acceleration, according to Equation 9.7.

---

23. ***REASONING AND SOLUTION*** A person is hanging motionless from a vertical rope over a swimming pool. She lets go of the rope and drops straight down. Consider the woman to be the system. When she lets go of the rope her angular momentum is zero. It is possible for the woman to curl into a ball; however, since her angular momentum is zero, she cannot change her angular velocity by exerting any internal torques. In order to begin spinning, an external torque must be applied to the woman. Therefore, it is not possible for the woman to curl into a ball and start spinning.

CHAPTER 9 | *ROTATIONAL DYNAMICS*

PROBLEMS

---

1. ***REASONING*** According to Equation 9.1, we have

$$\text{Magnitude of torque} = F\ell$$

where F is the magnitude of the applied force and ℓ is the lever arm. From the figure in the text, the lever arm is given by $\ell = (0.28\text{ m})\sin 50.0°$. Since both the magnitude of the torque and ℓ are known, Equation 9.1 can be solved for F.

SOLUTION Solving Equation 9.1 for F, we have

$$F = \frac{\text{Magnitude of torque}}{\ell} = \frac{45\text{ N}\cdot\text{m}}{(0.28\text{ m})\sin 50.0°} = \boxed{2.1\times10^2\text{ N}}$$

---

5. ***REASONING*** The torque on either wheel is given by Equation 9.1, $\tau = F\ell$, where F is the magnitude of the force and ℓ is the lever arm. Regardless of how the force is applied, the lever arm will be proportional to the radius of the wheel.

SOLUTION The ratio of the torque produced by the force in the truck to the torque produced in the car is

$$\frac{\tau_{\text{truck}}}{\tau_{\text{car}}} = \frac{F\ell_{\text{truck}}}{F\ell_{\text{car}}} = \frac{Fr_{\text{truck}}}{Fr_{\text{car}}} = \frac{r_{\text{truck}}}{r_{\text{car}}} = \frac{0.25\text{ m}}{0.19\text{ m}} = \boxed{1.3}$$

---

7. ***REASONING*** Since the meter stick does not move, it is in equilibrium. The forces and the torques, therefore, each add to zero. We can determine the location of the 6.00 N force, by using the condition that the sum of the torques must add to zero.

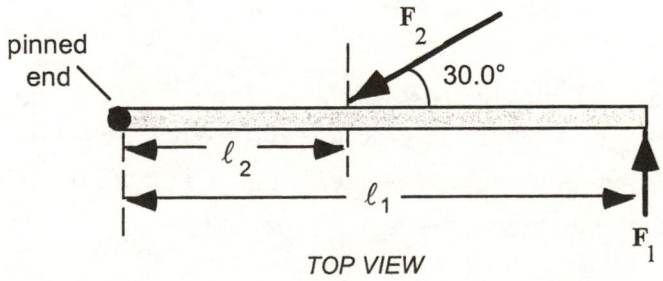

SOLUTION If we take counterclockwise torques as positive, then the torque of the first force about the pin is

$$\tau_1 = F_1\ell_1 = (2.00\text{ N})(1.00\text{ m}) = 2.00\text{ N}\cdot\text{m}$$

The torque due to the second force is

$$\tau_2 = -F_2\,(\sin 30.0°)\,\ell_2 = -(6.00\text{ N})(\sin 30.0°)\,\ell_2 = -(3.00\text{ N})\,\ell_2$$

The torque τ_2 is negative because the force F_2 tends to produce a clockwise rotation about the pinned end. Since the net torque is zero, we have

$$2.00\text{ N}\cdot\text{m} + [-(3.00\text{ N})\ell_2] = 0$$

Thus,

$$\ell_2 = \frac{2.00\text{ N}\cdot\text{m}}{3.00\text{ N}} = \boxed{0.667\text{ m}}$$

---

13. ***REASONING*** The drawing shows the bridge and the four forces that act on it: the upward force $\mathbf{F}_1$ exerted on the left end by the support, the force due to the weight $\mathbf{W}_h$ of the hiker, the weight $\mathbf{W}_b$ of the bridge, and the upward force $\mathbf{F}_2$ exerted on the right side by the support. Since the bridge is in equilibrium, the sum of the torques about any axis of rotation must be zero $(\Sigma\tau = 0)$, and the sum of the forces in the vertical direction must be zero $(\Sigma F_y = 0)$. These two conditions will allow us to determine the magnitudes of $\mathbf{F}_1$ and $\mathbf{F}_2$.

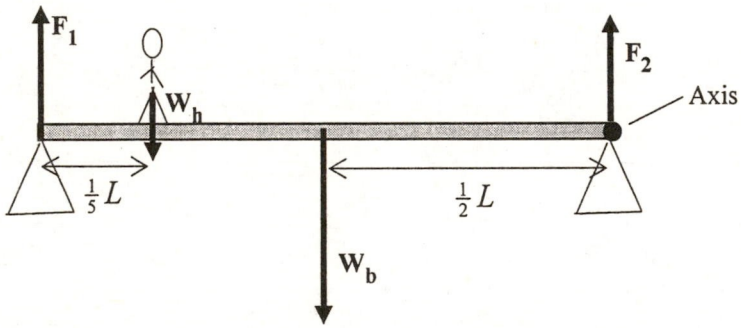

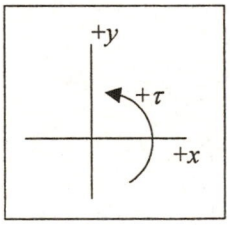

SOLUTION

a. We will begin by taking the axis of rotation about the right end of the bridge. The torque produced by $\mathbf{F}_2$ is zero, since its lever arm is zero. When we set the sum of the torques equal to zero, the resulting equation will have only one unknown, $\mathbf{F}_1$, in it. Setting the sum of the torques produced by the three forces equal to zero gives

$$\Sigma\tau = -F_1L + W_h\left(\tfrac{4}{5}L\right) + W_b\left(\tfrac{1}{2}L\right) = 0$$

Algebraically eliminating the length L of the bridge from this equation and solving for F_1 gives

$$F_1 = \tfrac{4}{5}W_h + \tfrac{1}{2}W_b = \tfrac{4}{5}(985\text{ N}) + \tfrac{1}{2}(3610\text{ N}) = \boxed{2590\text{ N}}$$

b. Since the bridge is in equilibrium, the sum of the forces in the vertical direction must be zero:

$$\Sigma F_y = F_1 - W_h - W_b + F_2 = 0$$

Solving for F_2 gives

$$F_2 = -F_1 + W_h + W_b = -2590\text{ N} + 985\text{ N} + 3610\text{ N} = \boxed{2010\text{ N}}$$

---

17. ***REASONING*** The figure at the right shows the door and the forces that act upon it. Since the door is uniform, the center of gravity, and, thus, the location of the weight **W**, is at the geometric center of the door.

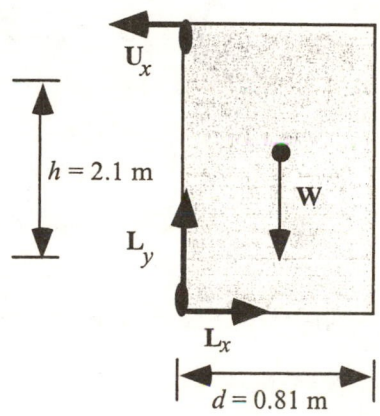

Let **U** represent the force applied to the door by the upper hinge, and **L** the force applied to the door by the lower hinge. Taking forces that point to the right and forces that point up as positive, we have

$$\Sigma F_x = L_x - U_x = 0 \quad \text{or} \quad L_x = U_x \qquad (1)$$

$$\Sigma F_y = L_y - W = 0 \quad \text{or} \quad L_y = W = 140\text{ N} \qquad (2)$$

Taking torques about an axis perpendicular to the plane of the door and through the lower hinge, with counterclockwise torques being positive, gives

$$\Sigma\tau = -W\left(\frac{d}{2}\right) + U_x h = 0 \qquad (3)$$

Equations (1), (2), and (3) can be used to find the force applied to the door by the hinges; Newton's third law can then be used to find the force applied by the door to the hinges.

SOLUTION

a. Solving Equation (3) for U_x gives

$$U_x = \frac{Wd}{2h} = \frac{(140\text{ N})(0.81\text{ m})}{2(2.1\text{ m})} = 27\text{ N}$$

From Equation (1), $L_x = 27$ N.

Therefore, the upper hinge exerts on the door a horizontal force of $\boxed{\text{27 N to the left}}$.

b. From Equation (1), $L_x = 27$ N, we conclude that the bottom hinge exerts on the door a horizontal force of $\boxed{\text{27 N to the right}}$.

c. From Newton's third law, the door applies a force of $\boxed{27\text{ N}}$ on the upper hinge. This force $\boxed{\text{points horizontally to the right}}$, according to Newton's third law.

d. Let **D** represent the force applied by the door to the lower hinge. Then, from Newton's third law, the force exerted on the lower hinge by the door has components $D_x = -27$ N and $D_y = -140$ N.

From the Pythagorean theorem,

$$D = \sqrt{D_x^2 + D_y^2} = \sqrt{(-27\text{ N})^2 + (-140\text{ N})^2} = \boxed{143\text{ N}}$$

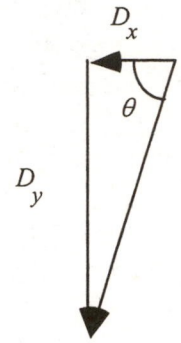

The angle θ is given by

$$\theta = \tan^{-1}\left(\frac{-140\text{ N}}{-27\text{ N}}\right) = 79°$$

The force is directed $\boxed{\text{79° below the horizontal}}$.

---

21. ***REASONING AND SOLUTION*** The figure below shows the massless board and the forces that act on the board due to the person and the scales.

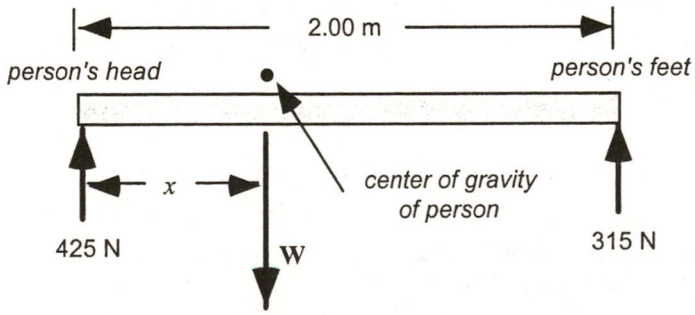

a. Applying Newton's second law to the vertical direction gives

$$315\text{ N} + 425\text{ N} - W = 0 \qquad \text{or} \qquad \mathbf{W} = \boxed{7.40\times10^2\text{ N, downward}}$$

b. Let x be the position of the center of gravity relative to the scale at the person's head. Taking torques about an axis through the contact point between the scale at the person's head and the board gives

$$(315\text{ N})(2.00\text{ m}) - (7.40 \times 10^2\text{ N})x = 0 \qquad \text{or} \qquad x = \boxed{0.851\text{ m}}$$

---

23. ***REASONING*** Since the man holds the ball motionless, the ball and the arm are in equilibrium. Therefore, the net force, as well as the net torque about any axis, must be zero.

SOLUTION Using Equation 9.1, the net torque about an axis through the elbow joint is

$$\Sigma\tau = M(0.0510\text{ m}) - (22.0\text{ N})(0.140\text{ m}) - (178\text{ N})(0.330\text{ m}) = 0$$

Solving this expression for M gives $M = \boxed{1.21\times10^3\text{ N}}$.

The net torque about an axis through the center of gravity is

$$\Sigma\tau = -(1210\text{ N})(0.0890\text{ m}) + F(0.140\text{ m}) - (178\text{ N})(0.190\text{ m}) = 0$$

Solving this expression for F gives $F = \boxed{1.01\times10^3\text{ N}}$. Since the forces must add to give a net force of zero, we know that the direction of **F** is $\boxed{\text{downward}}$.

---

33. ***REASONING AND SOLUTION***

a. The rim of the bicycle wheel can be treated as a hoop. Using the expression given in Table 9.1 in the text, we have

$$I_{\text{hoop}} = MR^2 = (1.20\text{ kg})(0.330\text{ m})^2 = \boxed{0.131\text{ kg}\cdot\text{m}^2}$$

b. Any one of the spokes may be treated as a long, thin rod that can rotate about one end. The expression in Table 9.1 gives

$$I_{\text{rod}} = \tfrac{1}{3}ML^2 = \tfrac{1}{3}(0.010\text{ kg})(0.330\text{ m})^2 = \boxed{3.6\times10^{-4}\text{ kg}\cdot\text{m}^2}$$

c. The total moment of inertia of the bicycle wheel is the sum of the moments of inertia of each constituent part. Therefore, we have

$$I = I_{\text{hoop}} + 50I_{\text{rod}} = 0.131\text{ kg}\cdot\text{m}^2 + 50(3.6\times10^{-4}\text{ kg}\cdot\text{m}^2) = \boxed{0.149\text{ kg}\cdot\text{m}^2}$$

---

37. ***REASONING AND SOLUTION***

a. From Table 9.1 in the text, the moment of inertia of the rod relative to an axis that is perpendicular to the rod at one end is given by

$$I_{\text{rod}} = \tfrac{1}{3}ML^2 = \tfrac{1}{3}(2.00\text{ kg})(2.00\text{ m})^2 = \boxed{2.67\text{ kg}\cdot\text{m}^2}$$

b. The moment of inertia of a point particle of mass m relative to a rotation axis located a perpendicular distance r from the particle is $I = mr^2$. Suppose that all the mass of the rod were located at a single point located a perpendicular distance R from the axis in part (a). If this point particle has the same moment of inertia as the rod in part (a), then the distance R, the radius of gyration, is given by

$$R = \sqrt{\frac{I}{m}} = \sqrt{\frac{2.67\text{ kg}\cdot\text{m}^2}{2.00\text{ kg}}} = \boxed{1.16\text{ m}}$$

---

39. ***REASONING*** The angular acceleration of the bicycle wheel can be calculated from Equation 8.4. Once the angular acceleration is known, Equation 9.7 can be used to find the net torque caused by the brake pads. The normal force can be calculated from the torque using Equation 9.1.

SOLUTION The angular acceleration of the wheel is, according to Equation 8.4,

$$\alpha = \frac{\omega - \omega_0}{t} = \frac{3.7\text{ rad/s} - 13.1\text{ rad/s}}{3.0\text{ s}} = -3.1\text{ rad/s}^2$$

If we assume that all the mass of the wheel is concentrated in the rim, we may treat the wheel as a hollow cylinder. From Table 9.1, we know that the moment of inertia of a hollow cylinder of mass m and radius r about an axis through its center is $I = mr^2$. The net torque that acts on the wheel due to the brake pads is, therefore,

$$\sum \tau = I\alpha = (mr^2)\alpha \qquad (1)$$

From Equation 9.1, the net torque that acts on the wheel due to the action of the two brake pads is

$$\sum \tau = -2f_{\text{k}}\ell \qquad (2)$$

where f_{k} is the kinetic frictional force applied to the wheel by each brake pad, and $\ell = 0.33\text{ m}$ is the lever arm between the axle of the wheel and the brake pad (see the drawing in the text). The factor of 2 accounts for the fact that there are two brake pads. The minus sign arises because the net torque must have the same sign as the angular acceleration. The kinetic frictional force can be written as (see Equation 4.8)

$$f_{\text{k}} = \mu_{\text{k}}F_{\text{N}} \qquad (3)$$

where μ_k is the coefficient of kinetic friction and F_N is the magnitude of the normal force applied to the wheel by each brake pad.

Combining Equations (1), (2), and (3) gives

$$-2(\mu_k F_N)\ell = (mr^2)\alpha$$

$$F_N = \frac{-mr^2\alpha}{2\mu_k \ell} = \frac{-(1.3\text{ kg})(0.33\text{ m})^2(-3.1\text{ rad/s}^2)}{2(0.85)(0.33\text{ m})} = \boxed{0.78\text{ N}}$$

---

43. ***REASONING*** The kinetic energy of the flywheel is given by Equation 9.9. The moment of inertia of the flywheel is the same as that of a solid disk, and, according to Table 9.1 in the text, is given by $I = \frac{1}{2}MR^2$. Once the moment of inertia of the flywheel is known, Equation 9.9 can be solved for the angular speed ω in rad/s. This quantity can then be converted to rev/min.

SOLUTION Solving Equation 9.9 for ω, we obtain,

$$\omega = \sqrt{\frac{2(\text{KE}_\text{R})}{I}} = \sqrt{\frac{2(\text{KE}_\text{R})}{\frac{1}{2}MR^2}} = \sqrt{\frac{4(1.2\times10^9\text{ J})}{(13\text{ kg})(0.30\text{ m})^2}} = 6.4\times10^4\text{ rad/s}$$

Converting this answer into rev/min, we find that

$$\omega = (6.4\times10^4\text{ rad/s})\left(\frac{1\text{ rev}}{2\pi\text{ rad}}\right)\left(\frac{60\text{ s}}{1\text{ min}}\right) = \boxed{6.1\times10^5\text{ rev/min}}$$

---

45. ***REASONING AND SOLUTION***

a. The tangential speed of each object is given by Equation 8.9, $v_T = r\omega$. Therefore,

For object 1: $v_{T1} = (2.00\text{ m})(6.00\text{ rad/s}) = \boxed{12.0\text{ m/s}}$

For object 2: $v_{T2} = (1.50\text{ m})(6.00\text{ rad/s}) = \boxed{9.00\text{ m/s}}$

For object 3: $v_{T3} = (3.00\text{ m})(6.00\text{ rad/s}) = \boxed{18.0\text{ m/s}}$

b. The total kinetic energy of this system can be calculated by computing the sum of the kinetic energies of each object in the system. Therefore,

$$\text{KE} = \tfrac{1}{2}m_1v_1^2 + \tfrac{1}{2}m_2v_2^2 + \tfrac{1}{2}m_3v_3^2$$

$$\text{KE} = \tfrac{1}{2}\left[(6.00\text{ kg})(12.0\text{ m/s})^2 + (4.00\text{ kg})(9.00\text{ m/s})^2 + (3.00\text{ kg})(18.0\text{ m/s})^2\right] = \boxed{1.08\times10^3\text{ J}}$$

c. The total moment of inertia of this system can be calculated by computing the sum of the moments of inertia of each object in the system. Therefore,

$$I = \sum mr^2 = m_1r_1^2 + m_2r_2^2 + m_3r_3^2$$

$$I = (6.00\text{ kg})(2.00\text{ m})^2 + (4.00\text{ kg})(1.50\text{ m})^2 + (3.00\text{ kg})(3.00\text{ m})^2 = \boxed{60.0\text{ kg}\cdot\text{m}^2}$$

d. The rotational kinetic energy of the system is, according to Equation 9.9,

$$\text{KE}_\text{R} = \tfrac{1}{2}I\omega^2 = \tfrac{1}{2}(60.0\text{ kg}\cdot\text{m}^2)(6.00\text{ rad/s})^2 = \boxed{1.08\times10^3\text{ J}}$$

This agrees, as it should, with the result for part (b).

---

49. ***REASONING AND SOLUTION*** The only force that does work on the cylinders as they move up the incline is the conservative force of gravity; hence, the total mechanical energy is conserved as the cylinders ascend the incline. We will let $h = 0$ on the horizontal plane at the bottom of the incline. Applying the principle of conservation of mechanical energy to the solid cylinder, we have

$$mgh_\text{s} = \tfrac{1}{2}mv_0^2 + \tfrac{1}{2}I_\text{s}\omega_0^2 \qquad (1)$$

where, from Table 9.1, $I_\text{s} = \tfrac{1}{2}mr^2$. In this expression, v_0 and ω_0 are the initial translational and rotational speeds, and h_s is the final height attained by the solid cylinder. Since the cylinder rolls without slipping, the rotational speed ω_0 and the translational speed v_0 are related according to Equation 8.12, $\omega_0 = v_0/r$. Then, solving Equation (1) for h_s, we obtain

$$h_\text{s} = \frac{3v_0^2}{4g}$$

Repeating the above for the hollow cylinder and using $I_\text{h} = mr^2$ we have

$$h_\text{h} = \frac{v_0^2}{g}$$

The height h attained by each cylinder is related to the distance s traveled along the incline and the angle θ of the incline by

$$s_s = \frac{h_s}{\sin\theta} \qquad \text{and} \qquad s_h = \frac{h_h}{\sin\theta}$$

Dividing these gives

$$\frac{s_s}{s_h} = \boxed{3/4}$$

---

53. ***REASONING*** Let the two disks constitute the system. Since there are no external torques acting on the system, the principle of conservation of angular momentum applies. Therefore we have $L_{\text{initial}} = L_{\text{final}}$, or

$$I_A\omega_A + I_B\omega_B = (I_A + I_B)\omega_{\text{final}}$$

This expression can be solved for the moment of inertia of disk B.

SOLUTION Solving the above expression for I_B, we obtain

$$I_B = I_A\left(\frac{\omega_{\text{final}} - \omega_A}{\omega_B - \omega_{\text{final}}}\right) = (3.4\ \text{kg}\cdot\text{m}^2)\left[\frac{-2.4\ \text{rad/s} - 7.2\ \text{rad/s}}{-9.8\ \text{rad/s} - (-2.4\ \text{rad/s})}\right] = \boxed{4.4\ \text{kg}\cdot\text{m}^2}$$

---

57. ***REASONING*** When the modules pull together, they do so by means of forces that are internal. These pulling forces, therefore, do not create a net external torque, and the angular momentum of the system is conserved. In other words, it remains constant. We will use the conservation of angular momentum to obtain a relationship between the initial and final angular speeds. Then, we will use Equation 8.9 ($v = r\omega$) to relate the angular speeds ω_0 and ω_f to the tangential speeds v_0 and v_f.

SOLUTION Let L be the initial length of the cable between the modules and ω_0 be the initial angular speed. Relative to the center-of-mass axis, the initial momentum of inertia of the two-module system is $I_0 = 2M(L/2)^2$, according to Equation 9.6. After the modules pull together, the length of the cable is $L/2$, the final angular speed is ω_f, and the momentum of inertia is $I_f = 2M(L/4)^2$. The conservation of angular momentum indicates that

$$\underbrace{I_f\omega_f}_{\substack{\text{Final angular}\\ \text{momentum}}} = \underbrace{I_0\omega_0}_{\substack{\text{Initial angular}\\ \text{momentum}}}$$

$$\left[2M\left(\frac{L}{4}\right)^2\right]\omega_f = \left[2M\left(\frac{L}{2}\right)^2\right]\omega_0$$

$$\omega_f = 4\omega_0$$

According to Equation 8.9, $\omega_f = v_f/(L/4)$ and $\omega_0 = v_0/(L/2)$. With these substitutions, the result that $\omega_f = 4\omega_0$ becomes

$$\frac{v_f}{L/4} = 4\left(\frac{v_0}{L/2}\right) \quad \text{or} \quad v_f = 2v_0 = 2(17\text{ m/s}) = \boxed{34\text{ m/s}}$$

---

61. ***REASONING*** The maximum torque will occur when the force is applied perpendicular to the diagonal of the square as shown. The lever arm ℓ is half the length of the diagonal. From the Pythagorean theorem, the lever arm is, therefore,

$$\ell = \frac{1}{2}\sqrt{(0.40\text{ m})^2 + (0.40\text{ m})^2} = 0.28\text{ m}$$

Since the lever arm is now known, we can use Equation 9.1 to obtain the desired result directly.

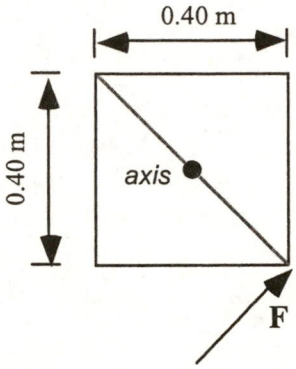

SOLUTION Equation 9.1 gives

$$\tau = F\ell = (15\text{ N})(0.28\text{ m}) = \boxed{4.2\text{ N}\cdot\text{m}}$$

---

65. ***REASONING*** The figure below shows eight particles, each one located at a different corner of an imaginary cube. As shown, if we consider an axis that lies along one edge of the cube, two of the particles lie on the axis, and for these particles $r = 0$. The next four particles closest to the axis have $r = \ell$, where ℓ is the length of one edge of the cube. The remaining two particles have $r = d$, where d is the length of the diagonal along any one of the faces. From the Pythagorean theorem, $d = \sqrt{\ell^2 + \ell^2} = \ell\sqrt{2}$.

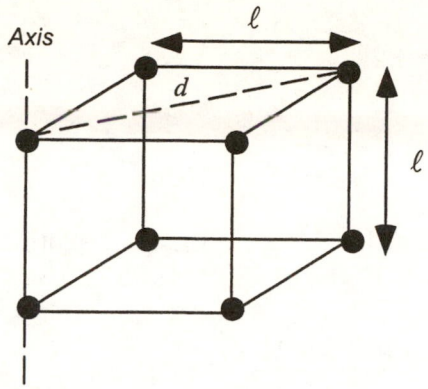

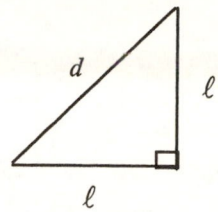

According to Equation 9.6, the moment of inertia of a system of particles is given by $I = \sum mr^2$.

SOLUTION Direct application of Equation 9.6 gives

$$I = \sum mr^2 = 4(m\ell^2) + 2(md^2) = 4(m\ell^2) + 2(2m\ell^2) = 8m\ell^2$$

or

$$I = 8(0.12\text{ kg})(0.25\text{ m})^2 = \boxed{0.060\text{ kg}\cdot\text{m}^2}$$

---

67. ***REASONING*** The drawing shows the forces acting on the board, which has a length L. The ground exerts the vertical normal force **V** on the lower end of the board. The maximum force of static friction has a magnitude of $\mu_s V$ and acts horizontally on the lower end of the board. The weight **W** acts downward at the board's center. The vertical wall applies a force **P** to the upper end of the board, this force being perpendicular to the wall since the wall is smooth (i.e., there is no friction along the wall). We take upward and to the right as our positive directions. Then, since the horizontal forces balance to zero, we have

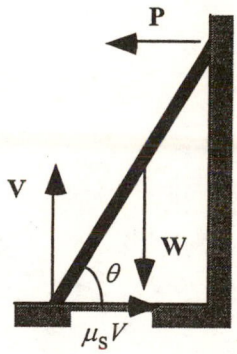

$$\mu_s V - P = 0 \tag{1}$$

The vertical forces also balance to zero giving

$$V - W = 0 \tag{2}$$

Using an axis through the lower end of the board, we express the fact that the torques balance to zero as

$$PL\sin\theta - W\left(\frac{L}{2}\right)\cos\theta = 0 \tag{3}$$

Equations (1), (2), and (3) may then be combined to yield an expression for θ.

SOLUTION Rearranging Equation (3) gives

$$\tan\theta = \frac{W}{2P} \tag{4}$$

But, $P = \mu_s V$ according to Equation (1), and $W = V$ according to Equation (2). Substituting these results into Equation (4) gives

$$\tan\theta = \frac{V}{2\mu_s V} = \frac{1}{2\mu_s}$$

Therefore,

$$\theta = \tan^{-1}\left(\frac{1}{2\mu_s}\right) = \tan^{-1}\left[\frac{1}{2(0.650)}\right] = \boxed{37.6°}$$

---

71. ***REASONING AND SOLUTION*** Consider the left board, which has a length L and a weight of (356 N)/2 = 178 N. Let $\mathbf{F}_V$ be the upward normal force exerted by the ground on the board. This force balances the weight, so $F_V = 178$ N. Let $\mathbf{f}_S$ be the force of static friction, which acts horizontally on the end of the board in contact with the ground. $\mathbf{f}_S$ points to the right. Since the board is in equilibrium, the net torque acting on the board through any axis must be zero. Measuring the torques with respect to an axis through the apex of the triangle formed by the boards, we have

$$+(178\text{ N})(\sin 30.0°)\left(\frac{L}{2}\right) + f_S(L\cos 30.0°) - F_V(L\sin 30.0°) = 0$$

or

$$44.5\text{ N} + f_s\cos 30.0° - F_V\sin 30.0° = 0$$

so that

$$f_s = \frac{(178\text{ N})(\sin 30.0°) - 44.5\text{ N}}{\cos 30.0°} = \boxed{51.4\text{ N}}$$

CHAPTER 10 | *SIMPLE HARMONIC MOTION AND ELASTICITY*

CONCEPTUAL QUESTIONS

---

7. ***REASONING AND SOLUTION*** The time required for a particle in simple harmonic motion to travel through one complete cycle (the period) is independent of the amplitude of the motion, even though at larger amplitudes the particle travels further. This is possible because, at larger amplitudes, the maximum speed of the particle is greater. Thus, even though the particle must cover larger distances at larger amplitudes, it does so with greater speeds.

---

13. ***REASONING AND SOLUTION*** The playground swing may be treated, to a good approximation, as a simple pendulum. The period of a simple pendulum is given by $T = 2\pi\sqrt{L/g}$. This expression for the period depends only on the length of the pendulum and the acceleration due to gravity; for angles less than 10° the period is independent of the amplitude of the motion. Therefore, if one person is pulled back 4° from the vertical while another person is pulled back 8° from the vertical, they will both have the same period. If they are released simultaneously, they will both come back to the starting points at the same time.

---

18. ***REASONING AND SOLUTION*** Pressure is defined in terms of the magnitude of the force per unit area, where the force is applied perpendicular to the surface.

Both sides of Equation 10.18, $F = S(\Delta X / L_0)A$ can be divided by the area A to give F/A on the left side. This F/A term cannot be called a pressure, such as the pressure that appears in Equation 10.19, because the force F in the "F/A" term is a shear force that is parallel to the surface, not perpendicular to it.

CHAPTER 10 | SIMPLE HARMONIC MOTION AND ELASTICITY

PROBLEMS

---

1. ***REASONING AND SOLUTION*** Using Equation 10.1, we first determine the spring constant:

$$k = \frac{F_x^{\text{Applied}}}{x} = \frac{89.0\text{ N}}{0.0191\text{ m}} = 4660\text{ N/m}$$

Again using Equation 10.1, we find that the force needed to compress the spring by 0.0508 m is

$$F_x^{\text{Applied}} = kx = (4660\text{ N/m})(0.0508\text{ m}) = \boxed{237\text{ N}}$$

---

5. ***REASONING AND SOLUTION*** According to Newton's second law, the force required to accelerate the trailer is $F_x = ma_x$ (Equation 4.2a). The displacement x of the spring is given by $F_x = -kx$ (Equation 10.2). Solving Equation 10.2 for x and using $F_x = ma_x$, we obtain

$$x = -\frac{F_x}{k} = -\frac{ma_x}{k} = -\frac{(92\text{ kg})\left(0.30\text{ m/s}^2\right)}{2300\text{ N/m}} = -0.012\text{ m}$$

The amount that the spring stretches is $\boxed{0.012\text{ m}}$.

---

7. ***REASONING AND SOLUTION*** The force that acts on the block is given by Newton's Second law, $F_x = ma_x$ (Equation 4.2a). Since the block has a constant acceleration, the acceleration is given by Equation 2.8 with $v_0 = 0\text{ m/s}$; that is, $a_x = 2d/t^2$, where d is the distance through which the block is pulled. Therefore, the force that acts on the block is given by

$$F_x = ma_x = \frac{2md}{t^2}$$

The force acting on the block is the restoring force of the spring. Thus, according to Equation 10.2, $F_x = -kx$, where k is the spring constant and x is the displacement. Solving Equation 10.2 for x and using the expression above for F_x, we obtain

$$x = -\frac{F_x}{k} = -\frac{2md}{kt^2} = -\frac{2(7.00\text{ kg})(4.00\text{ m})}{(415\text{ N/m})(0.750\text{ s})^2} = -0.240\text{ m}$$

The amount that the spring stretches is $\boxed{0.240\text{ m}}$.

---

11. ***REASONING*** When the ball is whirled in a horizontal circle of radius r at speed v, the centripetal force is provided by the restoring force of the spring. From Equation 5.3, the magnitude of the centripetal force is mv^2/r, while the magnitude of the restoring force is kx (see Equation 10.2). Thus,

$$\frac{mv^2}{r} = kx \qquad (1)$$

The radius of the circle is equal to $(L_0 + \Delta L)$, where L_0 is the unstretched length of the spring and ΔL is the amount that the spring stretches. Equation (1) becomes

$$\frac{mv^2}{L_0 + \Delta L} = k\Delta L \qquad (1')$$

If the spring were attached to the ceiling and the ball were allowed to hang straight down, motionless, the net force must be zero: $mg - kx = 0$, where $-kx$ is the restoring force of the spring. If we let Δy be the displacement of the spring in the vertical direction, then

$$mg = k\Delta y$$

Solving for Δy, we obtain

$$\Delta y = \frac{mg}{k} \qquad (2)$$

SOLUTION According to equation (1') above, the spring constant k is given by

$$k = \frac{mv^2}{\Delta L(L_0 + \Delta L)}$$

Substituting this expression for k into equation (2) gives

$$\Delta y = \frac{mg\Delta L(L_0 + \Delta L)}{mv^2} = \frac{g\Delta L(L_0 + \Delta L)}{v^2}$$

or

$$\Delta y = \frac{(9.80\text{ m/s}^2)(0.010\text{ m})(0.200\text{ m} + 0.010\text{ m})}{(3.00\text{ m/s})^2} = \boxed{2.29 \times 10^{-3}\text{ m}}$$

---

15. ***REASONING AND SOLUTION*** According to Equation 10.6, $\omega = 2\pi f$. The maximum speed and maximum acceleration of the atoms may be calculated from Equations 10.8 and 10.10, respectively.

a. Combining Equations 10.6 and 10.8, we obtain

$$v_{\max} = \omega A = (2\pi f)A = 2\pi(2.0\times10^{12}\ \text{Hz})(1.1\times10^{-11}\ \text{m}) = \boxed{140\ \text{m/s}}$$

b. Combining Equations 10.6 and 10.10, we obtain

$$a_{\max} = \omega^2 A = (4\pi^2 f^2)A = 4\pi^2(2.0\times10^{12}\ \text{Hz})^2(1.1\times10^{-11}\ \text{m}) = \boxed{1.7\times10^{15}\ \text{m/s}^2}$$

---

23. ***REASONING*** The angular frequency for simple harmonic motion is given by Equation 10.11 as $\omega = \sqrt{k/m}$. Since the frequency f is related to the angular frequency ω by $f = \omega/(2\pi)$ and f is related to the period T by $f = 1/T$, the period of the motion is given by

$$T = \frac{2\pi}{\omega} = 2\pi\sqrt{\frac{k}{m}}$$

SOLUTION

a. When $m_1 = m_2 = 3.0$ kg, we have that

$$T_1 = T_2 = 2\pi\sqrt{\frac{3.0\ \text{kg}}{120\ \text{N/m}}} = 0.99\ \text{s}$$

Both particles will pass through the position $x = 0$ m for the first time one-quarter of the way through one cycle, or

$$\Delta t = \tfrac{1}{4}T_1 = \tfrac{1}{4}T_2 = \frac{0.99\ \text{s}}{4} = \boxed{0.25\ \text{s}}$$

b. $T_1 = 0.99$ s, as in part (a) above, while

$$T_2 = 2\pi\sqrt{\frac{27.0\ \text{kg}}{120\ \text{N/m}}} = 3.0\ \text{s}$$

Each particle will pass through the position $x = 0$ m every odd-quarter of a cycle, $\tfrac{1}{4}T, \tfrac{3}{4}T, \tfrac{5}{4}T, \ldots$ Thus, the two particles will pass through $x = 0$ m when

3.0-kg particle $\quad t = \tfrac{1}{4}T_1, \tfrac{3}{4}T_1, \tfrac{5}{4}T_1, \ldots$

27.0-kg particle $\quad t = \tfrac{1}{4}T_2, \tfrac{3}{4}T_2, \tfrac{5}{4}T_2, \ldots$

Since $T_2 = 3T_1$, we see that both particles will be at $x = 0$ m simultaneously when $t = \frac{3}{4}T_1$, or $t = \frac{1}{4}T_2 = \frac{3}{4}T_1$. Thus,

$$t = \tfrac{3}{4}T_1 = \tfrac{3}{4}(0.99\text{ s}) = \boxed{0.75\text{ s}}$$

---

27. ***REASONING AND SOLUTION*** If we neglect air resistance, only the conservative forces of the spring and gravity act on the ball. Therefore, the principle of conservation of mechanical energy applies.

When the 2.00 kg object is hung on the end of the vertical spring, it stretches the spring by an amount y, where

$$y = \frac{F}{k} = \frac{mg}{k} = \frac{(2.00\text{ kg})(9.80\text{ m/s}^2)}{50.0\text{ N/m}} = 0.392\text{ m} \qquad (10.1)$$

This position represents the equilibrium position of the system with the 2.00-kg object suspended from the spring. The object is then pulled down another 0.200 m and released from rest ($v_0 = 0$ m/s). At this point the spring is stretched by an amount of $0.392\text{ m} + 0.200\text{ m} = 0.592\text{ m}$. This point represents the zero reference level ($h = 0$ m) for the gravitational potential energy.

***h* = 0 m:** The kinetic energy, the gravitational potential energy, and the elastic potential energy at the point of release are:

$$\text{KE} = \tfrac{1}{2}mv_0^2 = \tfrac{1}{2}m(0\text{ m/s})^2 = \boxed{0\text{ J}}$$

$$\text{PE}_{\text{gravity}} = mgh = mg(0\text{ m}) = \boxed{0\text{ J}}$$

$$\text{PE}_{\text{elastic}} = \tfrac{1}{2}ky_0^2 = \tfrac{1}{2}(50.0\text{ N/m})(0.592\text{ m})^2 = \boxed{8.76\text{ J}}$$

The total mechanical energy E_0 at the point of release is the sum of the three energies above: $E_0 = \boxed{8.76\text{ J}}$.

***h* = 0.200 m:** When the object has risen a distance of $h = 0.200$ m above the release point, the spring is stretched by an amount of $0.592\text{ m} - 0.200\text{ m} = 0.392\text{ m}$. Since the total mechanical energy is conserved, its value at this point is still $E = \boxed{8.76\text{ J}}$. The gravitational and elastic potential energies are:

$$\text{PE}_{\text{gravity}} = mgh = (2.00\text{ kg})(9.80\text{ m/s}^2)(0.200\text{ m}) = \boxed{3.92\text{ J}}$$

$$\text{PE}_{\text{elastic}} = \tfrac{1}{2}ky^2 = \tfrac{1}{2}(50.0\text{ N/m})(0.392\text{ m})^2 = \boxed{3.84\text{ J}}$$

Since $\text{KE} + \text{PE}_{\text{gravity}} + \text{PE}_{\text{elastic}} = E$,

$$\text{KE} = E - \text{PE}_{\text{gravity}} - \text{PE}_{\text{elastic}} = 8.76\text{ J} - 3.92\text{ J} - 3.84\text{ J} = \boxed{1.00\text{ J}}$$

***h* = 0.400 m:** When the object has risen a distance of $h = 0.400$ m above the release point, the spring is stretched by an amount of $0.592\text{ m} - 0.400\text{ m} = 0.192\text{ m}$. At this point, the total mechanical energy is still $E = \boxed{8.76\text{ J}}$. The gravitational and elastic potential energies are:

$$\text{PE}_{\text{gravity}} = mgh = (2.00\text{ kg})(9.80\text{ m/s}^2)(0.400\text{ m}) = \boxed{7.84\text{ J}}$$

$$\text{PE}_{\text{elastic}} = \tfrac{1}{2}ky^2 = \tfrac{1}{2}(50.0\text{ N/m})(0.192\text{ m})^2 = \boxed{0.92\text{ J}}$$

The kinetic energy is

$$\text{KE} = E - \text{PE}_{\text{gravity}} - \text{PE}_{\text{elastic}} = 8.76\text{ J} - 7.84\text{ J} - 0.92\text{ J} = \boxed{0\text{ J}}$$

The results are summarized in the table below:

h	KE	PE_{grav}	$\text{PE}_{\text{elastic}}$	E
0 m	0 J	0 J	8.76 J	8.76 J
0.200 m	1.00 J	3.92 J	3.84 J	8.76 J
0.400 m	0.00 J	7.84 J	0.92 J	8.76 J

---

31. ***REASONING*** The only force that acts on the block along the line of motion is the force due to the spring. Since the force due to the spring is a conservative force, the principle of conservation of mechanical energy applies. Initially, when the spring is unstrained, all of the mechanical energy is kinetic energy, $(1/2)mv_0^2$. When the spring is fully compressed, all of the mechanical energy is in the form of elastic potential energy, $(1/2)kx_{\text{max}}^2$, where x_{max}, the maximum compression of the spring, is the amplitude A. Therefore, the statement of energy conservation can be written as

$$\tfrac{1}{2}mv_0^2 = \tfrac{1}{2}kA^2$$

This expression may be solved for the amplitude A.

SOLUTION Solving for the amplitude A, we obtain

$$A = \sqrt{\frac{mv_0^2}{k}} = \sqrt{\frac{(1.00 \times 10^{-2}\ \text{kg})(8.00\ \text{m/s})^2}{124\ \text{N/m}}} = \boxed{7.18 \times 10^{-2}\ \text{m}}$$

---

33. ***REASONING AND SOLUTION*** The amount by which the spring stretches due to the weight of the 1.1 - kg object can be calculated using Equation 10.1 (with the variable x replaced by y), where the force F_y^{Applied} is equal to the weight of the object. The position of the object when the spring is stretched is the equilibrium position for the vertical harmonic motion of the object-spring system.

a. Solving Equation 10.1 for y with F_y^{Applied} equal to the weight of the object gives

$$y = \frac{F_y^{\text{Applied}}}{k} = \frac{mg}{k} = \frac{(1.1\ \text{kg})(9.80\ \text{m/s}^2)}{120\ \text{N/m}} = \boxed{9.0 \times 10^{-2}\ \text{m}}$$

b. The object is then pulled down another 0.20 m and released from rest ($v_0 = 0\,\text{m/s}$). At this point the spring is stretched by an amount of 0.090 m + 0.20 m = 0.29 m. We will let this point be the zero reference level ($h = 0\,\text{m}$) for the gravitational potential energy.

The kinetic energy, the gravitational potential energy, and the elastic potential energy at the point of release are:

$$\text{KE} = \tfrac{1}{2}mv_0^2 = \tfrac{1}{2}m(0\ \text{m/s})^2 = 0\ \text{J}$$

$$\text{PE}_{\text{gravity}} = mgh = mg(0\ \text{m}) = 0\ \text{J}$$

$$\text{PE}_{\text{elastic}} = \tfrac{1}{2}ky_0^2 = \tfrac{1}{2}(120\ \text{N/m})(0.29\ \text{m})^2 = 5.0\ \text{J}$$

The total mechanical energy E_0 is the sum of these three energies, so $E_0 = 5.0\ \text{J}$. When the object has risen a distance of $h = 0.20\ \text{m}$ above the release point, the spring is stretched by an amount of $0.29\ \text{m} - 0.20\ \text{m} = 0.090\ \text{m}$. Since the total mechanical energy is conserved, its value at this point is still 5.0 J. Thus,

$$E = \text{KE} + \text{PE}_{\text{gravity}} + \text{PE}_{\text{elastic}}$$

$$E = \tfrac{1}{2}mv^2 + mgh + \tfrac{1}{2}ky^2$$

$$5.0\ \text{J} = \tfrac{1}{2}(1.1\ \text{kg})v^2 + (1.1\ \text{kg})(9.80\ \text{m/s}^2)(0.20\ \text{m}) + \tfrac{1}{2}(120\ \text{N/m})(0.090\ \text{m})^2$$

Solving for v yields $v = \boxed{2.1 \text{ m/s}}$.

---

37. ***REASONING*** Using the principle of conservation of mechanical energy, the initial elastic potential energy stored in the elastic bands must be equal to the sum of the kinetic energy and the gravitational potential energy of the performer at the point of ejection:

$$\tfrac{1}{2}kx^2 = \tfrac{1}{2}mv_0^2 + mgh$$

where v_0 is the speed of the performer at the point of ejection and, from the figure at the right, $h = x \sin\theta$.

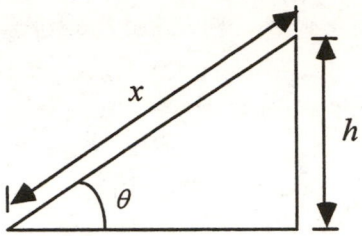

Thus,

$$\tfrac{1}{2}kx^2 = \tfrac{1}{2}mv_0^2 + mgx\sin\theta \qquad (1)$$

From the horizontal motion of the performer

$$v_{0x} = v_0 \cos\theta \qquad (2)$$

where

$$v_{0x} = \frac{s_x}{t} \qquad (3)$$

and $s_x = 26.8$ m. Combining equations (2) and (3) gives

$$v_0 = \frac{s_x}{t(\cos\theta)}$$

Equation (1) becomes:

$$\tfrac{1}{2}kx^2 = \tfrac{1}{2}m\frac{s_x^2}{t^2\cos^2\theta} + mgx\sin\theta$$

This expression can be solved for k, the spring constant of the firing mechanism.

SOLUTION Solving for k yields:

$$k = m\left(\frac{s_x}{xt\cos\theta}\right)^2 + \frac{2mg(\sin\theta)}{x}$$

$$k = (70.0\text{ kg})\left[\frac{26.8\text{ m}}{(3.00\text{ m})(2.14\text{ s})(\cos 40.0°)}\right]^2$$

$$+\frac{2(70.0\text{ kg})(9.80\text{ m/s}^2)(\sin 40.0°)}{3.00\text{ m}} = \boxed{2.37\times10^3\text{ N/m}}$$

---

47. ***REASONING AND SOLUTION*** If we assume that each block is subjected to one-fourth of the applied force F, then, according to Equation 10.18, the force on each block is given by

$$\frac{F}{4} = S\left(\frac{\Delta X}{L_0}\right)A$$

Solving for ΔX, we obtain

$$\Delta X = \frac{FL_0}{4SA} = \frac{(32\text{ N})(0.030\text{ m})}{4(2.6\times10^6\text{ N/m}^2)(1.2\times10^{-3}\text{ m}^2)} = \boxed{7.7\times10^{-5}\text{ m}}$$

---

49. ***REASONING AND SOLUTION*** Equation 10.20 gives the desired result. Solving for $\Delta V/V_0$ and taking the value for the bulk modulus B of aluminum from Table 10.3, we obtain

$$\frac{\Delta V}{V_0} = -\frac{\Delta P}{B} = -\frac{-1.01\times10^5\text{ Pa}}{7.1\times10^{10}\text{ N/m}^2} = \boxed{1.4\times10^{-6}}$$

---

53. ***REASONING AND SOLUTION*** The shearing stress is equal to the force per unit area applied to the rivet. Thus, when a shearing stress of 5.0×10^8 Pa is applied to each rivet, the force experienced by each rivet is

$$F = (\text{Stress})A = (\text{Stress})(\pi r^2) = (5.0\times10^8\text{ Pa})\left[\pi(5.0\times10^{-3}\text{ m})^2\right] = 3.9\times10^4\text{ N}$$

Therefore, the maximum tension T that can be applied to each beam, assuming that each rivet carries one-fourth of the total load, is $4F = \boxed{1.6\times10^5\text{ N}}$.

---

59. ***REASONING*** Our approach is straightforward. We will begin by writing Equation 10.17 $\left[F = Y\left(\frac{\Delta L}{L_0}\right)A\right]$ as it applies to the composite rod. In so doing, we will use subscripts for only those variables that have different values for the composite rod and the aluminum and tungsten sections. Thus, we note that the force applied to the end of the composite rod (see Figure 10.29) is also applied to each section of the rod, with the result that the magnitude F

of the force has no subscript. Similarly, the cross-sectional area A is the same for the composite rod and for the aluminum and tungsten sections. Next, we will express the change $\Delta L_{\text{Composite}}$ in the length of the composite rod as the sum of the changes in lengths of the aluminum and tungsten sections. Lastly, we will take into account that the initial length of the composite rod is twice the initial length of either of its two sections and thereby simply our equation algebraically to the point that we can solve it for the effective value of Young's modulus that applies to the composite rod.

SOLUTION Applying Equation 10.17 to the composite rod, we obtain

$$F = Y_{\text{Composite}}\left(\frac{\Delta L_{\text{Composite}}}{L_{0,\text{ Composite}}}\right)A \tag{1}$$

Since the change $\Delta L_{\text{Composite}}$ in the length of the composite rod is the sum of the changes in lengths of the aluminum and tungsten sections, we have $\Delta L_{\text{Composite}} = \Delta L_{\text{Aluminum}} + \Delta L_{\text{Tungsten}}$. Furthermore, the changes in length of each section can be expressed using Equation 10.17 $\left(\Delta L = \frac{FL_0}{YA}\right)$, so that

$$\Delta L_{\text{Composite}} = \Delta L_{\text{Aluminum}} + \Delta L_{\text{Tungsten}} = \frac{FL_{0,\text{ Aluminum}}}{Y_{\text{Aluminum}}A} + \frac{FL_{0,\text{ Tungsten}}}{Y_{\text{Tungsten}}A}$$

Substituting this result into Equation (1) gives

$$F = \left(\frac{Y_{\text{Composite}}A}{L_{0,\text{ Composite}}}\right)\Delta L_{\text{Composite}} = \left(\frac{Y_{\text{Composite}}A}{L_{0,\text{ Composite}}}\right)\left(\frac{FL_{0,\text{ Aluminum}}}{Y_{\text{Aluminum}}A} + \frac{FL_{0,\text{ Tungsten}}}{Y_{\text{Tungsten}}A}\right)$$

$$1 = Y_{\text{Composite}}\left(\frac{L_{0,\text{ Aluminum}}}{L_{0,\text{ Composite}}Y_{\text{Aluminum}}} + \frac{L_{0,\text{ Tungsten}}}{L_{0,\text{ Composite}}Y_{\text{Tungsten}}}\right)$$

In this result we now use the fact that $L_{0,\text{ Aluminum}}/L_{0,\text{ Composite}} = L_{0,\text{ Tungsten}}/L_{0,\text{ Composite}} = 1/2$ and obtain

$$1 = Y_{\text{Composite}}\left(\frac{1}{2Y_{\text{Aluminum}}} + \frac{1}{2Y_{\text{Tungsten}}}\right)$$

Solving for $Y_{\text{Composite}}$ shows that

$$Y_{\text{Composite}} = \frac{2Y_{\text{Tungsten}}Y_{\text{Aluminum}}}{Y_{\text{Tungsten}} + Y_{\text{Aluminum}}} = \frac{2\left(3.6\times10^{11}\text{ N/m}^2\right)\left(6.9\times10^{10}\text{ N/m}^2\right)}{\left(3.6\times10^{11}\text{ N/m}^2\right)+\left(6.9\times10^{10}\text{ N/m}^2\right)} = \boxed{1.2\times10^{11}\text{ N/m}^2}$$

The values for Y_{Tungsten} and Y_{Aluminum} have been taken from Table 10.1.

---

63. ***REASONING*** The strain in the wire is given by $\Delta L / L_0$. From Equation 10.17, the strain is therefore given by

$$\frac{\Delta L}{L_0} = \frac{F}{YA} \qquad (1)$$

where F must be equal to the magnitude of the centripetal force that keeps the stone moving in the circular path of radius R. Table 10.1 gives the value of Y for steel.

SOLUTION Combining Equation (1) with Equation 5.3 for the magnitude of the centripetal force, we obtain

$$\frac{\Delta L}{L_0} = \frac{F}{Y(\pi r^2)} = \frac{(mv^2/R)}{Y(\pi r^2)} = \frac{(8.0\text{ kg})(12\text{ m/s})^2/(4.0\text{ m})}{(2.0\times10^{11}\text{ Pa})\pi(1.0\times10^{-3}\text{ m})^2} = \boxed{4.6\times10^{-4}}$$

---

65. ***REASONING AND SOLUTION*** If we assume that the wire has a circular cross-section, then Equation 10.17 becomes

$$F = Y\left(\frac{\Delta L}{L_0}\right)(\pi r^2)$$

where F is the tension in the wire and r is the radius of the wire. Solving for r gives

$$r = \sqrt{\frac{FL_0}{\pi Y \Delta L}} \qquad (1)$$

Young's modulus for steel is given in Table 10.1 as $Y = 2.0 \times 10^{11}$ N/m^2. The change in length ΔL can be determined from the figure at the right. If L_0 represents the length of the wire when it is horizontal, and L its length when stretched, then the figure shows that

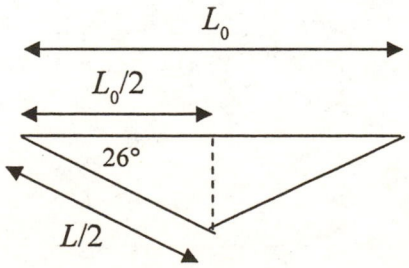

$$\cos 26° = \frac{L_0/2}{L/2} \quad \text{or} \quad L = \frac{L_0}{\cos 26°} = 1.11L_0 \qquad (2)$$

Therefore,

$$\Delta L = L - L_0 = 0.11L_0$$

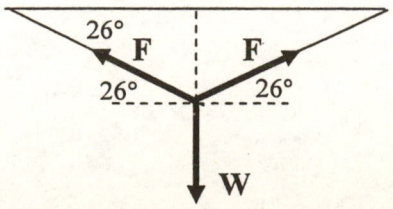

The tension in the wire can be determined from the free-body diagram at the right. Applying Newton's second law to the forces in the vertical direction yields

$$2F\sin 26° - W = 0 \quad \text{or} \quad F = \frac{W}{2\sin 26°} \tag{3}$$

Substituting Equations (2) and (3) into Equation (1) yields

$$r = \sqrt{\frac{W L_0}{2\pi(\sin 26°)Y(0.11L_0)}} = \sqrt{\frac{(96\text{ N})L_0}{2\pi(\sin 26°)(2.0\times 10^{11}\text{ N/m}^2)(0.11L_0)}} = \boxed{4.0\times 10^{-5}\text{ m}}$$

---

67. ***REASONING AND SOLUTION*** According to Equations 10.6 and 10.11, $2\pi f = \sqrt{k/m}$. According to the data in the problem, the frequency of vibration of either spring is

$$f = \frac{5.0\text{ cycles}}{3.0\text{ s}} = 1.67\text{ Hz}$$

Squaring both sides of the equation $2\pi f = \sqrt{k/m}$ and solving for k, we obtain

$$k = 4\pi^2 f^2 m = 4\pi^2(1.67\text{ Hz})^2(320\text{ kg}) = \boxed{3.5\times 10^4\text{ N/m}}$$

---

69. ***REASONING AND SOLUTION*** Recall that the relationship between frequency f and period T is $f = 1/T$. Then, according to Equations 10.6 and 10.16, the period of the simple pendulum is given by

$$T = 2\pi\sqrt{\frac{L}{g}}$$

where L is the length of the pendulum. Solving for g and noting that the period is $T = (280\text{ s})/100 = 2.8$ s, we obtain

$$g = \frac{4\pi^2 L}{T^2} = \frac{4\pi^2(1.2\text{ m})}{(2.8\text{ s})^2} = \boxed{6.0\text{ m/s}^2}$$

---

71. ***REASONING AND SOLUTION*** The amount of compression can be obtained from Equation 10.17,

$$F = Y\left(\frac{\Delta L}{L_0}\right)A$$

where F is the magnitude of the force on the stand due to the weight of the statue. Solving for ΔL gives

$$\Delta L = \frac{FL_0}{YA} = \frac{mgL_0}{YA} = \frac{(3500\text{ kg})(9.80\text{ m/s}^2)(1.8\text{ m})}{(2.3\times10^{10}\text{ N/m}^2)(7.3\times10^{-2}\text{ m}^2)} = \boxed{3.7\times10^{-5}\text{ m}}$$

---

75. ***REASONING*** The frequency of vibration of the spring is related to the added mass m by Equations 10.6 and 10.11:

$$f = \frac{1}{2\pi}\sqrt{\frac{k}{m}} \qquad (1)$$

The spring constant can be determined from Equation 10.1.

SOLUTION Since the spring stretches by 0.018 m when a 2.8-kg object is suspended from its end, the spring constant is, according to Equation 10.1,

$$k = \frac{F_x^{\text{Applied}}}{x} = \frac{mg}{x} = \frac{(2.8\text{ kg})(9.80\text{ m/s}^2)}{0.018\text{ m}} = 1.52\times10^3\text{ N/m}$$

Solving Equation (1) for m, we find that the mass required to make the spring vibrate at 3.0 Hz is

$$m = \frac{k}{4\pi^2 f^2} = \frac{1.52\times10^3\text{ N/m}}{4\pi^2(3.0\text{ Hz})^2} = \boxed{4.3\text{ kg}}$$

---

79. ***REASONING AND SOLUTION*** The natural frequency of the suspension system is given by Equation 10.11:

$$\omega = \sqrt{\frac{k}{m}} = \sqrt{\frac{1.50\times10^6\text{ N/m}}{215\text{ kg}}} = 83.5\text{ rad/s}$$

Thus, the wheel will resonate when its angular speed is 83.5 rad/s. This corresponds to a linear speed of

$$v = r\omega = (0.400\text{ m})(83.5\text{ rad/s}) = \boxed{33.4\text{ m/s}}$$

---

81. ***REASONING*** Equation 10.20 can be used to find the fractional change in volume of the brass sphere when it is exposed to the Venusian atmosphere. Once the fractional change in volume is known, it can be used to calculate the fractional change in radius.

SOLUTION According to Equation 10.20, the fractional change in volume is

$$\frac{\Delta V}{V_0} = -\frac{\Delta P}{B} = -\frac{8.9\times 10^6 \text{ Pa}}{6.7\times 10^{10} \text{ Pa}} = -1.33\times 10^{-4}$$

where we have taken the value for the bulk modulus B of brass from Table 10.3. The initial volume of the sphere is $V_0 = \frac{4}{3}\pi r_0^3$. If we assume that the change in the radius of the sphere is very small relative to the initial radius, we can think of the sphere's change in volume as the addition or subtraction of a spherical shell of volume ΔV, whose radius is r_0 and whose thickness is Δr. Then, the change in volume of the sphere is equal to the volume of the shell and is given by $\Delta V = \left(4\pi r_0^2\right)\Delta r$. Combining the expressions for V_0 and ΔV, and solving for $\Delta r / r_0$, we have

$$\frac{\Delta r}{r_0} = \left(\frac{1}{3}\right)\frac{\Delta V}{V_0}$$

Therefore,

$$\frac{\Delta r}{r_0} = \frac{1}{3}\left(-1.33\times 10^{-4}\right) = \boxed{-4.4\times 10^{-5}}$$

CHAPTER 11 | *FLUIDS*

CONCEPTUAL QUESTIONS

---

9. ***REASONING AND SOLUTION*** The *gauge pressure* of a fluid enclosed in a container is the amount by which the container pressure *exceeds* atmospheric pressure. The actual pressure of the fluid, taking atmospheric pressure into account, is called the absolute pressure. On earth, at sea level, the atmosphere exerts a pressure of 1.0×10^5 Pa on every surface with which it is in contact.

An exam question asks, "Why does the cork fly out with a loud 'pop' when a bottle of champagne is opened?" To answer this question, a student replies, "Because the gas pressure in the bottle is about 0.3×10^5 Pa." The cork flies out because the absolute pressure within the bottle exceeds the atmospheric pressure on the outside. Assuming that the student recognizes this fact, the student must be referring to a gauge pressure of 0.3×10^5 Pa. Then, the absolute pressure within the bottle would be 1.0×10^5 Pa plus 0.3×10^5 Pa and would indeed exceed atmospheric pressure.

---

16. ***REASONING AND SOLUTION*** The plastic cup containing the two blocks floats in the water because there is sufficient buoyant force to balance the total weight of the cup and the blocks. According to Archimedes' principle, the magnitude of the buoyant force is equal to the weight of the water that the cup displaces. Thus, the weight of the displaced water consists of the following three parts:

W_{plastic} = the weight of the cup itself
W_{wooden} = the weight of the wooden block
W_{lead} = the weight of the lead block

When the wooden block is removed from the cup and placed in the water, it floats. Thus, its weight W_{wooden} is still balanced by a buoyant force, which has a magnitude equal to the weight of the water that the wooden block displaces. Thus, the weight of water displaced by the floating block is W_{wooden}, just as it is when the block is in the cup. As a result, the level of the water in the pot does not change because the same amount of water is being displaced in both cases.

When the lead block (instead of the wooden block) is removed from the cup and placed in the water, it sinks to the bottom. The reason it sinks is that the magnitude of the buoyant force acting on it is less than its weight W_{lead}. When the lead block is in the cup, however, its weight is balanced by a buoyant force that has a magnitude equal to W_{lead}. Thus, the lead block outside the cup causes less water to be displaced than does the lead block in the cup. The result is that the level of the water in the pot is higher when the lead block is in the cup than when it is outside the cup. In other words, the level falls when the lead block is removed from the cup and put into the water.

---

26. ***REASONING AND SOLUTION*** As discussed in the text, the dynamic lift on airplane wings occurs because the air travels faster over the curved upper surface of the wing than it does over the flatter lower surface (see Figure 11.34). According to Bernoulli's equation, the pressure above the wing is lower (higher air speed), while the pressure below the wing is higher (lower air speed). The net upward pressure on the wing causes an upward force that lifts the plane.

The airport in Phoenix, Arizona, has occasionally been closed to large planes because of unusually low air density. According to Equation 11.12, the net upward pressure on the wing of a plane is given by $P_{\text{bottom}} - P_{\text{top}} = (1/2)\rho_{\text{air}}(v_{\text{top}}^2 - v_{\text{bottom}}^2)$ where P_{bottom} and P_{top} are the air pressures on the bottom and top surfaces of the wing, while v_{top} and v_{bottom} are the air speeds at the top and bottom wing surfaces. When the air density is low, the quantity $(v_{\text{top}}^2 - v_{\text{bottom}}^2)$ must be greater than it is under normal conditions to provide the same minimum upward force to lift the plane. The speeds, v_{top} and v_{bottom}, depend on the speed of the plane relative to the ground. If the runway is not exceptionally long, the plane may not have sufficient distance over which to increase its speed so that the quantity $(v_{\text{top}}^2 - v_{\text{bottom}}^2)$ is large enough for take-off.

CHAPTER 11 | *FLUIDS*

PROBLEMS

---

1. ***REASONING AND SOLUTION*** Equation 11.1 can be solved for the mass m. In order to use Equation 11.1, however, we must know the volume of the ice. Since the lake is circular, the volume of the ice is equal to the circular area of the lake $(A = \pi r^2)$ multiplied by the thickness d of the ice.

$$m = \rho V = \rho(\pi r^2)d = (917\ \text{kg/m}^3)\pi(480\ \text{m})^2(0.010\ \text{m}) = \boxed{6.6\times10^6\ \text{kg}}$$

---

5. ***REASONING*** Equation 11.1 can be used to find the volume occupied by 1.00 kg of silver. Once the volume is known, the area of a sheet of silver of thickness d can be found from the fact that the volume is equal to the area of the sheet times its thickness.

SOLUTION Solving Equation 11.1 for V and using a value of $\rho = 10\ 500\ \text{kg/m}^3$ for the density of silver (see Table 11.1), we find that the volume of 1.00 kg of silver is

$$V = \frac{m}{\rho} = \frac{1.00\ \text{kg}}{10\ 500\ \text{kg/m}^3} = 9.52\times10^{-5}\ \text{m}^3$$

The area of the silver, is, therefore,

$$A = \frac{V}{d} = \frac{9.52\times10^{-5}\ \text{m}^3}{3.00\times10^{-7}\ \text{m}} = \boxed{317\ \text{m}^2}$$

---

9. ***REASONING*** The total mass of the solution is the sum of the masses of its constituents. Therefore,

$$\rho_s V_s = \rho_w V_w + \rho_g V_g \qquad (1)$$

where the subscripts s, w, and g refer to the solution, the water, and the ethylene glycol, respectively. The volume of the water can be written as $V_w = V_s - V_g$. Making this substitution for V_w, Equation (1) above can be rearranged to give

$$\frac{V_g}{V_s} = \frac{\rho_s - \rho_w}{\rho_g - \rho_w} \qquad (2)$$

Equation (2) can be used to calculate the relative volume of ethylene glycol in the solution.

SOLUTION The density of ethylene glycol is given in the problem. The density of water is given in Table 11.1 as $1.000\times10^3\ \text{kg/m}^3$. The specific gravity of the solution is given as 1.0730. Therefore, the density of the solution is

$$\rho_s = (\text{specific gravity of solution})\times\rho_w$$

$$= (1.0730)(1.000\times10^3\ \text{kg/m}^3) = 1.0730\times10^3\ \text{kg/m}^3$$

Substituting the values for the densities into Equation (2), we obtain

$$\frac{V_g}{V_s} = \frac{\rho_s - \rho_w}{\rho_g - \rho_w} = \frac{1.0730\times10^3\ \text{kg/m}^3 - 1.000\times10^3\ \text{kg/m}^3}{1116\ \text{kg/m}^3 - 1.000\times10^3\ \text{kg/m}^3} = 0.63$$

Therefore, the volume percentage of ethylene glycol is $\boxed{63\%}$.

---

11. ***REASONING*** Since the inside of the box is completely evacuated, there is no air to exert an upward force on the lid from the inside. Furthermore, since the weight of the lid is negligible, there is only one force that acts on the lid; the downward force caused by the air pressure on the outside of the lid. In order to pull the lid off the box, one must supply a force that is at least equal in magnitude and opposite in direction to the force exerted on the lid by the outside air.

SOLUTION According to Equation 11.3, pressure is defined as $P = F/A$; therefore, the magnitude of the force on the lid due to the air pressure is

$$F = (0.85\times10^5\ \text{N/m}^2)(1.3\times10^{-2}\ \text{m}^2) = \boxed{1.1\times10^3\ \text{N}}$$

---

17. ***REASONING AND SOLUTION*** Both the cylinder and the hemisphere have circular cross sections. According to Equation 11.3, the pressure exerted on the ground by the hemisphere is

$$P = \frac{W_h}{A_h} = \frac{W_h}{\pi r_h^2}$$

where W_h and r_h are the weight and radius of the hemisphere. Similarly, the pressure exerted on the ground by the cylinder is

$$P = \frac{W_c}{A_c} = \frac{W_c}{\pi r_c^2}$$

where W_c and r_c are the weight and radius of the cylinder. Since each object exerts the same pressure on the ground, we can equate the right-hand sides of the expressions above to obtain

$$\frac{W_h}{\pi r_h^2} = \frac{W_c}{\pi r_c^2}$$

Solving for r_h^2, we obtain

$$r_h^2 = r_c^2 \frac{W_h}{W_c} \qquad (1)$$

The weight of the hemisphere is

$$W_h = \rho g V_h = \rho g\left[\tfrac{1}{2}\left(\tfrac{4}{3}\pi r_h^3\right)\right] = \tfrac{2}{3}\rho g \pi r_h^3$$

where ρ and V_h are the density and volume of the hemisphere, respectively. The weight of the cylinder is

$$W_c = \rho g V_c = \rho g \pi r_c^2 h$$

where ρ and V_c are the density and volume of the cylinder, respectively, and h is the height of the cylinder. Substituting these expressions for the weights into Equation (1) gives

$$r_h^2 = r_c^2 \frac{W_h}{W_c} = r_c^2 \frac{\frac{2}{3}\rho g \pi r_h^3}{\rho g \pi r_c^2 h}$$

Solving for r_h gives

$$r_h = \tfrac{3}{2}h = \tfrac{3}{2}(0.500\text{ m}) = \boxed{0.750\text{ m}}$$

---

19. ***REASONING*** Since the faucet is closed, the water in the pipe may be treated as a static fluid. The gauge pressure P_2 at the faucet on the first floor is related to the gauge pressure P_1 at the faucet on the second floor by Equation 11.4, $P_2 = P_1 + \rho g h$.

SOLUTION

a. Solving Equation 11.4 for P_1, we find the gauge pressure at the second-floor faucet is

$$P_1 = P_2 - \rho g h = 1.90\times10^5\text{ Pa} - (1.00\times10^3\text{ kg/m}^3)(9.80\text{ m/s}^2)(6.50\text{ m}) = \boxed{1.26\times10^5\text{ Pa}}$$

b. If the second faucet were placed at a height h above the first-floor faucet so that the gauge pressure P_1 at the second faucet were zero, then no water would flow from the second faucet, even if it were open. Solving Equation 11.4 for h when P_1 equals zero, we obtain

$$h = \frac{P_2 - P_1}{\rho g} = \frac{1.90 \times 10^5 \text{ Pa} - 0}{(1.00 \times 10^3 \text{ kg/m}^3)(9.80 \text{ m/s}^2)} = \boxed{19.4 \text{ m}}$$

---

23. ***REASONING AND SOLUTION*** Using Equation 11.4, we have

$$P_{\text{heart}} - P_{\text{brain}} = \rho g h = (1.000 \times 10^3 \text{ kg/m}^3)(9.80 \text{ m/s}^2)(12 \text{ m}) = \boxed{1.2 \times 10^5 \text{ Pa}}$$

---

27. ***REASONING*** Let the length of the tube be denoted by L, and let the length of the liquid be denoted by ℓ. When the tube is whirled in a circle at a constant angular speed about an axis through one end, the liquid collects at the other end and experiences a centripetal force given by (see Equation 8.11, and use $F = ma$) $F = mr\omega^2 = mL\omega^2$.

Since there is no air in the tube, this is the only radial force experienced by the liquid, and it results in a pressure of

$$P = \frac{F}{A} = \frac{mL\omega^2}{A}$$

where A is the cross-sectional area of the tube. The mass of the liquid can be expressed in terms of its density ρ and volume V: $m = \rho V = \rho A \ell$. The pressure may then be written as

$$P = \frac{\rho A \ell L \omega^2}{A} = \rho \ell L \omega^2 \qquad (1)$$

If the tube were completely filled with liquid and allowed to hang vertically, the pressure at the bottom of the tube (that is, where $h = L$) would be given by

$$P = \rho g L \qquad (2)$$

SOLUTION According to the statement of the problem, the quantities calculated by Equations (1) and (2) are equal, so that $\rho \ell L \omega^2 = \rho g L$. Solving for ω gives

$$\omega = \sqrt{\frac{g}{\ell}} = \sqrt{\frac{9.80 \text{ m/s}^2}{0.0100 \text{ m}}} = \boxed{31.3 \text{ rad/s}}$$

---

31. ***REASONING*** According to Equation 11.4, the initial pressure at the bottom of the pool is $P_0 = (P_{\text{atm}})_0 + \rho gh$, while the final pressure is $P_f = (P_{\text{atm}})_f + \rho gh$. Therefore, the change in pressure at the bottom of the pool is

$$\Delta P = P_f - P_0 = \left[(P_{\text{atm}})_f + \rho gh\right] - \left[(P_{\text{atm}})_0 + \rho gh\right] = (P_{\text{atm}})_f - (P_{\text{atm}})_0$$

According to Equation 11.3, $F = PA$, the change in the force at the bottom of the pool is

$$\Delta F = (\Delta P)A = \left[(P_{\text{atm}})_f - (P_{\text{atm}})_0\right]A$$

SOLUTION Direct substitution of the data given in the problem into the expression above yields

$$\Delta F = (765 \text{ mm Hg} - 755 \text{ mm Hg})(12 \text{ m})(24 \text{ m})\left(\frac{133 \text{ Pa}}{1.0 \text{ mm Hg}}\right) = \boxed{3.8 \times 10^5 \text{ N}}$$

Note that the conversion factor 133 Pa = 1.0 mm Hg is used to convert mm Hg to Pa.

---

37. ***REASONING*** The pressure P' exerted on the bed of the truck by the plunger is $P' = P - P_{\text{atm}}$. According to Equation 11.3, $F = P'A$, so the force exerted on the bed of the truck can be expressed as $F = (P - P_{\text{atm}})A$. If we assume that the plunger remains perpendicular to the floor of the load bed, the torque that the plunger creates about the axis shown in the figure in the text is

$$\tau = F\ell = (P - P_{\text{atm}})A\ell = (P - P_{\text{atm}})(\pi r^2)\ell$$

SOLUTION Direct substitution of the numerical data into the expression above gives

$$\tau = (3.54 \times 10^6 \text{ Pa} - 1.01 \times 10^5 \text{ Pa})\pi(0.150 \text{ m})^2(3.50 \text{ m}) = \boxed{8.50 \times 10^5 \text{ N} \cdot \text{m}}$$

---

39. ***REASONING*** According to Archimedes principle, the buoyant force that acts on the block is equal to the weight of the water that is displaced by the block. The block displaces an amount of water V, where V is the volume of the block. Therefore, the weight of the water displaced by the block is $W = mg = (\rho_{\text{water}}V)g$.

SOLUTION The buoyant force that acts on the block is, therefore,

$$F = \rho_{\text{water}}Vg = (1.00 \times 10^3 \text{ kg/m}^3)(0.10 \text{ m} \times 0.20 \text{ m} \times 0.30 \text{ m})(9.80 \text{ m/s}^2) = \boxed{59 \text{ N}}$$

---

43. ***REASONING AND SOLUTION*** Under water, the weight of the person with empty lungs is $W_{\text{empty}} = W - \rho_{\text{water}} g V_{\text{empty}}$, where W is the weight of the person in air and V_{empty} is the volume of the empty lungs. Similarly, when the person's lungs are partially full under water, the weight of the person is $W_{\text{full}} = W - \rho_{\text{water}} g V_{\text{full}}$. Subtracting the second equation from the first equation and rearranging gives

$$V_{\text{full}} - V_{\text{empty}} = \frac{W_{\text{empty}} - W_{\text{full}}}{\rho_{\text{water}} g} = \frac{40.0\text{ N} - 20.0\text{ N}}{\left(1.00\times10^{3}\text{ kg/m}^{3}\right)\left(9.80\text{ m/s}^{2}\right)} = \boxed{2.04\times10^{-3}\text{ m}^{3}}$$

---

47. ***REASONING*** The height of the cylinder that is in the oil is given by $h_{\text{oil}} = V_{\text{oil}} / (\pi r^2)$, where V_{oil} is the volume of oil displaced by the cylinder and r is the radius of the cylinder. We must, therefore, find the volume of oil displaced by the cylinder. After the oil is poured in, the buoyant force that acts on the cylinder is equal to the sum of the weight of the water displaced by the cylinder and the weight of the oil displaced by the cylinder. Therefore, the magnitude of the buoyant force is given by $F = \rho_{\text{water}} g V_{\text{water}} + \rho_{\text{oil}} g V_{\text{oil}}$. Since the cylinder floats in the fluid, the net force that acts on the cylinder must be zero. Therefore, the buoyant force that supports the cylinder must be equal to the weight of the cylinder, or

$$\rho_{\text{water}} g V_{\text{water}} + \rho_{\text{oil}} g V_{\text{oil}} = mg$$

where m is the mass of the cylinder. Substituting values into the expression above leads to

$$V_{\text{water}} + (0.725) V_{\text{oil}} = 7.00\times10^{-3}\text{ m}^{3} \qquad (1)$$

From the figure in the text, $V_{\text{cylinder}} = V_{\text{water}} + V_{\text{oil}}$. Substituting values into the expression for V_{cylinder} gives

$$V_{\text{water}} + V_{\text{oil}} = 8.48\times10^{-3}\text{ m}^{3} \qquad (2)$$

Subtracting Equation (1) from Equation (2) yields $V_{\text{oil}} = 5.38\times10^{-3}\text{ m}^{3}$.

SOLUTION The height of the cylinder that is in the oil is, therefore,

$$h_{\text{oil}} = \frac{V_{\text{oil}}}{\pi r^2} = \frac{5.38\times10^{-3}\text{ m}^{3}}{\pi(0.150\text{ m})^2} = \boxed{7.6\times10^{-2}\text{ m}}$$

---

51. ***REASONING AND SOLUTION*** The mass flow rate Q_{mass} is the amount of fluid mass that flows per unit time. Therefore,

$$Q_{\text{mass}} = \frac{m}{t} = \frac{\rho V}{t} = \frac{(1030\ \text{kg/m}^3)(9.5\times10^{-4}\ \text{m}^3)}{6.0\ \text{h}}\left(\frac{1.0\ \text{h}}{3600\ \text{s}}\right) = \boxed{4.5\times10^{-5}\ \text{kg/s}}$$

---

55. ***REASONING AND SOLUTION***

a. The volume flow rate is given by Equation 11.10. Assuming that the line has a circular cross section, $A = \pi r^2$, we have

$$Q = Av = (\pi r^2)v = \pi\ (0.0065\ \text{m})^2(1.2\ \text{m/s}) = \boxed{1.6\times10^{-4}\ \text{m}^3/\text{s}}$$

b. The volume flow rate calculated in part (a) above is the flow rate for all twelve holes. Therefore, the volume flow rate through one of the twelve holes is

$$Q_{\text{hole}} = \frac{Q}{12} = (\pi r_{\text{hole}}^2)v_{\text{hole}}$$

Solving for v_{hole} we have

$$v_{\text{hole}} = \frac{Q}{12\pi r_{\text{hole}}^2} = \frac{1.6\times10^{-4}\ \text{m}^3/\text{s}}{12\pi(4.6\times10^{-4}\ \text{m})^2} = \boxed{2.0\times10^{1}\ \text{m/s}}$$

---

57. ***REASONING AND SOLUTION***

a. Using Equation 11.12, the form of Bernoulli's equation with $y_1 = y_2$, we have

$$P_1 - P_2 = \tfrac{1}{2}\rho\left(v_2^2 - v_1^2\right) = \frac{1.29\ \text{kg/m}^3}{2}\left[(15\ \text{m/s})^2 - (0\ \text{m/s})^2\right] = \boxed{150\ \text{Pa}}$$

b. The pressure inside the roof is greater than the pressure on the outside. Therefore, there is a net outward force on the roof. If the wind speed is sufficiently high, some roofs are "blown outward."

---

59. ***REASONING AND SOLUTION*** Let the speed of the air below and above the wing be given by v_1 and v_2, respectively. According to Equation 11.12, the form of Bernoulli's equation with $y_1 = y_2$, we have

$$P_1 - P_2 = \tfrac{1}{2}\rho\left(v_2^2 - v_1^2\right) = \frac{1.29\ \text{kg/m}^3}{2}\left[(251\ \text{m/s})^2 - (225\ \text{m/s})^2\right] = 7.98\times10^{3}\ \text{Pa}$$

From Equation 11.3, the lifting force is, therefore,

$$F = \left(P_1 - P_2\right)A = (7.98 \times 10^3 \text{ Pa})(24.0 \text{ m}^2) = \boxed{1.92 \times 10^5 \text{ N}}$$

---

63. ***REASONING*** Since the pressure difference is known, Bernoulli's equation can be used to find the speed v_2 of the gas in the pipe. Bernoulli's equation also contains the unknown speed v_1 of the gas in the Venturi meter; therefore, we must first express v_1 in terms of v_2. This can be done by using Equation 11.9, the equation of continuity.

SOLUTION

a. From the equation of continuity (Equation 11.9) it follows that $v_1 = \left(A_2 / A_1\right)v_2$. Therefore,

$$v_1 = \frac{0.0700 \text{ m}^2}{0.0500 \text{ m}^2} v_2 = (1.40)v_2$$

Substituting this expression into Bernoulli's equation (Equation 11.12), we have

$$P_1 + \tfrac{1}{2}\rho(1.40\ v_2)^2 = P_2 + \tfrac{1}{2}\rho v_2^2$$

Solving for v_2, we obtain

$$v_2 = \sqrt{\frac{2(P_2 - P_1)}{\rho\left[(1.40)^2 - 1\right]}} = \sqrt{\frac{2(120 \text{ Pa})}{(1.30 \text{ kg/m}^3)\left[(1.40)^2 - 1\right]}} = \boxed{14 \text{ m/s}}$$

b. According to Equation 11.10, the volume flow rate is

$$Q = A_2 v_2 = (0.0700 \text{ m}^2)(14 \text{ m/s}) = \boxed{0.98 \text{ m}^3/\text{s}}$$

---

69. ***REASONING AND SOLUTION*** Bernoulli's equation (Equation 11.12) is

$$P_1 + \tfrac{1}{2}\rho v_1^2 = P_2 + \tfrac{1}{2}\rho v_2^2$$

If we let the right hand side refer to the air above the plate, and the left hand side refer to the air below the plate, then $v_1 = 0$, since the air below the plate is stationary. We wish to find v_2 for the situation illustrated in part *b* of the figure shown in the text. Solving the equation above for v_2 (with $v_2 = v_{2b}$ and $v_1 = 0$) gives

$$v_{2b} = \sqrt{\frac{2(P_1 - P_2)}{\rho}} \tag{1}$$

In Equation (1), P_1 is atmospheric pressure and P_2 must be determined. We must first consider the situation in part *a* of the text figure.

The figure at the right shows the forces that act on the rectangular plate in part *a* of the text drawing. F_1 is the force exerted on the plate from the air below the plate, and F_2 is the force exerted on the plate from the air above the plate. Applying Newton's second law, we have (taking "up" to be the positive direction),

$$F_1 - F_2 - mg = 0$$

$$F_1 - F_2 = mg$$

Thus, the difference in pressures exerted by the air on the plate in part *a* of the drawing is

$$P_1 - P_2 = \frac{F_1 - F_2}{A} = \frac{mg}{A} \tag{2}$$

where A is the area of the plate. From Bernoulli's equation (Equation 11.12) we have, with $v_2 = v_{2a}$ and $v_{1a} = 0$,

$$P_1 - P_2 = \tfrac{1}{2}\rho v_{2a}^2 \tag{3}$$

where v_{2a} is the speed of the air along the top of the plate in part *a* of the text drawing. Combining Equations (2) and (3) we have

$$\frac{mg}{A} = \tfrac{1}{2}\rho v_{2a}^2 \tag{4}$$

The figure below, on the left, shows the forces that act on the plate in part *b* of the text drawing. The notation is the same as that used when the plate was horizontal (part *a* of the text figure). The figure at the right below shows the same forces resolved into components along the plate and perpendicular to the plate.

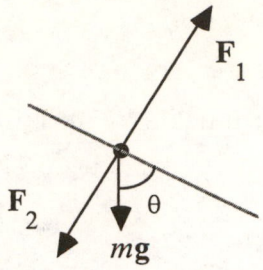

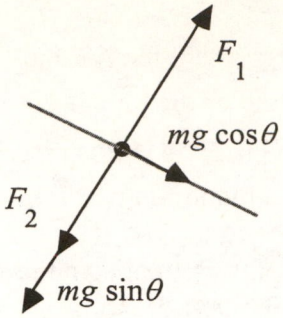

Applying Newton's second law we have $F_1 - F_2 - mg\sin\theta = 0$, or

$$F_1 - F_2 = mg\sin\theta$$

Thus, the difference in pressures exerted by the air on the plate in part *b* of the text figure is

$$P_1 - P_2 = \frac{F_1 - F_2}{A} = \frac{mg\sin\theta}{A}$$

Using Equation (4) above,

$$P_1 - P_2 = \tfrac{1}{2}\rho v_{2a}^2 \sin\theta$$

Thus, Equation (1) becomes

$$v_{2b} = \sqrt{\frac{2\left(\tfrac{1}{2}\rho v_{2a}^2 \sin\theta\right)}{\rho}}$$

Therefore,

$$v_{2b} = \sqrt{v_{2a}^2 \sin\ \theta} = \sqrt{(11.0\ \text{m/s})^2 \sin 30.0°} = \boxed{7.78\ \text{m/s}}$$

---

71. ***REASONING AND SOLUTION***

a. If the water behaves as an ideal fluid, and since the pipe is horizontal and has the same radius throughout, the speed and pressure of the water are the same at all points in the pipe. Since the right end of the pipe is open to the atmosphere, the pressure at the right end is atmospheric pressure; therefore, the pressure at the left end is also atmospheric pressure, or $\boxed{1.01\times10^5\ \text{Pa}}$.

b. If the water is treated as a viscous fluid, the volume flow rate Q is described by Poiseuille's law (Equation 11.14):

$$Q = \frac{\pi R^4 (P_2 - P_1)}{8\eta L}$$

Let P_1 represent the pressure at the right end of the pipe, and let P_2 represent the pressure at the left end of the pipe. Solving for P_2 (with P_1 equal to atmospheric pressure), we obtain

$$P_2 = \frac{8\eta LQ}{\pi R^4} + P_1$$

Therefore,

$$P_2 = \frac{8(1.00\times10^{-3}\ \text{Pa}\cdot\text{s})(1.3\ \text{m})(9.0\times10^{-3}\ \text{m}^3/\text{s})}{\pi(6.4\times10^{-3}\ \text{m})^4} + 1.013\times10^5\ \text{Pa} = \boxed{1.19\times10^5\ \text{Pa}}$$

---

75. ***REASONING AND SOLUTION***

a. Using Stoke's law, the viscous force is

$$F = 6\pi\eta Rv = 6\pi(1.00\times10^{-3}\ \text{Pa}\cdot\text{s})(5.0\times10^{-4}\ \text{m})(3.0\ \text{m/s}) = \boxed{2.8\times10^{-5}\ \text{N}}$$

b. When the sphere reaches its terminal speed, the net force on the sphere is zero, so that the magnitude of **F** must be equal to the magnitude of $m\mathbf{g}$, or $F = mg$. Therefore, $6\pi\eta Rv_T = mg$, where v_T is the terminal speed of the sphere. Solving for v_T, we have

$$v_T = \frac{mg}{6\pi\eta R} = \frac{(1.0\times10^{-5}\ \text{kg})(9.80\ \text{m/s}^2)}{6\pi(1.00\times10^{-3}\ \text{Pa}\cdot\text{s})(5.0\times10^{-4}\ \text{m})} = \boxed{1.0\times10^1\ \text{m/s}}$$

---

79. ***REASONING*** The buoyant force exerted on the balloon by the air must be equal in magnitude to the weight of the balloon and its contents (load and hydrogen). The magnitude of the buoyant force is given by $\rho_{\text{air}}Vg$. Therefore,

$$\rho_{\text{air}}Vg = W_{\text{load}} + \rho_{\text{hydrogen}}Vg$$

where, since the balloon is spherical, $V = (4/3)\pi r^3$. Making this substitution for V and solving for r, we obtain

$$r = \left[\frac{3W_{\text{load}}}{4\pi g(\rho_{\text{air}} - \rho_{\text{hydrogen}})}\right]^{1/3}$$

SOLUTION Direct substitution of the data given in the problem yields

$$r = \left[\frac{3(5750\text{ N})}{4\pi(9.80\text{ m/s}^2)\,(1.29\text{ kg/m}^3 - 0.0899\text{ kg/m}^3)}\right]^{1/3} = \boxed{4.89\text{ m}}$$

---

83. ***REASONING AND SOLUTION*** Using Equation 11.10 for the volume flow rate, we have $Q = Av$, where A is the cross-sectional area of the pipe and v is the speed of the fluid. Since the pipe has a circular cross section, $A = \pi r^2$. Therefore, solving Equation 11.10 for v, we have

$$v = \frac{Q}{A} = \frac{Q}{\pi r^2} = \frac{1.50\text{ m}^3/\text{s}}{\pi(0.500\text{ m})^2} = \boxed{1.91\text{ m/s}}$$

---

85. ***REASONING AND SOLUTION***

a. The pressure at the level of house A is given by Equation 11.4 as $P = P_{\text{atm}} + \rho gh$. Now the height h consists of the 15.0 m and the diameter d of the tank. We first calculate the radius of the tank, from which we can infer d. Since the tank is spherical, its full mass is given by $M = \rho V = \rho[(4/3)\pi r^3]$. Therefore,

$$r^3 = \frac{3M}{4\pi\rho} \quad \text{or} \quad r = \left(\frac{3M}{4\pi\rho}\right)^{1/3} = \left[\frac{3(5.25\times10^5\text{ kg})}{4\pi(1.000\times10^3\text{ kg/m}^3)}\right]^{1/3} = 5.00\text{ m}$$

Therefore, the diameter of the tank is 10.0 m, and the height h is given by

$$h = 10.0\text{ m} + 15.0\text{ m} = 25.0\text{ m}$$

According to Equation 11.4, the gauge pressure in house A is, therefore,

$$P - P_{\text{atm}} = \rho gh = (1.000\times10^3\text{ kg/m}^3)(9.80\text{ m/s}^2)(25.0\text{ m}) = \boxed{2.45\times10^5\text{ Pa}}$$

b. The pressure at house B is $P = P_{\text{atm}} + \rho gh$, where

$$h = 15.0\text{ m} + 10.0\text{ m} - 7.30\text{ m} = 17.7\text{ m}$$

According to Equation 11.4, the gauge pressure in house B is

$$P - P_{\text{atm}} = \rho gh = (1.000\times10^3\text{ kg/m}^3)(9.80\text{ m/s}^2)(17.7\text{ m}) = \boxed{1.73\times10^5\text{ Pa}}$$

---

89. ***REASONING AND SOLUTION*** The weight of the coin in air is equal to the sum of the weights of the silver and copper that make up the coin:

$$W_{\text{air}} = m_{\text{coin}} g = \left(m_{\text{silver}} + m_{\text{copper}}\right) g = \left(\rho_{\text{silver}} V_{\text{silver}} + \rho_{\text{copper}} V_{\text{copper}}\right) g \tag{1}$$

The weight of the coin in water is equal to its weight in air minus the buoyant force exerted on the coin by the water. Therefore,

$$W_{\text{water}} = W_{\text{air}} - \rho_{\text{water}} (V_{\text{silver}} + V_{\text{copper}}) g \tag{2}$$

Solving Equation (2) for the sum of the volumes gives

$$V_{\text{silver}} + V_{\text{copper}} = \frac{W_{\text{air}} - W_{\text{water}}}{\rho_{\text{water}} g} = \frac{m_{\text{coin}} g - W_{\text{water}}}{\rho_{\text{water}} g}$$

or

$$V_{\text{silver}} + V_{\text{copper}} = \frac{(1.150 \times 10^{-2}\ \text{kg})(9.80\ \text{m/s}^2) - 0.1011\ \text{N}}{(1.000 \times 10^3\ \text{kg/m}^3)(9.80\ \text{m/s}^2)} = 1.184 \times 10^{-6}\ \text{m}^3$$

From Equation (1) we have

$$V_{\text{silver}} = \frac{m_{\text{coin}} - \rho_{\text{copper}} V_{\text{copper}}}{\rho_{\text{silver}}} = \frac{1.150 \times 10^{-2}\ \text{kg} - (8890\ \text{kg/m}^3) V_{\text{copper}}}{10\ 500\ \text{kg/m}^3}$$

or

$$V_{\text{silver}} = 1.095 \times 10^{-6}\ \text{m}^3 - (0.8467) V_{\text{copper}} \tag{3}$$

Substitution of Equation (3) into Equation (2) gives

$$W_{\text{water}} = W_{\text{air}} - \rho_{\text{water}} \left[1.095 \times 10^{-6}\ \text{m}^3 - (0.8467) V_{\text{copper}} + V_{\text{copper}}\right] g$$

Solving for V_{copper} gives $V_{\text{copper}} = 5.806 \times 10^{-7}\ \text{m}^3$. Substituting into Equation (3) gives

$$V_{\text{silver}} = 1.095 \times 10^{-6}\ \text{m}^3 - (0.8467)(5.806 \times 10^{-7}\ \text{m}^3) = 6.034 \times 10^{-7}\ \text{m}^3$$

From Equation 11.1, the mass of the silver is

$$m_{\text{silver}} = \rho_{\text{silver}} V_{\text{silver}} = (10\ 500\ \text{kg/m}^3)(6.034 \times 10^{-7}\ \text{m}^3) = \boxed{6.3 \times 10^{-3}\ \text{kg}}$$

91. ***REASONING*** According to Equation 11.4, the pressure P_{mercury} at a point 7.10 cm below the ethyl alcohol-mercury interface is

$$P_{\text{mercury}} = P_{\text{interface}} + \rho_{\text{mercury}} g h_{\text{mercury}} \tag{1}$$

where $P_{\text{interface}}$ is the pressure at the alcohol-mercury interface, and $h_{\text{mercury}} = 0.0710\ \text{m}$. The pressure at the interface is

$$P_{\text{interface}} = P_{\text{atm}} + \rho_{\text{ethyl}} g h_{\text{ethyl}} \tag{2}$$

Equation (2) can be used to find the pressure at the interface. This value can then be used in Equation (1) to determine the pressure 7.10 cm below the interface.

SOLUTION Direct substitution of the numerical data into Equation (2) yields

$$P_{\text{interface}} = 1.01 \times 10^5\ \text{Pa} + (806\ \text{kg/m}^3)(9.80\ \text{m/s}^2)(1.10\ \text{m}) = 1.10 \times 10^5\ \text{Pa}$$

Therefore, the pressure 7.10 cm below the ethyl alcohol-mercury interface is

$$P_{\text{mercury}} = 1.10 \times 10^5\ \text{Pa} + (13\ 600\ \text{kg/m}^3)(9.80\ \text{m/s}^2)(0.0710\ \text{m}) = \boxed{1.19 \times 10^5\ \text{Pa}}$$

CHAPTER 12 | *TEMPERATURE AND HEAT*

CONCEPTUAL QUESTIONS

---

8. ***REASONING AND SOLUTION*** A hot steel ring fits snugly over a cold brass cylinder. The temperatures of the ring and the cylinder are, respectively, above and below room temperature. As the assembly equilibrates to room temperature, the diameter of the steel ring contracts while the diameter of the brass cylinder expands. The ring and the cylinder, therefore, fit much more snugly than they did initially. It becomes nearly impossible to pull the ring off the cylinder once the assembly has equilibrated at room temperature.

---

15. ***REASONING AND SOLUTION*** Two identical mugs contain hot coffee from the same pot. One mug is full, while the other is only one-quarter full. According to Equation 12.4, the amount of heat that must be removed from either cup to bring it to room temperature is $Q = cm\Delta T$, where m is the mass of coffee in the cup, c is the specific heat capacity of the coffee and ΔT is the temperature difference between the initial temperature of the coffee and room temperature. Since the mug that is full contains a larger mass m of coffee, more heat Q must be removed from the full mug than from the mug that is only one-quarter full. Therefore, the mug that is full remains warmer longer.

---

24. ***REASONING AND SOLUTION*** A bottle of carbonated soda is left outside in subfreezing temperatures. The temperature of the soda is "subfreezing," but the soda remains in the liquid state. Because the bottle is sealed, the pressure above the soda remains too high for the soda to freeze. When the soda is brought inside and opened, the pressure drops to atmospheric pressure. Because the temperature of the soda is below the freezing point at atmospheric pressure, the soda freezes.

CHAPTER 12 | *TEMPERATURE AND HEAT*

PROBLEMS

---

1. ***REASONING AND SOLUTION*** The temperature of –273.15 °C is 273.15 Celsius degrees *below* the ice point of 0 °C. This number of Celsius degrees corresponds to

$$(273.15\ \text{C°})\left(\frac{(9/5)\ \text{F°}}{1\ \text{C°}}\right) = 491.67\ \text{F°}$$

Subtracting 491.67 F° from the ice point of 32.00 °F on the Fahrenheit scale gives a Fahrenheit temperature of $\boxed{-459.67\ °\text{F}}$

---

5. ***REASONING AND SOLUTION*** The difference between these two averages, expressed in Fahrenheit degrees, is

$$98.6\ °\text{F} - 98.2\ °\text{F} = 0.4\ \text{F°}$$

Since 1 C° is equal to $\frac{9}{5}$ F°, we can make the following conversion

$$(0.4\ \text{F°})\left(\frac{1\ \text{C°}}{(9/5)\ \text{F°}}\right) = \boxed{0.2\ \text{C°}}$$

---

7. ***REASONING*** The temperature at any pressure can be determined from the equation of the graph in Figure 12.4. Since the gas pressure is 5.00×10^3 Pa when the temperature is 0.00 °C, and the pressure is zero when the temperature is –273.15 °C, the slope of the line is given by

$$\frac{\Delta P}{\Delta T} = \frac{5.00 \times 10^3\ \text{Pa} - 0.00\ \text{Pa}}{0.00\ °\text{C} - (-273.15\ °\text{C})} = 18.3\ \text{Pa/C°}$$

Therefore, the equation of the line is $\Delta P = (18.3\ \text{Pa/C°})\Delta T$, or

$$P - P_0 = (18.3\ \text{Pa/C°})(T - T_0)$$

where $P_0 = 5.00 \times 10^3$ Pa when $T_0 = 0.00$ °C.

SOLUTION Solving the equation above for T, we obtain

$$T = \frac{P - P_0}{18.3 \text{ Pa/C°}} + T_0 = \frac{2.00 \times 10^3 \text{ Pa} - 5.00 \times 10^3 \text{ Pa}}{18.3 \text{ Pa/C°}} + 0.00 \text{ °C} = \boxed{-164 \text{ °C}}$$

---

11. ***REASONING AND SOLUTION*** The steel in the bridge expands according to Equation 12.2, $\Delta L = \alpha L_0 \Delta T$. Solving for L_0 and using the value for the coefficient of thermal expansion of steel given in Table 12.1, we find that the approximate length of the Golden Gate bridge is

$$L_0 = \frac{\Delta L}{\alpha \Delta T} = \frac{0.53 \text{ m}}{\left[12 \times 10^{-6} \text{ (C°)}^{-1}\right](32 \text{ °C} - 2 \text{ °C})} = \boxed{1500 \text{ m}}$$

---

15. ***REASONING AND SOLUTION*** Assuming that the rod expands linearly with heat, we first calculate the quantity $\Delta L / \Delta T$ using the data given in the problem.

$$\frac{\Delta L}{\Delta T} = \frac{8.47 \times 10^{-4} \text{ m}}{100.0 \text{ °C} - 25.0 \text{ °C}} = 1.13 \times 10^{-5} \text{ m/C°}$$

Therefore, when the rod is cooled from 25.0 °C, it will shrink by

$$\begin{aligned} \Delta L &= (1.13 \times 10^{-5} \text{ m/C°})\ \Delta T \\ &= (1.13 \times 10^{-5} \text{ m/C°})\ (0.00 \text{ °C} - 25.0 \text{ °C}) = \boxed{-2.82 \times 10^{-4} \text{ m}} \end{aligned}$$

---

19. ***REASONING AND SOLUTION*** Recall that $\omega = 2\pi / T$, Equation 10.6, where ω is the angular frequency of the pendulum and T is the period. Using this fact and Equation 10.16, we know that the period of the pendulum before the temperature rise is given by $T_1 = 2\pi\sqrt{L_0 / g}$, where L_0 is the length of the pendulum. After the temperature has risen, the period becomes (using Equation 12.2), $T_2 = 2\pi\sqrt{\left(L_0 + \alpha L_0 \Delta T\right)/g}$. Dividing these expressions, solving for T_2, and taking the coefficient of thermal expansion of brass from Table 12.1, we find that

$$T_2 = T_1\sqrt{1 + \alpha \Delta T} = (2.0000 \text{ s})\sqrt{1 + (19 \times 10^{-6} \text{ /C°})\ (140 \text{ C°})} = \boxed{2.0027 \text{ s}}$$

---

23. ***REASONING*** The change in length of the wire is the sum of the change in length of each of the two segments: $\Delta L = \Delta L_{\text{al}} + \Delta L_{\text{st}}$. Using Equation 12.2 to express the changes in length, we have

$$\alpha L_0 \Delta T = \alpha_{\text{al}} L_{0\text{al}} \Delta T + \alpha_{\text{st}} L_{0\text{st}} \Delta T$$

Dividing both sides by L_0 and algebraically canceling ΔT gives

$$\alpha = \alpha_{\text{al}}\left(\frac{L_{0\text{al}}}{L_0}\right) + \alpha_{\text{st}}\left(\frac{L_{0\text{st}}}{L_0}\right)$$

The length of the steel segment of the wire is given by $L_{0\text{st}} = L_0 - L_{0\text{al}}$. Making this substitution leads to

$$\alpha = \alpha_{\text{al}}\left(\frac{L_{0\text{al}}}{L_0}\right) + \alpha_{\text{st}}\left(\frac{L_0 - L_{0\text{al}}}{L_0}\right)$$

$$= \alpha_{\text{al}}\left(\frac{L_{0\text{al}}}{L_0}\right) + \alpha_{\text{st}}\left(\frac{L_0}{L_0}\right) - \alpha_{\text{st}}\left(\frac{L_{0\text{al}}}{L_0}\right)$$

This expression can be solved for the desired quantity, $L_{0\text{al}}/L_0$.

SOLUTION Solving for the ratio $(L_{0\text{al}}/L_0)$ and taking values for the coefficients of thermal expansion for aluminum and steel from Table 12.1 gives

$$\frac{L_{0\text{al}}}{L_0} = \frac{\alpha - \alpha_{\text{st}}}{\alpha_{\text{al}} - \alpha_{\text{st}}} = \frac{19\times10^{-6}(\text{C}^\circ)^{-1} - 12\times10^{-6}(\text{C}^\circ)^{-1}}{23\times10^{-6}(\text{C}^\circ)^{-1} - 12\times10^{-6}(\text{C}^\circ)^{-1}} = \boxed{0.6}$$

---

29. ***REASONING*** The increase ΔV in volume is given by Equation 12.3 as $\Delta V = \beta V_0 \Delta T$, where β is the coefficient of volume expansion, V_0 is the initial volume, and ΔT is the increase in temperature. The lead and quartz objects experience the same change in volume. Therefore, we can use Equation 12.3 to express the two volume changes and set them equal. We will solve the resulting equation for ΔT_{Quartz}.

SOLUTION Recognizing that the lead and quartz objects experience the same change in volume and expressing that change with Equation 12.3, we have

$$\underbrace{\beta_{\text{Lead}} V_0 \Delta T_{\text{Lead}}}_{\Delta V_{\text{Lead}}} = \underbrace{\beta_{\text{Quartz}} V_0 \Delta T_{\text{Quartz}}}_{\Delta V_{\text{Quartz}}}$$

In this result V_0 is the initial volume of each object. Solving for ΔT_{Quartz} and taking values for the coefficients of volume expansion for lead and quartz from Table 12.1 gives

$$\Delta T_{\text{Quartz}} = \frac{\beta_{\text{Lead}} \Delta T_{\text{Lead}}}{\beta_{\text{Quartz}}} = \frac{\left[87\times10^{-6}(\text{C}^\circ)^{-1}\right](4.0\ \text{C}^\circ)}{1.5\times10^{-6}(\text{C}^\circ)^{-1}} = \boxed{230\ \text{C}^\circ}$$

---

31. ***REASONING AND SOLUTION*** Both the gasoline and the tank expand as the temperature increases. The coefficients of volumetric expansion β_g and β_s for gasoline and steel are available in Table 12.1. According to Equation 12.3, the volume expansion of the gasoline is

$$\Delta V_g = \beta_g V_0 \Delta T = \left[950\times10^{-6}\ (\text{C}°)^{-1}\right](20.0\ \text{gal})(18\ \text{C}°) = 0.34\ \text{gal}$$

while the volume of the steel tank expands by an amount

$$\Delta V_s = \beta_s V_0 \Delta T = \left[36\times10^{-6}\,(\text{C}°)^{-1}\right](20.0\ \text{gal})(18\ \text{C}°) = 0.013\ \text{gal}$$

The amount of gasoline which spills out is

$$\Delta V_g - \Delta V_s = \boxed{0.33\ \text{gal}}$$

---

35. ***REASONING*** In order to keep the water from expanding as its temperature increases from 15 to 25 °C, the atmospheric pressure must be increased to compress the water as it tries to expand. The magnitude of the pressure change ΔP needed to compress a substance by an amount ΔV is, according to Equation 10.20, $\Delta P = B(\Delta V / V_0)$. The ratio $\Delta V / V_0$ is, according to Equation 12.3, $\Delta V / V_0 = \beta \Delta T$. Combining these two equations yields

$$\Delta P = B\beta\Delta T$$

SOLUTION Taking the value for the coefficient of volumetric expansion β for water from Table 12.1, we find that the change in atmospheric pressure that is required to keep the water from expanding is

$$\Delta P = (2.2\times10^9\ \text{N/m}^2)\left[207\times10^{-6}\ (\text{C}°)^{-1}\right](25\ °\text{C} - 15\ °\text{C})$$

$$= \left(4.6\times10^6\ \text{Pa}\right)\left(\frac{1\ \text{atm}}{1.01\times10^5\ \text{Pa}}\right) = \boxed{45\ \text{atm}}$$

---

37. ***REASONING*** The cavity that contains the liquid in either Pyrex thermometer expands according to Equation 12.3, $\Delta V_g = \beta_g V_0 \Delta T$. On the other hand, the volume of mercury expands by an amount $\Delta V_m = \beta_m V_0 \Delta T$, while the volume of alcohol expands by an amount $\Delta V_a = \beta_a V_0 \Delta T$. Therefore, the net change in volume for the mercury thermometer is

$$\Delta V_m - \Delta V_g = (\beta_m - \beta_g) V_0 \Delta T$$

while the net change in volume for the alcohol thermometer is

$$\Delta V_a - \Delta V_g = (\beta_a - \beta_g)V_0 \Delta T$$

In each case, this volume change is related to a movement of the liquid into a cylindrical region of the thermometer with volume $\pi r^2 h$, where r is the radius of the region and h is the height of the region. For the mercury thermometer, therefore,

$$h_m = \frac{(\beta_m - \beta_g)V_0 \Delta T}{\pi r^2}$$

Similarly, for the alcohol thermometer

$$h_a = \frac{(\beta_a - \beta_g)V_0 \Delta T}{\pi r^2}$$

These two expressions can be combined to give the ratio of the heights, h_a/h_m.

SOLUTION Taking the values for the coefficients of volumetric expansion for methyl alcohol, Pyrex glass, and mercury from Table 12.1, we divide the two expressions for the heights of the liquids in the thermometers and find that

$$\frac{h_a}{h_m} = \frac{\beta_a - \beta_g}{\beta_m - \beta_g} = \frac{1200 \times 10^{-6}\ (\text{C}°)^{-1} - 9.9 \times 10^{-6}\ (\text{C}°)^{-1}}{182 \times 10^{-6}\ (\text{C}°)^{-1} - 9.9 \times 10^{-6}\ (\text{C}°)^{-1}} = 6.9$$

Therefore, the degree marks are $\boxed{\text{6.9 times further apart}}$ on the alcohol thermometer than on the mercury thermometer.

---

39. ***REASONING*** According to Equation 12.4, the heat required to warm the pool can be calculated from $Q = cm\Delta T$. The specific heat capacity c of water is given in Table 12.2. In order to use Equation 12.4, we must first determine the mass of the water in the pool. Equation 11.1 indicates that the mass can be calculated from $m = \rho V$, where ρ is the density of water and V is the volume of water in the pool.

SOLUTION Combining these two expressions, we have $Q = c\rho V \Delta T$, or

$$Q = [4186\ \text{J/(kg}\cdot\text{C°)}](1.00 \times 10^3\ \text{kg/m}^3)(12.0\ \text{m} \times 9.00\ \text{m} \times 1.5\ \text{m})\ (27\ °\text{C} - 15\ °\text{C}) = 8.14 \times 10^9\ \text{J}$$

Using the fact that $1\ \text{kWh} = 3.6 \times 10^6\ \text{J}$, the cost of using electrical energy to heat the water in the pool at a cost of \$0.10 per kWh is

$$(8.14\times10^9\ \text{J})\left(\frac{\$0.10}{3.6\times10^6\ \text{J}}\right)=\boxed{\$230.}$$

---

45. ***REASONING*** Let the system be comprised only of the metal forging and the oil. Then, according to the principle of energy conservation, the heat lost by the forging equals the heat gained by the oil, or $Q_{\text{metal}} = Q_{\text{oil}}$. According to Equation 12.4, the heat lost by the forging is $Q_{\text{metal}} = c_{\text{metal}}m_{\text{metal}}(T_{0\text{metal}} - T_{\text{eq}})$, where T_{eq} is the final temperature of the system at thermal equilibrium. Similarly, the heat gained by the oil is given by $Q_{\text{oil}} = c_{\text{oil}}m_{\text{oil}}(T_{\text{eq}} - T_{0\text{oil}})$.

SOLUTION

$$Q_{\text{metal}} = Q_{\text{oil}}$$

$$c_{\text{metal}}\,m_{\text{metal}}\,(T_{0\text{metal}} - T_{\text{eq}}) = c_{\text{oil}}\,m_{\text{oil}}\,(T_{\text{eq}} - T_{0\text{oil}})$$

Solving for $T_{0\text{metal}}$, we have

$$T_{0\text{metal}} = \frac{c_{\text{oil}}\,m_{\text{oil}}\,(T_{\text{eq}} - T_{0\text{oil}})}{c_{\text{metal}}\,m_{\text{metal}}} + T_{\text{eq}}$$

or

$$T_{0\text{metal}} = \frac{[2700\ \text{J/(kg}\cdot\text{C°)}](710\ \text{kg})(47\ °\text{C} - 32\ °\text{C})}{[430\ \text{J/(kg}\cdot\text{C°)}](75\ \text{kg})} + 47\ °\text{C} = \boxed{940\ °\text{C}}$$

---

49. ***REASONING AND SOLUTION*** According to Equation 12.4, the total heat per kilogram required to raise the temperature of the water is

$$\frac{Q}{m} = c\Delta T = [4186\ \text{J/(kg}\cdot\text{C°)}]\ (32.0\ °\text{C}) = 1.34\times10^5\ \text{J/kg}$$

The mass flow rate, $\Delta m/\Delta t$, is given by Equations 11.7 and 11.10 as $\Delta m/\Delta t = \rho A v = \rho Q_{\text{V}}$, where ρ and Q_{V} are the density and volume flow rate, respectively. We have, therefore,

$$\frac{\Delta m}{\Delta t} = (1.000\times10^3\ \text{kg/m}^3)(5.0\times10^{-6}\ \text{m}^3/\text{s}) = 5.0\times10^{-3}\ \text{kg/s}$$

Therefore, the minimum power rating of the heater must be

$$(1.34\times10^5\ \text{J/kg})(5.0\times10^{-3}\ \text{kg/s}) = 6.7\times10^2\ \text{J/s} = \boxed{6.7\times10^2\ \text{W}}$$

---

51. ***REASONING*** Heat Q_1 must be added to raise the temperature of the aluminum in its solid phase from 130 °C to its melting point at 660 °C. According to Equation 12.4, $Q_1 = cm\Delta T$. The specific heat c of aluminum is given in Table 12.2. Once the solid aluminum is at its melting point, additional heat Q_2 must be supplied to change its phase from solid to liquid. The additional heat required to melt or liquefy the aluminum is $Q_2 = mL_f$, where L_f is the latent heat of fusion of aluminum. Therefore, the total amount of heat which must be added to the aluminum in its solid phase to liquefy it is

$$Q_{\text{total}} = Q_1 + Q_2 = m(c\Delta T + L_f)$$

SOLUTION Substituting values, we obtain

$$Q_{\text{total}} = (0.45\text{ kg})\left\{[9.00\times10^2\text{ J/(kg}\cdot\text{C°)}](660\text{ °C} - 130\text{ °C}) + 4.0\times10^5\text{ J/kg}\right\} = \boxed{3.9\times10^5\text{ J}}$$

---

55. ***REASONING*** From the principle of conservation of energy, the heat lost by the coin must be equal to the heat gained by the liquid nitrogen. The heat lost by the silver coin is, from Equation 12.4, $Q = c_{\text{coin}}m_{\text{coin}}\Delta T_{\text{coin}}$ (see Table 12.2 for the specific heat capacity of silver). If the liquid nitrogen is at its boiling point, –195.8 °C, then the heat gained by the nitrogen will cause it to change phase from a liquid to a vapor. The heat gained by the liquid nitrogen is $Q = m_{\text{nitrogen}}L_v$, where m_{nitrogen} is the mass of liquid nitrogen that vaporizes, and L_v is the latent heat of vaporization for nitrogen (see Table 12.3).

SOLUTION

$$Q_{\substack{\text{lost by}\\ \text{coin}}} = Q_{\substack{\text{gained by}\\ \text{nitrogen}}}$$

$$c_{\text{coin}}m_{\text{coin}}\Delta T_{\text{coin}} = m_{\text{nitrogen}}L_v$$

Solving for the mass of the nitrogen that vaporizes

$$m_{\text{nitrogen}} = \frac{c_{\text{coin}}m_{\text{coin}}\Delta T_{\text{coin}}}{L_v}$$

$$= \frac{[235\text{ J/(kg}\cdot\text{C°)}](1.5\times10^{-2}\text{ kg})[25\text{ °C} - (-195.8\text{ °C})]}{2.00\times10^5\text{ J/kg}} = \boxed{3.9\times10^{-3}\text{ kg}}$$

---

59. ***REASONING*** According to the statement of the problem, the initial state of the system is comprised of the ice and the steam. From the principle of energy conservation, the heat lost by the steam equals the heat gained by the ice, or $Q_{\text{steam}} = Q_{\text{ice}}$. When the ice and the steam are brought together, the steam immediately begins losing heat to the ice. An amount

$Q_{1(\text{lost})}$ is released as the temperature of the steam drops from 130 °C to 100 °C, the boiling point of water. Then an amount of heat $Q_{2(\text{lost})}$ is released as the steam condenses into liquid water at 100 °C. The remainder of the heat lost by the "steam" $Q_{3(\text{lost})}$ is the heat that is released as the water at 100 °C cools to the equilibrium temperature of $T_{\text{eq}} = 50.0\ °\text{C}$. According to Equation 12.4, $Q_{1(\text{lost})}$ and $Q_{3(\text{lost})}$ are given by

$$Q_{1(\text{lost})} = c_{\text{steam}} m_{\text{steam}} (T_{\text{steam}} - 100.0\ °\text{C}) \quad \text{and} \quad Q_{3(\text{lost})} = c_{\text{water}} m_{\text{steam}} (100.0\ °\text{C} - T_{\text{eq}})$$

$Q_{2(\text{lost})}$ is given by $Q_{2(\text{lost})} = m_{\text{steam}} L_{\text{v}}$, where L_{v} is the latent heat of vaporization of water. The total heat lost by the steam has three effects on the ice. First, a portion of this heat $Q_{1(\text{gained})}$ is used to raise the temperature of the ice to its melting point at 0.00 °C. Then, an amount of heat $Q_{2(\text{gained})}$ is used to melt the ice completely (we know this because the problem states that after thermal equilibrium is reached the liquid phase is present at 50.0 °C). The remainder of the heat $Q_{3(\text{gained})}$ gained by the "ice" is used to raise the temperature of the resulting liquid at 0.0 °C to the final equilibrium temperature. According to Equation 12.4, $Q_{1(\text{gained})}$ and $Q_{3(\text{gained})}$ are given by

$$Q_{1(\text{gained})} = c_{\text{ice}} m_{\text{ice}} (0.00\ °\text{C} - T_{\text{ice}}) \quad \text{and} \quad Q_{3(\text{gained})} = c_{\text{water}} m_{\text{ice}} (T_{\text{eq}} - 0.00\ °\text{C})$$

$Q_{2(\text{gained})}$ is given by $Q_{2(\text{gained})} = m_{\text{ice}} L_{\text{f}}$, where L_{f} is the latent heat of fusion of ice.

SOLUTION According to the principle of energy conservation, we have

$$Q_{\text{steam}} = Q_{\text{ice}}$$

$$Q_{1(\text{lost})} + Q_{2(\text{lost})} + Q_{3(\text{lost})} = Q_{1(\text{gained})} + Q_{2(\text{gained})} + Q_{3(\text{gained})}$$

or

$$c_{\text{steam}} m_{\text{steam}} (T_{\text{steam}} - 100.0\ °\text{C}) + m_{\text{steam}} L_{\text{v}} + c_{\text{water}} m_{\text{steam}} (100.0\ °\text{C} - T_{\text{eq}})$$

$$= c_{\text{ice}} m_{\text{ice}} (0.00\ °\text{C} - T_{\text{ice}}) + m_{\text{ice}} L_{\text{f}} + c_{\text{water}} m_{\text{ice}} (T_{\text{eq}} - 0.00\ °\text{C})$$

Values for specific heats are given in Table 12.2, and values for the latent heats are given in Table 12.3. Solving for the ratio of the masses gives

$$\frac{m_{\text{steam}}}{m_{\text{ice}}} = \frac{c_{\text{ice}} (0.00\ °\text{C} - T_{\text{ice}}) + L_{\text{f}} + c_{\text{water}} (T_{\text{eq}} - 0.00\ °\text{C})}{c_{\text{steam}} (T_{\text{steam}} - 100.0\ °\text{C}) + L_{\text{v}} + c_{\text{water}} (100.0\ °\text{C} - T_{\text{eq}})}$$

$$= \frac{\left[2.00\times10^3\ \text{J/(kg}\cdot\text{C°)}\right]\left[0.0\ \text{°C}-(-10.0\text{°C})\right]+33.5\times10^4\ \text{J/kg}+\left[4186\ \text{J/(kg}\cdot\text{C°)}\right](50.0\ \text{°C}-0.0\ \text{°C})}{\left[2020\ \text{J/(kg}\cdot\text{C°)}\right](130\ \text{°C}-100.0\ \text{°C})+22.6\times10^5\ \text{J/kg}+\left[4186\ \text{J/(kg}\cdot\text{C°)}\right](100.0\ \text{°C}-50.0\ \text{°C})}$$

or

$$\frac{m_{\text{steam}}}{m_{\text{ice}}} = \boxed{0.223}$$

---

63. ***REASONING*** The system is comprised of the unknown material, the glycerin, and the aluminum calorimeter. From the principle of energy conservation, the heat gained by the unknown material is equal to the heat lost by the glycerin and the calorimeter. The heat gained by the unknown material is used to melt the material and then raise its temperature from the initial value of –25.0 °C to the final equilibrium temperature of $T_{\text{eq}} = 20.0\ \text{°C}$.

SOLUTION

$$Q_{\substack{\text{gained by}\\ \text{unknown}}} = Q_{\substack{\text{lost by}\\ \text{glycerine}}} + Q_{\substack{\text{lost by}\\ \text{calorimeter}}}$$

$$m_{\text{u}}L_{\text{f}} + c_{\text{u}}m_{\text{u}}\Delta T_{\text{u}} = c_{\text{gl}}m_{\text{gl}}\Delta T_{\text{gl}} + c_{\text{al}}m_{\text{al}}\Delta T_{\text{al}}$$

Taking values for the specific heat capacities of glycerin and aluminum from Table 12.2, we have

$$(0.10\ \text{kg})L_{\text{f}} + [160\ \text{J/(kg}\cdot\text{C°)}](0.10\ \text{kg})(45.0\ \text{C°}) = [2410\ \text{J/(kg}\cdot\text{C°)}](0.100\ \text{kg})(7.0\ \text{C°}) + [9.0\times10^2\ \text{J/(kg}\cdot\text{C°)}](0.150\ \text{kg})(7.0\ \text{C°})$$

Solving for L_{f} yields,

$$L_{\text{f}} = \boxed{1.9\times10^4\ \text{J/kg}}$$

---

65. ***REASONING*** In order to melt, the bullet must first heat up to 327.3 °C (its melting point) and then undergo a phase change. According to Equation 12.4, the amount of heat necessary to raise the temperature of the bullet to 327.3 °C is $Q = cm(327.3\ \text{°C} - 30.0\ \text{°C})$, where m is the mass of the bullet. The amount of heat required to melt the bullet is given by $Q_{\text{melt}} = mL_{\text{f}}$, where L_{f} is the latent heat of fusion of lead.

The lead bullet melts completely when it comes to a sudden halt; all of the kinetic energy of the bullet is converted into heat; therefore,

$$\text{KE} = Q + Q_{\text{melt}}$$

$$\frac{1}{2}mv^2 = cm(327.3\ \text{°C} - 30.0\ \text{°C}) + mL_{\text{f}}$$

The value for the specific heat c of lead is given in Table 12.2, and the value for the latent heat of fusion L_f of lead is given in Table 12.3. This expression can be solved for v, the minimum speed of the bullet for such an event to occur.

SOLUTION Solving for v, we find that the minimum speed of the lead bullet is

$$v = \sqrt{2L_f + 2c\,(327.3\ °\text{C} - 30.0\ °\text{C})}$$

$$v = \sqrt{2(2.32 \times 10^4\ \text{J/kg}) + 2[128\ \text{J/(kg}\cdot\text{C°)}]\,(327.3\ °\text{C} - 30.0\ °\text{C})} = \boxed{3.50 \times 10^2\ \text{m/s}}$$

---

67. ***REASONING AND SOLUTION*** From inspection of the graph that accompanies this problem, a pressure of 3.5×10^6 Pa corresponds to a temperature of 0 °C. Therefore, liquid carbon dioxide exists in equilibrium with its vapor phase at $\boxed{0\ °\text{C}}$ when the vapor pressure is 3.5×10^6 Pa.

---

75. ***REASONING*** We must first find the equilibrium temperature T_{eq} of the iced tea. Once this is known, we can use the vapor pressure curve that accompanies Problem 68 to find the partial pressure of water vapor at that temperature and then estimate the relative humidity using Equation 12.6.

According to the principle of energy conservation, when the ice is mixed with the tea, the heat lost by the tea is gained by the ice, or $Q_{tea} = Q_{ice}$. The heat gained by the ice is used to melt the ice at 0.0 °C; the remainder of the heat is used to bring the water at 0.0 °C up to the final equilibrium temperature T_{eq}.

SOLUTION

$$Q_{tea} = Q_{ice}$$

$$c_{water}\,m_{tea}\,(30.0\ °\text{C} - T_{eq}) = m_{ice}L_f + c_{water}\,m_{ice}\,(T_{eq} - 0.00\ °\text{C})$$

The specific heat capacity of water is given in Table 12.2, and the latent heat of fusion L_f of water is given in Table 12.3. Solving for T_{eq}, we have

$$T_{eq} = \frac{c_{water}\,m_{tea}\,(30.0\ °\text{C}) - m_{ice}L_f}{c_{water}\,(m_{tea} + m_{ice})}$$

$$= \frac{[4186\ \text{J/(kg}\cdot\text{C°)}]\,(0.300\ \text{kg})(30.0\ °\text{C}) - (0.0670\ \text{kg})(33.5 \times 10^4\ \text{J/kg})}{[4186\ \text{J/(kg}\cdot\text{C°)}]\,(0.300\ \text{kg} + 0.0670\ \text{kg})} = 9.91\ °\text{C}$$

According to the vapor pressure curve that accompanies Problem 68, at a temperature of 9.91 °C, the equilibrium vapor pressure is approximately 1250 Pa. At 30 °C, the equilibrium vapor pressure is approximately 4400 Pa. Therefore, according to Equation 12.6, the percent relative humidity is approximately

$$\text{Percent relative humidity} = \left(\frac{1250\ \text{Pa}}{4400\ \text{Pa}}\right) \times 100 = \boxed{28\%}$$

---

79. ***REASONING*** From the conservation of energy, the heat lost by the mercury is equal to the heat gained by the water. As the mercury loses heat, its temperature decreases; as the water gains heat, its temperature rises to its boiling point. Any remaining heat gained by the water will then be used to vaporize the water.

According to Equation 12.4, the heat lost by the mercury is $Q_{\text{mercury}} = (cm\Delta T)_{\text{mercury}}$. The heat required to vaporize the water is, from Equation 12.5, $Q_{\text{vap}} = (m_{\text{vap}}L_{\text{v}})_{\text{water}}$. Thus, the total amount of heat gained by the water is $Q_{\text{water}} = (cm\Delta T)_{\text{water}} + (m_{\text{vap}}L_{\text{v}})_{\text{water}}$.

SOLUTION

$$Q_{\substack{\text{lost by}\\ \text{mercury}}} = Q_{\substack{\text{gained by}\\ \text{water}}}$$

$$(cm\Delta T)_{\text{mercury}} = (cm\Delta T)_{\text{water}} + (m_{\text{vap}}L_{\text{v}})_{\text{water}}$$

where $\Delta T_{\text{mercury}} = (205\ °\text{C} - 100.0\ °\text{C})$ and $\Delta T_{\text{water}} = (100.0\ °\text{C} - 80.0\ °\text{C})$. The specific heats of mercury and water are given in Table 12.2, and the latent heat of vaporization of water is given in Table 12.3. Solving for the mass of the water that vaporizes gives

$$m_{\text{vap}} = \frac{c_{\text{mercury}}m_{\text{mercury}}\Delta T_{\text{mercury}} - c_{\text{water}}m_{\text{water}}\Delta T_{\text{water}}}{(L_{\text{v}})_{\text{water}}}$$

$$= \frac{[139\ \text{J/(kg}\cdot\text{C°)}](2.10\ \text{kg})(105\ \text{C°}) - [4186\ \text{J/(kg}\cdot\text{C°)}](0.110\ \text{kg})(20.0\ \text{C°})}{22.6\times10^{5}\ \text{J/kg}}$$

$$= \boxed{9.49\times10^{-3}\ \text{kg}}$$

---

83. ***REASONING AND SOLUTION*** The cider will expand according to Equation 12.3; therefore, the change in volume of the cider is

$$\Delta V = \beta V_0 \Delta T = \left[280\times10^{-6}\ (\text{C°})^{-1}\right](1.0\ \text{gal})(22\ \text{C°}) = 6.2\times10^{-3}\ \text{gal}$$

At a cost of two dollars per gallon, this amounts to

$$\left(6.2\times10^{-3}\ \text{gal}\right)\left(\frac{\$\ 2.00}{1\ \text{gal}}\right)=\$\ 0.01 \quad \text{or} \quad \boxed{\text{one penny}}$$

---

85. ***REASONING AND SOLUTION*** From the vapor pressure curve that accompanies Problem 68, it is seen that the partial pressure of water vapor in the atmosphere at 10 °C is about 1400 Pa, and that the equilibrium vapor pressure at 30 °C is about 4200 Pa. The relative humidity is, from Equation 12.6,

$$\begin{matrix}\text{Percent}\\ \text{relative}\\ \text{humidity}\end{matrix}=\left(\frac{1400\ \text{Pa}}{4200\ \text{Pa}}\right)\times 100=\boxed{33\%}$$

---

91. ***REASONING AND SOLUTION*** As the rock falls through a distance h, its initial potential energy $m_{\text{rock}}gh$ is converted into kinetic energy. This kinetic energy is then converted into heat when the rock is brought to rest in the pail. If we ignore the heat absorbed by the pail, the principle of conservation of energy indicates that

$$m_{\text{rock}}gh = c_{\text{rock}}m_{\text{rock}}\Delta T + c_{\text{water}}m_{\text{water}}\Delta T$$

where we have used Equation 12.4 to express the heat absorbed by the rock and the water. Table 12.2 gives the specific heat capacity of the water. Solving for ΔT yields

$$\Delta T = \frac{m_{\text{rock}}gh}{c_{\text{rock}}m_{\text{rock}} + c_{\text{water}}m_{\text{water}}}$$

Substituting values yields

$$\Delta T = \frac{(0.20\ \text{kg})(9.80\ \text{m/s}^2)(15\ \text{m})}{[1840\ \text{J/(kg}\cdot\text{C°)}](0.20\ \text{kg}) + [4186\ \text{J/(kg}\cdot\text{C°)}](0.35\ \text{kg})} = \boxed{0.016\ \text{C°}}$$

CHAPTER 13 | *THE TRANSFER OF HEAT*

CONCEPTUAL QUESTIONS

---

5. ***REASONING AND SOLUTION*** A piece of Styrofoam and a piece of wood are sandwiched together to form a layered slab. The two pieces have the same thickness and cross-sectional area. The exposed surfaces have constant temperatures. The temperature of the exposed Styrofoam surface is greater than the temperature of the exposed wood surface. The rate of heat flow through either layer can be determined from Equation 13.1: $Q/t = kA\Delta T/L$, where k is the thermal conductivity of the layer, A and L are the cross-sectional area and thickness of the layer, respectively, and ΔT is the temperature difference between the ends of the layer. Since heat is not trapped within the sandwich, the rate at which heat flows through the sandwich, Q/t, must be uniform throughout both layers. Therefore, $(kA\Delta T/L)_{\text{Styrofoam}} = (kA\Delta T/L)_{\text{wood}}$. Since both layers have the same cross-sectional area and thickness, A and L are the same for both layers. Therefore, $k_{\text{Styrofoam}}\Delta T_{\text{Styrofoam}} = k_{\text{wood}}\Delta T_{\text{wood}}$. From Table 13.1, we see that the thermal conductivity of Styrofoam is less than the thermal conductivity of wood; therefore, the temperature difference between the two ends of the wood layer must be smaller than the temperature difference between the two ends of the Styrofoam layer. From this, we can conclude that the temperature at the Styrofoam-wood interface must be closer to the lower temperature of the exposed wood surface.

---

13. ***REASONING AND SOLUTION*** One car has a metal body, while another car has a plastic body. On a cold winter day, these cars are parked side by side. The metal car feels colder to the touch of your bare hand even though both cars are at the same temperature. This is because your fingers are sensitive to the rate at which heat is transferred to or from them, rather than to the temperature itself. The metal car feels colder than the plastic car at the same temperature, because heat flows from your bare hand into the metal car more readily than it flows into the plastic car. The flow occurs into the metal more readily, because the thermal conductivity of the metal is greater than that of the plastic.

---

16. ***REASONING AND SOLUTION***

a. The highly reflective paint reduces the ability of the so-called "radiator" to deliver heat into the room via the mechanism of radiation. This is because the paint allows the surface of the device to reflect more radiation than it otherwise would. Being a better reflector means that the device has become a poorer absorber of radiation, and poor absorbers are also poor emitters. Since the painted "radiator" loses less heat by the mechanism of radiation, it becomes hotter than it would if it were unpainted.

b. Since painting the device reduces its ability to radiate electromagnetic waves, we dismiss radiation as the primary mechanism by which "radiators" deliver heat. We also dismiss conduction, since air is not a good conductor of heat. That leaves convection. "Radiators" indeed function primarily via convection. The fact that the paint enables the device to become hotter for a given supply of hot water or steam is beneficial for convection. The hotter the device becomes, the more effectively it can generate the convection currents that distribute the heat around the room.

CHAPTER 13 | *THE TRANSFER OF HEAT*

PROBLEMS

---

1. ***REASONING*** The heat conducted through the iron poker is given by Equation 13.1, $Q = (kA\,\Delta T)t / L$. If we assume that the poker has a circular cross-section, then its cross-sectional area is $A = \pi r^2$. Table 13.1 gives the thermal conductivity of iron as $79\ \text{J}/(\text{s}\cdot\text{m}\cdot\text{C}°)$.

SOLUTION The amount of heat conducted from one end of the poker to the other in 5.0 s is, therefore,

$$Q = \frac{(k\,A\,\Delta T)t}{L} = \frac{[79\ \text{J}/(\text{s}\cdot\text{m}\cdot\text{C}°)]\pi(5.0\times10^{-3}\ \text{m})^2(502\ °\text{C} - 26\ °\text{C})(5.0\ \text{s})}{1.2\ \text{m}} = \boxed{12\ \text{J}}$$

---

5. ***REASONING*** The heat transferred in a time t is given by Equation 13.1, $Q = (k\,A\,\Delta T)t / L$. If the same amount of heat per second is conducted through the two plates, then $(Q/t)_{\text{al}} = (Q/t)_{\text{st}}$. Using Equation 13.1, this becomes

$$\frac{k_{\text{al}} A\,\Delta T}{L_{\text{al}}} = \frac{k_{\text{st}} A\,\Delta T}{L_{\text{st}}}$$

This expression can be solved for L_{st}.

SOLUTION Solving for L_{st} gives

$$L_{\text{st}} = \frac{k_{\text{st}}}{k_{\text{al}}} L_{\text{al}} = \frac{14\ \text{J}/(\text{s}\cdot\text{m}\cdot\text{C}°)}{240\ \text{J}/(\text{s}\cdot\text{m}\cdot\text{C}°)}(0.035\ \text{m}) = \boxed{2.0\times10^{-3}\ \text{m}}$$

---

7. ***REASONING AND SOLUTION*** Values for the thermal conductivities of Styrofoam and air are given in Table 11.1. The conductance of an 0.080 mm thick sample of Styrofoam of cross-sectional area A is

$$\frac{k_{\text{s}} A}{L_{\text{s}}} = \frac{[0.010\ \text{J}/(\text{s}\cdot\text{m}\cdot\text{C}°)]\,A}{0.080\times10^{-3}\ \text{m}} = [125\ \text{J}/(\text{s}\cdot\text{m}^2\cdot\text{C}°)]\,A$$

The conductance of a 3.5 mm thick sample of air of cross-sectional area A is

$$\frac{k_a A}{L_a} = \frac{[0.0256\ \text{J/(s}\cdot\text{m}\cdot\text{C°)}]\ A}{3.5\times10^{-3}\ \text{m}} = [7.3\ \text{J/(s}\cdot\text{m}^2\cdot\text{C°)}]\ A$$

Dividing the conductance of Styrofoam by the conductance of air for samples of the same cross-sectional area A, gives

$$\frac{[125\ \text{J/(s}\cdot\text{m}^2\cdot\text{C°)}]\ A}{[7.3\ \text{J/(s}\cdot\text{m}^2\cdot\text{C°)}]\ A} = 17$$

Therefore, the body can adjust the conductance of the tissues beneath the skin by $\boxed{\text{a factor of 17}}$.

---

11. ***REASONING*** The heat lost per second due to conduction through the glass is given by Equation 13.1 as $Q/t = (kA\Delta T)/L$. In this expression, we have no information for the thermal conductivity k, the cross-sectional area A, or the length L. Nevertheless, we can apply the equation to the initial situation and again to the situation where the outside temperature has fallen. This will allow us to eliminate the unknown variables from the calculation.

SOLUTION Applying Equation 13.1 to the initial situation and to the situation after the outside temperature has fallen, we obtain

$$\left(\frac{Q}{t}\right)_{\text{Initial}} = \frac{kA\left(T_{\text{In}} - T_{\text{Out, initial}}\right)}{L} \quad \text{and} \quad \left(\frac{Q}{t}\right)_{\text{Colder}} = \frac{kA\left(T_{\text{In}} - T_{\text{Out, colder}}\right)}{L}$$

Dividing these two equations to eliminate the common variables gives

$$\frac{(Q/t)_{\text{Colder}}}{(Q/t)_{\text{Initial}}} = \frac{\dfrac{kA\left(T_{\text{In}} - T_{\text{Out, colder}}\right)}{L}}{\dfrac{kA\left(T_{\text{In}} - T_{\text{Out, initial}}\right)}{L}} = \frac{T_{\text{In}} - T_{\text{Out, colder}}}{T_{\text{In}} - T_{\text{Out, initial}}}$$

Remembering that twice as much heat is lost per second when the outside is colder, we find

$$\frac{2(Q/t)_{\text{Initial}}}{(Q/t)_{\text{Initial}}} = 2 = \frac{T_{\text{In}} - T_{\text{Out, colder}}}{T_{\text{In}} - T_{\text{Out, initial}}}$$

Solving for the colder outside temperature gives

$$T_{\text{Out, colder}} = 2T_{\text{Out, initial}} - T_{\text{In}} = 2(5.0\ °\text{C}) - (25\ °\text{C}) = \boxed{-15\ °\text{C}}$$

15. ***REASONING*** If the cylindrical rod were made of solid copper, the amount of heat it would conduct in a time t is, according to Equation 13.1, $Q_{\text{copper}} = (k_{\text{copper}} A_2 \, \Delta T / L)t$. Similarly, the amount of heat conducted by the lead-copper combination is the sum of the heat conducted through the copper portion of the rod and the heat conducted through the lead portion:

$$Q_{\text{combination}} = \left[k_{\text{copper}} (A_2 - A_1)\Delta T / L + k_{\text{lead}} A_1 \Delta T / L \right] t .$$

Since the lead-copper combination conducts one-half the amount of heat than does the solid copper rod, $Q_{\text{combination}} = \frac{1}{2} Q_{\text{copper}}$, or

$$\frac{k_{\text{copper}}(A_2 - A_1)\Delta T}{L} + \frac{k_{\text{lead}} A_1 \Delta T}{L} = \frac{1}{2}\left(\frac{k_{\text{copper}} A_2 \, \Delta T}{L} \right)$$

This expression can be solved for A_1 / A_2, the ratio of the cross-sectional areas. Since the cross-sectional area of a cylinder is circular, $A = \pi r^2$. Thus, once the ratio of the areas is known, the ratio of the radii can be determined.

SOLUTION Solving for the ratio of the areas, we have

$$\frac{A_1}{A_2} = \frac{k_{\text{copper}}}{2\left(k_{\text{copper}} - k_{\text{lead}}\right)}$$

The cross-sectional areas are circular so that $A_1 / A_2 = (\pi r_1^2)/(\pi r_2^2) = (r_1 / r_2)^2$; therefore,

$$\frac{r_1}{r_2} = \sqrt{\frac{k_{\text{copper}}}{2(k_{\text{copper}} - k_{\text{lead}})}} = \sqrt{\frac{390 \text{ J/(s}\cdot\text{m}\cdot\text{C°)}}{2[390 \text{ J/(s}\cdot\text{m}\cdot\text{C°)} - 35 \text{ J/(s}\cdot\text{m}\cdot\text{C°)}]}} = \boxed{0.74}$$

where we have taken the thermal conductivities of copper and lead from Table 13.1.

---

17. ***REASONING AND SOLUTION*** Solving the Stefan-Boltzmann law, Equation 13.2, for the time t, and using the fact that $Q_{\text{blackbody}} = Q_{\text{bulb}}$, we have

$$t_{\text{blackbody}} = \frac{Q_{\text{blackbody}}}{\sigma T^4 A} = \frac{Q_{\text{bulb}}}{\sigma T^4 A} = \frac{P_{\text{bulb}} \, t_{\text{bulb}}}{\sigma T^4 A}$$

where P_{bulb} is the power rating of the light bulb. Therefore,

$$t_{\text{blackbody}} = \frac{(100.0\ \text{J/s})\ (3600\ \text{s})}{\left[5.67\times10^{-8}\ \text{J/(s}\cdot\text{m}^2\cdot\text{K}^4)\right]\ (303\ \text{K})^4\ \left[(6\ \text{sides})(0.0100\ \text{m})^2/\text{side}\right]}$$

$$\times\left(\frac{1\ \text{h}}{3600\ \text{s}}\right)\left(\frac{1\ \text{d}}{24\ \text{h}}\right) = \boxed{14.5\ \text{d}}$$

---

21. ***REASONING AND SOLUTION*** The net power generated by the stove is given by Equation 13.3, $P_{\text{net}} = e\sigma A(T^4 - T_0^4)$. Solving for T gives

$$T = \left(\frac{P_{\text{net}}}{e\sigma A} + T_0^4\right)^{1/4}$$

$$= \left\{\frac{7300\ \text{W}}{(0.900)[5.67\times10^{-8}\ \text{J}/(\text{s}\cdot\text{m}^2\cdot\text{K}^4)](2.00\ \text{m}^2)} + (302\ \text{K})^4\right\}^{1/4} = \boxed{532\ \text{K}}$$

---

25. ***REASONING*** The total radiant power emitted by an object that has a Kelvin temperature T, surface area A, and emissivity e can be found by rearranging Equation 13.2, the Stefan-Boltzmann law: $Q = e\sigma T^4 At$. The emitted power is $P = Q/t = e\sigma T^4 A$. Therefore, when the original cylinder is cut perpendicular to its axis into N smaller cylinders, the ratio of the power radiated by the pieces to that radiated by the original cylinder is

$$\frac{P_{\text{pieces}}}{P_{\text{original}}} = \frac{e\sigma T^4 A_2}{e\sigma T^4 A_1} \qquad (1)$$

where A_1 is the surface area of the original cylinder, and A_2 is the sum of the surface areas of all N smaller cylinders. The surface area of the original cylinder is the sum of the surface area of the ends and the surface area of the cylinder body; therefore, if L and r represent the length and cross-sectional radius of the original cylinder, with $L = 10r$,

$$A_1 = (\text{area of ends}) + (\text{area of cylinder body})$$

$$= 2(\pi r^2) + (2\pi r)L = 2(\pi r^2) + (2\pi r)(10r) = 22\pi r^2$$

When the original cylinder is cut perpendicular to its axis into N smaller cylinders, the total surface area A_2 is

$$A_2 = N2(\pi r^2) + (2\pi r)L = N2(\pi r^2) + (2\pi r)(10r) = (2N + 20)\pi r^2$$

Substituting the expressions for A_1 and A_2 into Equation (1), we obtain the following expression for the ratio of the power radiated by the N pieces to that radiated by the original cylinder

$$\frac{P_{\text{pieces}}}{P_{\text{original}}} = \frac{e\sigma T^4 A_2}{e\sigma T^4 A_1} = \frac{(2N+20)\pi r^2}{22\pi r^2} = \frac{N+10}{11}$$

SOLUTION Since the total radiant power emitted by the N pieces is twice that emitted by the original cylinder, $P_{\text{pieces}} / P_{\text{original}} = 2$, we have $(N + 10)/11 = 2$. Solving this expression for N gives $N = 12$. Therefore, $\boxed{\text{there are 12 smaller cylinders}}$.

---

27. ***REASONING AND SOLUTION*** According to Equation 13.1, the heat per second lost is

$$\frac{Q}{t} = \frac{kA\,\Delta T}{L} = \frac{[0.040\ \text{J/(s}\cdot\text{m}\cdot\text{C}^\circ)]\,(1.6\ \text{m}^2)(25\ \text{C}^\circ)}{2.0\times10^{-3}\ \text{m}} = \boxed{8.0\times10^2\ \text{J/s}}$$

where the value for the thermal conductivity k of wool has been taken from Table 13.1.

---

31. ***REASONING AND SOLUTION*** The power radiated per square meter by the car when it has reached a temperature T is given by the Stefan-Boltzmann law, Equation 13.2, $P_{\text{radiated}} / A = e\sigma T^4$, where $P_{\text{radiated}} = Q/t$. Solving for T we have

$$T = \left[\frac{(P_{\text{radiated}} / A)}{e\sigma}\right]^{1/4} = \left\{\frac{560\ \text{W/m}^2}{(1.00)\left[5.67\times10^{-8}\ \text{J/(s}\cdot\text{m}^2\cdot\text{K}^4)\right]}\right\}^{1/4} = \boxed{320\ \text{K}}$$

---

35. ***REASONING*** The rate at which heat is conducted along either rod is given by Equation 13.1, $Q/t = (kA\,\Delta T)/L$. Since both rods conduct the same amount of heat per second, we have

$$\frac{k_s A_s\ \Delta T}{L_s} = \frac{k_i A_i\ \Delta T}{L_i} \tag{1}$$

Since the same temperature difference is maintained across both rods, we can algebraically cancel the ΔT terms. Because both rods have the same mass, $m_s = m_i$; in terms of the densities of silver and iron, the statement about the equality of the masses becomes $\rho_s(L_s A_s) = \rho_i(L_i A_i)$, or

$$\frac{A_s}{A_i} = \frac{\rho_i L_i}{\rho_s L_s} \tag{2}$$

Equations (1) and (2) may be combined to find the ratio of the lengths of the rods. Once the ratio of the lengths is known, Equation (2) can be used to find the ratio of the cross-sectional areas of the rods. If we assume that the rods have circular cross sections, then each has an area of $A = \pi r^2$. Hence, the ratio of the cross-sectional areas can be used to find the ratio of the radii of the rods.

SOLUTION

a. Solving Equation (1) for the ratio of the lengths and substituting the right hand side of Equation (2) for the ratio of the areas, we have

$$\frac{L_s}{L_i} = \frac{k_s A_s}{k_i A_i} = \frac{k_s(\rho_i L_i)}{k_i(\rho_s L_s)} \quad \text{or} \quad \left(\frac{L_s}{L_i}\right)^2 = \frac{k_s \rho_i}{k_i \rho_s}$$

Solving for the ratio of the lengths, we have

$$\frac{L_s}{L_i} = \sqrt{\frac{k_s \rho_i}{k_i \rho_s}} = \sqrt{\frac{[420\ \text{J/(s}\cdot\text{m}\cdot\text{C°)}](7860\ \text{kg/m}^3)}{[79\ \text{J/(s}\cdot\text{m}\cdot\text{C°)}](10\ 500\ \text{kg/m}^3)}} = \boxed{2.0}$$

b. From Equation (2) we have

$$\frac{\pi r_s^2}{\pi r_i^2} = \frac{\rho_i L_i}{\rho_s L_s} \quad \text{or} \quad \left(\frac{r_s}{r_i}\right)^2 = \frac{\rho_i L_i}{\rho_s L_s}$$

Solving for the ratio of the radii, we have

$$\frac{r_s}{r_i} = \sqrt{\frac{\rho_i}{\rho_s}\left(\frac{L_i}{L_s}\right)} = \sqrt{\frac{7860\ \text{kg/m}^3}{10\ 500\ \text{kg/m}^3}\left(\frac{1}{2.0}\right)} = \boxed{0.61}$$

CHAPTER 14 | *THE IDEAL GAS LAW AND KINETIC THEORY*

CONCEPTUAL QUESTIONS

---

4. ***REASONING AND SOLUTION*** According to the ideal gas law (Equation 14.1), $PV = nRT$. Therefore, the pressure of a gas confined to a fixed volume is directly proportional to the Kelvin temperature. Therefore, when the temperature is increased, the pressure increases proportionally.

The gas above the liquid in a can of hair spray is at a relatively high pressure. If the temperature of the can is increased to a sufficiently high temperature, the pressure of the confined gas could increase to a value larger than the walls of the can will sustain. The can would then explode. Therefore, the label on hair spray cans usually contains the warning "Do not store at high temperatures."

---

10. ***REASONING AND SOLUTION*** The kinetic theory of gases assumes that a gas molecule rebounds with the same speed after colliding with the wall of a container. From the impulse-momentum theorem, Equation 7.4, we know that $\mathbf{F}\Delta t = m(\mathbf{v}_f - \mathbf{v}_0)$, where the magnitude of **F** is the magnitude of the force on the wall. This implies that in a time interval t, a gas molecule of mass m, moving with velocity **v** before the collision and –**v** after the collision, exerts a force of magnitude $F = 2mv/t$ on the walls of the container, since the magnitude of $\mathbf{v}_f - \mathbf{v}_0$ is $2v$. If the speed of the gas molecule after the collision is less than that before the collision, the force exerted on the wall of the container will be less than $F = 2mv/t$, since the magnitude of $\mathbf{v}_f - \mathbf{v}_0$ is then less than $2v$. Therefore, the pressure of the gas will be less than that predicted by kinetic theory.

---

14. ***REASONING AND SOLUTION*** The atoms in a container of helium have the same translational rms speed as the molecules in a container of argon. Equation 14.6 relates the average translational kinetic energy per particle in an ideal gas to the Kelvin temperature: $\frac{1}{2}mv_{\text{rms}}^2 = \frac{3}{2}kT$. Since the mass of an argon atom is greater than the mass of a helium atom, then, from Equation 14.6, the average translational kinetic energy per atom of the argon atoms is greater than the average translational kinetic energy per atom of the helium atoms. Therefore, the temperature of the argon atoms is greater than that of the helium atoms.

CHAPTER 14 | *THE IDEAL GAS LAW AND KINETIC THEORY*

PROBLEMS

---

1. ***REASONING AND SOLUTION*** Since hemoglobin has a molecular mass of 64 500 u, the mass per mole of hemoglobin is 64 500 g/mol. The number of hemoglobin molecules per mol is Avogadro's number, or 6.022×10^{23} mol^{-1}. Therefore, one molecule of hemoglobin has a mass (in kg) of

$$\left(\frac{64\,500 \text{ g/mol}}{6.022\times10^{23} \text{ mol}^{-1}}\right)\left(\frac{1 \text{ kg}}{1000 \text{ g}}\right) = \boxed{1.07\times10^{-22} \text{ kg}}$$

---

5. ***REASONING AND SOLUTION*** The number n of moles contained in a sample is equal to the number N of atoms in the sample divided by the number N_A of atoms per mole (Avogadro's number):

$$n = \frac{N}{N_A} = \frac{30.1\times10^{23}}{6.022\times10^{23} \text{ mol}^{-1}} = 5.00 \text{ mol}$$

Since the sample has a mass of 135 g, the mass per mole is

$$\frac{135 \text{ g}}{5.00 \text{ mol}} = 27.0 \text{ g/mol}$$

The mass per mole (in g/mol) of a substance has the same numerical value as the atomic mass of the substance. Therefore, the atomic mass is 27.0 u. The periodic table of the elements reveals that the unknown element is $\boxed{\text{aluminum}}$.

---

7. ***REASONING*** The number n of moles of water molecules in the glass is equal to the mass m of water divided by the mass per mole. According to Equation 11.1, the mass of water is equal to its density ρ times its volume V. Thus, we have

$$n = \frac{m}{\text{Mass per mole}} = \frac{\rho V}{\text{Mass per mole}}$$

The volume of the cylindrical glass is $V = \pi r^2 h$, where r is the radius of the cylinder and h is its height. The number of moles of water can be written as

$$n = \frac{\rho V}{\text{Mass per mole}} = \frac{\rho\left(\pi r^2 h\right)}{\text{Mass per mole}}$$

SOLUTION The molecular mass of water (H_2O) is 2(1.00794 u) + (15.9994 u) = 18.0 u. The mass per mole of H_2O is 18.0 g/mol. The density of water (see Table 11.1) is 1.00×10^3 kg/m^3 or 1.00 g/cm^3.

$$n = \frac{\rho\left(\pi r^2 h\right)}{\text{Mass per mole}} = \frac{\left(1.00\ \text{g/cm}^3\right)(\pi)(4.50\ \text{cm})^2(12.0\ \text{cm})}{18.0\ \text{g/mol}} = \boxed{42.4\ \text{mol}}$$

---

9. ***REASONING AND SOLUTION*** The number of moles of air that is pumped into the tire is $\Delta n = n_f - n_i$. According to the ideal gas law (Equation 14.1), $n = PV/(RT)$; therefore,

$$\Delta n = n_f - n_i = \frac{P_f V_f}{RT_f} - \frac{P_i V_i}{RT_i} = \frac{V}{RT}\left(P_f - P_i\right)$$

Substitution of the data given in the problem leads to

$$\Delta n = \frac{4.1 \times 10^{-4}\ \text{m}^3}{[8.31\ \text{J/(mol}\cdot\text{K)}](296\ \text{K})}\left[\left(6.2 \times 10^5\ \text{Pa}\right) - \left(4.8 \times 10^5\ \text{Pa}\right)\right] = \boxed{2.3 \times 10^{-2}\ \text{mol}}$$

---

15. ***REASONING*** According to the ideal gas law (Equation 14.1), $PV = nRT$. Since n, the number of moles, is constant, $n_1 R = n_2 R$. Thus, according to Equation 14.1, we have

$$\frac{P_1 V_1}{T_1} = \frac{P_2 V_2}{T_2}$$

SOLUTION Solving for T_2, we have

$$T_2 = \left(\frac{P_2}{P_1}\right)\left(\frac{V_2}{V_1}\right) T_1 = \left(\frac{48.5\ P_1}{P_1}\right)\left[\frac{V_1/16}{V_1}\right](305\ \text{K}) = \boxed{925\ \text{K}}$$

---

19. ***REASONING AND SOLUTION***

a. Since the heat gained by the gas in one tank is equal to the heat lost by the gas in the other tank, $Q_1 = Q_2$, or (letting the subscript 1 correspond to the neon in the left tank, and letting 2 correspond to the neon in the right tank) $cm_1\Delta T_1 = cm_2\Delta T_2$,

$$cm_1(T - T_1) = cm_2(T_2 - T)$$

$$m_1(T - T_1) = m_2(T_2 - T)$$

Solving for T gives

$$T = \frac{m_2T_2 + m_1T_1}{m_2 + m_1} \qquad (1)$$

The masses m_1 and m_2 can be found by first finding the number of moles n_1 and n_2. From the ideal gas law, $PV = nRT$, so

$$n_1 = \frac{P_1V_1}{RT_1} = \frac{(5.0\times10^5\ \text{Pa})(2.0\ \text{m}^3)}{[8.31\ \text{J/(mol}\cdot\text{K)}](220\ \text{K})} = 5.5\times10^2\ \text{mol}$$

This corresponds to a mass $m_1 = (5.5\times10^2\ \text{mol})(20.179\ \text{g/mol}) = 1.1\times10^4\ \text{g} = 1.1\times10^1\ \text{kg}$. Similarly, $n_2 = 2.4 \times 10^2$ mol and $m_2 = 4.9 \times 10^3$ g = 4.9 kg. Substituting these mass values into Equation (1) yields

$$T = \frac{(4.9\ \text{kg})(580\ \text{K}) + (1.1\times10^1\ \text{kg})(220\ \text{K})}{(4.9\ \text{kg}) + (1.1\times10^1\ \text{kg})} = \boxed{3.3\times10^2\ \text{K}}$$

b. From the ideal gas law,

$$P = \frac{nRT}{V} = \frac{[(5.5\times10^2\ \text{mol}) + (2.4\times10^2\ \text{mol})][8.31\ \text{J/(mol}\cdot\text{K)}](3.3\times10^2\ \text{K})}{2.0\ \text{m}^3 + 5.8\ \text{m}^3} = \boxed{2.8\times10^5\text{Pa}}$$

---

23. ***REASONING*** The graph that accompanies Problem 68 in Chapter 12 can be used to determine the equilibrium vapor pressure of water in the air when the temperature is 30.0 °C (303 K). Equation 12.6 can then be used to find the partial pressure of water in the air at this temperature. Using this pressure, the ideal gas law can then be used to find the number of moles of water vapor per cubic meter.

SOLUTION According to the graph that accompanies Problem 68 in Chapter 12, the equilibrium vapor pressure of water vapor at 30.0 °C is approximately 4250 Pa. According to Equation 12.6,

$$\begin{matrix}\text{Partial}\\ \text{pressure of}\\ \text{water vapor}\end{matrix} = \begin{pmatrix}\text{Percent relative}\\ \text{humidity}\end{pmatrix}\begin{pmatrix}\text{Equilibrium vapor pressure of}\\ \text{water at the existing temperature}\end{pmatrix} \times \frac{1}{100\%}$$

$$= \frac{(55\%)(4250\text{ Pa})}{100\%} = 2.34 \times 10^3\text{ Pa}$$

The ideal gas law then gives the number of moles of water vapor per cubic meter of air as

$$\frac{n}{V} = \frac{P}{RT} = \frac{(2.34 \times 10^3\text{ Pa})}{[8.31\text{ J/(mol}\cdot\text{K)}](303\text{ K})} = \boxed{0.93\text{ mol/m}^3} \tag{14.1}$$

---

27. ***REASONING*** According to the ideal gas law (Equation 14.1), $PV = nRT$. Since n, the number of moles of the gas, is constant, $n_1R = n_2R$. Therefore, $P_1V_1/T_1 = P_2V_2/T_2$, where $T_1 = 273\text{ K}$ and T_2 is the temperature we seek. Since the beaker is cylindrical, the volume V of the gas is equal to Ad, where A is the cross-sectional area of the cylindrical volume and d is the height of the region occupied by the gas, as measured from the bottom of the beaker. With this substitution for the volume, the expression obtained from the ideal gas law becomes

$$\frac{P_1d_1}{T_1} = \frac{P_2d_2}{T_2} \tag{1}$$

where the pressures P_1 and P_2 are equal to the sum of the atmospheric pressure and the pressure caused by the mercury in each case. These pressures can be determined using Equation 11.4. Once the pressures are known, Equation (1) can be solved for T_2.

SOLUTION Using Equation 11.4, we obtain the following values for the pressures P_1 and P_2. Note that the initial height of the mercury is $h_1 = \frac{1}{2}(1.520\text{ m}) = 0.760\text{ m}$, while the final height of the mercury is $h_2 = \frac{1}{4}(1.520\text{ m}) = 0.380\text{ m}$.

$$P_1 = P_0 + \rho g h_1 = (1.01\times10^5\text{ Pa}) + \left[(1.36\times10^4\text{kg/m}^3)(9.80\text{ m/s}^2)(0.760\text{ m})\right] = 2.02\times10^5\text{ Pa}$$

$$P_2 = P_0 + \rho g h_2 = (1.01\times10^5\text{ Pa}) + \left[(1.36\times10^4\text{kg/m}^3)(9.80\text{ m/s}^2)(0.380\text{ m})\right] = 1.52\times10^5\text{ Pa}$$

In these pressure calculations, the density of mercury is $\rho = 1.36\times10^4\text{ kg/m}^3$. In Equation (1) we note that $d_1 = 0.760\text{ m}$ and $d_2 = 1.14\text{ m}$. Solving Equation (1) for T_2 and substituting values, we obtain

$$T_2 = \left(\frac{P_2 d_2}{P_1 d_1}\right) T_1 = \left[\frac{(1.52\times10^5\ \text{Pa})(1.14\ \text{m})}{(2.02\times10^5\ \text{Pa})(0.760\ \text{m})}\right](273\ \text{K}) = \boxed{308\ \text{K}}$$

---

29. ***REASONING AND SOLUTION*** Using the expressions for $\overline{v^2}$ and $(\bar{v})^2$ given in the statement of the problem, we obtain:

a. $$\overline{v^2} = \tfrac{1}{3}(v_1^2 + v_2^2 + v_3^2) = \tfrac{1}{3}\left[(3.0\ \text{m/s})^2 + (7.0\ \text{m/s})^2 + (9.0\ \text{m/s})^2\right] = \boxed{46.3\ \text{m}^2/\text{s}^2}$$

b. $$(\bar{v})^2 = \left[\tfrac{1}{3}(v_1 + v_2 + v_3)\right]^2 = \left[\tfrac{1}{3}(3.0\ \text{m/s} + 7.0\ \text{m/s} + 9.0\ \text{m/s})\right]^2 = \boxed{40.1\ \text{m}^2/\text{s}^2}$$

$\overline{v^2}$ and $(\bar{v})^2$ are *not* equal, because they are two different physical quantities.

---

35. ***REASONING AND SOLUTION*** Since we are treating the air as an ideal gas, $PV = nRT$, so that $U = \frac{5}{2}nRT = \frac{5}{2}PV = \frac{5}{2}(7.7\times10^6\ \text{Pa})(5.6\times10^5\ \text{m}^3) = 1.1\times10^{13}\ \text{J}$. The number of joules of energy consumed per day by one house is

$$30.0\ \text{kW}\cdot\text{h} = \left(30.0\times10^3\ \frac{\text{J}\cdot\text{h}}{\text{s}}\right)\left(\frac{3600\ \text{s}}{1\ \text{h}}\right) = 1.08\times10^8\ \text{J}$$

The number of homes that could be served for one day by 1.1×10^{13} J of energy is

$$(1.1\times10^{13}\ \text{J})\left(\frac{1\ \text{home}}{1.08\times10^8\ \text{J}}\right) = \boxed{1.0\times10^5\ \text{homes}}$$

---

37. ***REASONING AND SOLUTION*** The average force exerted by one electron on the screen is, from the impulse-momentum theorem (Equation 7.4), $\overline{F} = \Delta p/\Delta t = m\Delta v/\Delta t$. Therefore, in a time Δt, N electrons exert an average force $\overline{F} = Nm\Delta v/\Delta t = (N/\Delta t)m\Delta v$. Since the pressure on the screen is the average force per unit area (Equation 10.19), we have

$$P = \frac{\overline{F}}{A} = \frac{(N/\Delta t)m\Delta v}{A}$$

$$= \frac{(6.2\times10^{16}\ \text{electrons/s})(9.11\times10^{-31}\ \text{kg})(8.4\times10^7\ \text{m/s} - 0\ \text{m/s})}{1.2\times10^{-7}\ \text{m}^2} = \boxed{4.0\times10^1\ \text{Pa}}$$

---

41. ***REASONING AND SOLUTION*** According to Fick's law of diffusion (Equation 14.8), the mass of ethanol that diffuses through the cylinder in one hour (3600 s) is

$$m = \frac{(DA\,\Delta C)t}{L} = \frac{(12.4\times10^{-10}\ \text{m}^2/\text{s})(4.00\times10^{-4}\ \text{m}^2)(1.50\ \text{kg/m}^3)(3600\ \text{s})}{(0.0200\ \text{m})} = \boxed{1.34\times10^{-7}\ \text{kg}}$$

---

43. ***REASONING*** Since mass is conserved, the mass flow rate is the same at all points, as described by the equation of continuity (Equation 11.8). Therefore, the mass flow rate at which CCl_4 enters the tube is the same as that at point A. The concentration difference of CCl_4 between point A and the left end of the tube, ΔC, can be calculated by using Fick's law of diffusion (Equation 14.8). The concentration of CCl_4 at point A can be found from $C_A = C_{\text{left end}} - \Delta C$.

SOLUTION

a. As discussed above in the reasoning, the mass flow rate of CCl_4 as it passes point A is the same as the mass flow rate at which CCl_4 enters the left end of the tube; therefore, the mass flow rate of CCl_4 at point A is $\boxed{5.00\times10^{-13}\ \text{kg/s}}$.

b. Solving Fick's law for ΔC, we obtain

$$\Delta C = \frac{mL}{DAt} = \frac{(m/t)L}{DA}$$

$$= \frac{(5.00\times10^{-13}\ \text{kg/s})(5.00\times10^{-3}\ \text{m})}{(20.0\times10^{-10}\ \text{m}^2/\text{s})(3.00\times10^{-4}\ \text{m}^2)} = 4.2\times10^{-3}\ \text{kg/m}^3$$

Then,

$$C_A = C_{\text{left end}} - \Delta C = (1.00\times10^{-2}\ \text{kg/m}^3) - (4.2\times10^{-3}\ \text{kg/m}^3) = \boxed{5.8\times10^{-3}\ \text{kg/m}^3}$$

---

47. ***REASONING AND SOLUTION*** According to the ideal gas law (Equation 14.1), the total number of moles n of fresh air in the sample is

$$n = \frac{PV}{RT} = \frac{(1.0\times10^5\ \text{Pa})(5.0\times10^{-4}\ \text{m}^3)}{[8.31\ \text{J/(mole}\cdot\text{K)}]\ (310\ \text{K})} = 1.94\times10^{-2}\ \text{mol}$$

The total number of molecules in the sample is nN_A, where N_A is Avogadro's number. Since the sample contains approximately 21% oxygen, the total number of oxygen molecules in the sample is $(0.21)nN_A$ or

$$(0.21)(1.94\times10^{-2}\ \text{mol})(6.022\times10^{23}\ \text{mol}^{-1})=\boxed{2.5\times10^{21}}$$

---

51. ***REASONING*** The internal energy of the neon at any Kelvin temperature T is given by Equation 14.7, $U=(3/2)nRT$. Therefore, when the temperature of the neon increases from an initial temperature T_i to a final temperature T_f, the internal energy of the neon increases by an amount

$$\Delta U = U_f - U_i = \frac{3}{2}nR\,(T_f - T_i)$$

In order to use this equation, we must first determine n. Since the neon is confined to a tank, the number of moles n is constant, and we can use the information given concerning the initial conditions to determine an expression for the quantity nR. According to the ideal gas law (Equation 14.1),

$$nR = \frac{P_iV_i}{T_i}$$

These two expressions can be combined to obtain an equation in terms of the variables that correspond to the data given in the problem statement.

SOLUTION Combining the two expressions and substituting the given values yields

$$\Delta U = \frac{3}{2}\left(\frac{P_iV_i}{T_i}\right)(T_f - T_i)$$

$$= \frac{3}{2}\left[\frac{(1.01\times10^5\ \text{Pa})(680\ \text{m}^3)}{(293.2\ \text{K})}\right](294.3\ \text{K} - 293.2\ \text{K}) = \boxed{3.9\times10^5\ \text{J}}$$

---

53. ***REASONING AND SOLUTION*** Since the density of aluminum is $2700\ \text{kg/m}^3$, the number of atoms of aluminum per cubic meter is, using the given data,

$$N/V = \left(\frac{2700\times10^3\ \text{g/m}^3}{26.9815\ \text{g/mol}}\right)\left(6.022\times10^{23}\ \text{mol}^{-1}\right) = 6.0\times10^{28}\ \text{atoms/m}^3$$

Assuming that the volume of the solid contains many small cubes, with one atom at the center of each, then there are $(6.0\times10^{28})^{1/3}$ atoms/m $= 3.9\times10^9$ atoms/m along each edge of a 1.00-m^3 cube. Therefore, the spacing between the centers of neighboring atoms is

$$d = \frac{1}{3.9\times10^9/\text{m}} = \boxed{2.6\times10^{-10}\ \text{m}}$$

CHAPTER 15 | *THERMODYNAMICS*

CONCEPTUAL QUESTIONS

---

1. ***REASONING AND SOLUTION*** The plunger of a bicycle tire pump is pushed down rapidly with the end of the pump sealed so that no air escapes. Since the compression occurs rapidly, there is no time for heat to flow into or out of the system. Therefore, to a very good approximation, the process may be treated as an adiabatic compression that is described by Equation 15.4:

$$W = (3/2)nR(T_i - T_f)$$

The person who pushes the plunger down does work on the system, therefore W is negative. It follows that the term $(T_i - T_f)$ must also be negative. Thus, the final temperature T_f must be greater than the initial temperature T_i. This increase in temperature is evidenced by the fact that the pump becomes warm to the touch.

Alternate Explanation:
Since the compression occurs rapidly, there is no time for heat to flow into or out of the system. Therefore, to a very good approximation, the process may be treated as an adiabatic compression. According to the first law of thermodynamics, the change in the internal energy is $\Delta U = Q - W = -W$, since $Q = 0$ for adiabatic processes. Since work is done on the system, W is negative; therefore the change in the internal energy, ΔU, is positive. The work done by the person pushing the plunger is manifested as an increase in the internal energy of the air in the pump. The internal energy of an ideal gas is proportional to the Kelvin temperature. Since the internal energy of the gas increases, the temperature of the air in the pump must also increase. This increase in temperature is evidenced by the fact that the pump becomes warm to the touch.

---

8. ***REASONING AND SOLUTION*** A material contracts when it is heated. To determine the molar specific heat capacities, we first calculate the heat Q needed to raise the temperature of the material by an amount ΔT. From the first law of thermodynamics, $Q = \Delta U + W$. When the heating occurs at constant pressure, the work done is given by Equation 15.2: $W = P\Delta V = P(V_f - V_i)$. When the volume is constant, $\Delta V = 0$. Therefore, we have:

$$Q_P = \Delta U + P\left(V_f - V_i\right) \quad \text{and} \quad Q_V = \Delta U$$

Equation 15.6 indicates that the molar heat capacities will be given by $C = Q/\left(n\Delta T\right)$. Therefore

$$C_P = \frac{\Delta U + P\left(V_f - V_i\right)}{n\Delta T} \quad \text{and} \quad C_V = \frac{\Delta U}{n\Delta T}$$

Since the material contracts when it is heated, V_f is less than V_i. Therefore the term $P(V_f - V_i)$ is negative. Hence, the numerator of C_P is smaller than the numerator of C_V. Therefore, C_V is larger than C_P.

---

16. ***REASONING AND SOLUTION*** A refrigerator is kept in a garage that is not heated in the cold winter or air-conditioned in the hot summer. In order to make ice cubes, the refrigerator uses electrical energy to provide the work to remove heat from the interior of the freezer and deposit the heat outside of the refrigerator. In the summer, the "hot" reservoir will be at a higher temperature than it is in the winter. Therefore, more work will be required to remove heat from the interior of the freezer in the summer, and the refrigerator will use more electrical energy. Hence, it will cost more for the refrigerator to make a kilogram of ice cubes in the summer.

CHAPTER 15 | *THERMODYNAMICS*

PROBLEMS

1. ***REASONING*** Energy in the form of work leaves the system, while energy in the form of heat enters. More energy leaves than enters, so we expect the internal energy of the system to decrease, that is, we expect the change ΔU in the internal energy to be negative. The first law of thermodynamics will confirm our expectation. As far as the environment is concerned, we note that when the system loses energy, the environment gains it, and when the system gains energy the environment loses it. Therefore, the change in the internal energy of the environment must be opposite to that of the system.

SOLUTION
a. The system gains heat so Q is positive, according to our convention. The system does work, so W is also positive, according to our convention. Applying the first law of thermodynamics from Equation 15.1, we find for the system that

$$\Delta U = Q - W = (77\ \text{J}) - (164\ \text{J}) = \boxed{-87\ \text{J}}$$

As expected, this value is negative, indicating a decrease.

b. The change in the internal energy of the environment is opposite to that of the system, so that $\boxed{\Delta U_{\text{environment}} = +87\ \text{J}}$.

---

5. ***REASONING*** Since the change in the internal energy and the heat released in the process are given, the first law of thermodynamics (Equation 15.1) can be used to find the work done. Since we are told how much work is required to make the car go one mile, we can determine how far the car can travel. When the gasoline burns, its internal energy decreases and heat flows into the surroundings; therefore, both ΔU and Q are negative.

SOLUTION According to the first law of thermodynamics, the work that is done when one gallon of gasoline is burned in the engine is

$$W = Q - \Delta U = -1.00 \times 10^8\ \text{J} - (-1.19 \times 10^8\ \text{J}) = 0.19 \times 10^8\ \text{J}$$

Since 6.0×10^5 J of work is required to make the car go one mile, the car can travel

$$0.19 \times 10^8\ \text{J} \left(\frac{1\ \text{mile}}{6.0 \times 10^5\ \text{J}} \right) = \boxed{32\ \text{miles}}$$

---

9. ***REASONING*** According to Equation 15.2, $W = P\Delta V$, the average pressure $\overline{P}$ of the expanding gas is equal to $\overline{P} = W / \Delta V$, where the work W done by the gas on the bullet can be found from the work-energy theorem (Equation 6.3). Assuming that the barrel of the gun is cylindrical with radius r, the volume of the barrel is equal to its length L multiplied by the area (πr^2) of its cross section. Thus, the change in volume of the expanding gas is $\Delta V = L\pi r^2$.

SOLUTION The work done by the gas on the bullet is given by Equation 6.3 as

$$W = \tfrac{1}{2}m(v_{\text{final}}^2 - v_{\text{initial}}^2) = \tfrac{1}{2}(2.6\times10^{-3}\text{ kg})[(370\text{ m/s})^2 - 0] = 180\text{ J}$$

The average pressure of the expanding gas is, therefore,

$$\overline{P} = \frac{W}{\Delta V} = \frac{180\text{ J}}{(0.61\text{ m})\pi(2.8\times10^{-3}\text{ m})^2} = \boxed{1.2\times10^7\text{ Pa}}$$

---

13. ***REASONING*** The work done in an isobaric process is given by Equation 15.2, $W = P\Delta V$; therefore, the pressure is equal to $P = W / \Delta V$. In order to use this expression, we must first determine a numerical value for the work done; this can be calculated using the first law of thermodynamics (Equation 15.1), $\Delta U = Q - W$.

SOLUTION Solving Equation 15.1 for the work W, we find

$$W = Q - \Delta U = 1500\text{ J}-(+4500\text{ J}) = -3.0\times10^3\text{ J}$$

Therefore, the pressure is

$$P = \frac{W}{\Delta V} = \frac{-3.0\times10^3\text{ J}}{-0.010\text{ m}^3} = \boxed{3.0\times10^5\text{ Pa}}$$

The change in volume ΔV, which is the final volume minus the initial volume, is negative because the final volume is 0.010 m^3 *less* than the initial volume.

---

15. ***REASONING AND SOLUTION*** The first law of thermodynamics states that $\Delta U = Q - W$. The work W involved in an isobaric process is, according to Equation 15.2, $W = P\Delta V$. Combining these two expressions leads to $\Delta U = Q - P\Delta V$. Solving for Q gives

$$Q = \Delta U + P\Delta V \qquad (1)$$

Since this is an expansion, $\Delta V > 0$, so $P\Delta V > 0$. From the ideal gas law, $PV = nRT$, we have $P\Delta V = nR\Delta T$. Since $P\Delta V > 0$, it follows that $nR\Delta T > 0$. The internal energy of an ideal gas is directly proportional to its Kelvin temperature T. Therefore, since $nR\Delta T > 0$, it follows that $\Delta U > 0$. Since both terms on the right hand side of Equation (1) are positive, the left hand side of Equation (1) must also be positive. Thus, Q is positive. By the convention described in the text, this means that

$$\boxed{\text{heat can only flow into an ideal gas during an isobaric expansion}}$$

---

19. ***REASONING*** According to the first law of thermodynamics (Equation 15.1), $\Delta U = Q - W$. For a monatomic ideal gas (Equation 14.7), $U = \frac{3}{2}nRT$. Therefore, for the process in question, the change in the internal energy is $\Delta U = \frac{3}{2}nR\Delta T$. Combining the last expression for ΔU with Equation 15.1 yields

$$\tfrac{3}{2}nR\Delta T = Q - W$$

This expression can be solved for ΔT.

SOLUTION

a. The heat is $Q = +1200\text{ J}$, since it is absorbed by the system. The work is $W = +2500\text{ J}$, since it is done *by* the system. Solving the above expression for ΔT and substituting the values for the data given in the problem statement, we have

$$\Delta T = \frac{Q - W}{\frac{3}{2}nR} = \frac{1200\text{ J} - 2500\text{ J}}{\frac{3}{2}(0.50\text{ mol})[8.31\text{ J}/(\text{mol}\cdot\text{K})]} = \boxed{-2.1\times10^{2}\text{ K}}$$

b. Since $\Delta T = T_{\text{final}} - T_{\text{initial}}$ is negative, T_{initial} must be greater than T_{final}; this change represents a $\boxed{\text{decrease}}$ in temperature.

Alternatively, one could deduce that the temperature decreases from the following physical argument. Since the system loses more energy in doing work than it gains in the form of heat, the internal energy of the system decreases. Since the internal energy of an ideal gas depends only on the temperature, a decrease in the internal energy must correspond to a decrease in the temperature.

---

23. ***REASONING*** When the expansion is isothermal, the work done can be calculated from Equation (15.3): $W = nRT\ln\left(V_{\text{f}}/V_{\text{i}}\right)$. When the expansion is adiabatic, the work done can be calculated from Equation 15.4: $W = \frac{3}{2}nR(T_{\text{i}} - T_{\text{f}})$.

Since the gas does the same amount of work whether it expands adiabatically or isothermally, we can equate the right hand sides of these two equations. We also note that since the initial temperature is the same for both cases, the temperature T in the isothermal expansion is the same as the initial temperature T_i for the adiabatic expansion. We then have

$$nRT_i \ln\left(\frac{V_f}{V_i}\right) = \tfrac{3}{2}nR(T_i - T_f) \quad \text{or} \quad \ln\left(\frac{V_f}{V_i}\right) = \frac{\tfrac{3}{2}(T_i - T_f)}{T_i}$$

SOLUTION Solving for the ratio of the volumes gives

$$\frac{V_f}{V_i} = e^{\frac{3}{2}(T_i - T_f)/T_i} = e^{\frac{3}{2}(405\text{ K} - 245\text{ K})/(405\text{ K})} = \boxed{1.81}$$

---

27. ***REASONING*** According to Equation 15.5, $P_iV_i^\gamma = P_fV_f^\gamma$. The ideal gas law states that $V = nRT/P$ for both the initial and final conditions. Thus, we have

$$P_i\left(\frac{nRT_i}{P_i}\right)^\gamma = P_f\left(\frac{nRT_f}{P_f}\right)^\gamma \quad \text{or} \quad \frac{P_f}{P_i} = \left(\frac{T_i}{T_f}\right)^{\gamma/(1-\gamma)}$$

Since the ratio of the temperatures is known, the last expression can be solved for the final pressure P_f.

SOLUTION Since $T_i/T_f = 1/2$, and $\gamma = 5/3$, we find that

$$P_f = P_i\left(\frac{T_i}{T_f}\right)^{\gamma/(1-\gamma)} = \left(1.50\times10^5\text{ Pa}\right)\left(\frac{1}{2}\right)^{(5/3)/[1-(5/3)]} = \boxed{8.49\times10^5\text{ Pa}}$$

---

29. ***REASONING AND SOLUTION***

a. The final temperature of the adiabatic process is given by solving Equation 15.4 for T_f.

$$T_f = T_i - \frac{W}{\tfrac{3}{2}nR} = 393\text{ K} - \frac{825\text{ J}}{\tfrac{3}{2}(1.00\text{ mol})[8.31\text{ J/(mol}\cdot\text{K)}]} = \boxed{327\text{ K}}$$

b. According to Equation 15.5 for the adiabatic expansion of an ideal gas, $P_iV_i^\gamma = P_fV_f^\gamma$. Therefore,

$$V_f^\gamma = V_i^\gamma\left(\frac{P_i}{P_f}\right)$$

From the ideal gas law, $PV = nRT$; therefore, the ratio of the pressures is given by

$$\frac{P_i}{P_f} = \left(\frac{T_i}{T_f}\right)\left(\frac{V_f}{V_i}\right)$$

Combining the previous two equations gives

$$V_f^{\gamma} = V_i^{\gamma}\left(\frac{T_i}{T_f}\right)\left(\frac{V_f}{V_i}\right)$$

Solving for V_f we obtain

$$\frac{V_f^{\gamma}}{V_f} = \left(\frac{T_i}{T_f}\right)\left(\frac{V_i^{\gamma}}{V_i}\right) \qquad \text{or} \qquad V_f^{(\gamma-1)} = \left(\frac{T_i}{T_f}\right)V_i^{(\gamma-1)}$$

$$V_f = \left[\left(\frac{T_i}{T_f}\right)V_i^{(\gamma-1)}\right]^{1/(\gamma-1)} = V_i\left(\frac{T_i}{T_f}\right)^{1/(\gamma-1)}$$

Therefore,

$$V_f = V_i\left(\frac{T_i}{T_f}\right)^{1/(\gamma-1)} = (0.100\ \text{m}^3)\left(\frac{393\ \text{K}}{327\ \text{K}}\right)^{1/(2/3)} = (0.100\ \text{m}^3)\left(\frac{393\ \text{K}}{327\ \text{K}}\right)^{3/2} = \boxed{0.132\ \text{m}^3}$$

---

31. ***REASONING AND SOLUTION*** According to the first law of thermodynamics (Equation 15.1), $\Delta U = U_f - U_i = Q - W$. Since the internal energy of this gas is doubled by the addition of heat, the initial and final internal energies are U and $2U$, respectively. Therefore,

$$\Delta U = U_f - U_i = 2U - U = U$$

Equation 15.1 for this situation then becomes $U = Q - W$. Solving for Q gives

$$Q = U + W \tag{1}$$

The initial internal energy of the gas can be calculated from Equation 14.7:

$$U = \tfrac{3}{2}nRT = \tfrac{3}{2}(2.5\ \text{mol})\left[8.31\ \text{J/(mol}\cdot\text{K)}\right](350\ \text{K}) = 1.1\times10^4\ \text{J}$$

a. If the process is carried out isochorically (i.e., at constant volume), then W = 0, and the heat required to double the internal energy is

$$Q = U + W = U + 0 = \boxed{1.1\times10^4\ \text{J}}$$

b. If the process is carried out isobarically (i.e., at constant pressure), then $W = P\Delta V$, and Equation (1) above becomes

$$Q = U + W = U + P\Delta V \tag{2}$$

From the ideal gas law, $PV = nRT$, we have that $P\Delta V = nR\Delta T$, and Equation (2) becomes

$$Q = U + nR\Delta T \tag{3}$$

The internal energy of an ideal gas is directly proportional to its Kelvin temperature. Since the internal energy of the gas is doubled, the final Kelvin temperature will be twice the initial Kelvin temperature, or $\Delta T = 350$ K. Substituting values into Equation (3) gives

$$Q = 1.1\times10^4 \text{ J} + (2.5 \text{ mol})[8.31 \text{ J/(mol}\cdot\text{K)}](350 \text{ K}) = \boxed{1.8\times10^4 \text{ J}}$$

---

35. ***REASONING AND SOLUTION*** The amount of heat required to change the temperature of the gas is given by Equation 15.6, where C_P is given by Equation 15.7.

$$Q = C_P n\Delta T = \tfrac{5}{2}Rn\Delta T = \tfrac{5}{2}[8.31 \text{ J}/(\text{mol}\cdot\text{K})]\ (1.5 \text{ mol})\ (77 \text{ K}) = \boxed{2400 \text{ J}}$$

---

41. ***REASONING AND SOLUTION*** The efficiency of a heat engine is defined by Equation 15.11 as $e = |W|/|Q_H|$, where $|W|$ is the magnitude of the work done and $|Q_H|$ is the magnitude of the heat input. The principle of energy conservation requires that $|Q_H| = |W| + |Q_C|$, where $|Q_C|$ is the magnitude of the heat rejected to the cold reservoir (Equation 15.12). Combining Equations 15.11 and 15.12 gives

$$e = \frac{|W|}{|W| + |Q_C|} = \frac{16\ 600 \text{ J}}{16\ 600 \text{ J} + 9700 \text{ J}} = \boxed{0.631}$$

---

45. ***REASONING*** The efficiency of either engine is given by Equation 15.13, $e = 1 - (|Q_C|/|Q_H|)$. Since engine A receives three times more input heat, produces five times more work, and rejects two times more heat than engine B, it follows that $|Q_{HA}| = 3|Q_{HB}|$, $|W_A| = 5|W_B|$, and $|Q_{CA}| = 2|Q_{CB}|$. As required by the principle of energy conservation for engine A (Equation 15.12),

$$\underbrace{|Q_{HA}|}_{3|Q_{HB}|} = \underbrace{|Q_{CA}|}_{2|Q_{CB}|} + \underbrace{|W_A|}_{5|W_B|}$$

Thus,

$$3|Q_{HB}| = 2|Q_{CB}| + 5|W_B| \tag{1}$$

Since engine B also obeys the principle of energy conservation (Equation 15.12),

$$|Q_{HB}| = |Q_{CB}| + |W_B| \tag{2}$$

Substituting $|Q_{HB}|$ from Equation (2) into Equation (1) yields

$$3(|Q_{CB}| + |W_B|) = 2|Q_{CB}| + 5|W_B|$$

Solving for $|W_B|$ gives

$$|W_B| = \tfrac{1}{2}|Q_{CB}|$$

Therefore, Equation (2) predicts for engine B that

$$|Q_{HB}| = |Q_{CB}| + |W_B| = \frac{3}{2}|Q_{CB}|$$

SOLUTION

a. Substituting $|Q_{CA}| = 2|Q_{CB}|$ and $|Q_{HA}| = 3|Q_{HB}|$ into Equation 15.13 for engine A, we have

$$e_A = 1 - \frac{|Q_{CA}|}{|Q_{HA}|} = 1 - \frac{2|Q_{CB}|}{3|Q_{HB}|} = 1 - \frac{2|Q_{CB}|}{3\left(\frac{3}{2}|Q_{CB}|\right)} = 1 - \frac{4}{9} = \boxed{\frac{5}{9}}$$

b. Substituting $|Q_{HB}| = \frac{3}{2}|Q_{CB}|$ into Equation 15.13 for engine B, we have

$$e_B = 1 - \frac{|Q_{CB}|}{|Q_{HB}|} = 1 - \frac{|Q_{CB}|}{\frac{3}{2}|Q_{CB}|} = 1 - \frac{2}{3} = \boxed{\frac{1}{3}}$$

---

49. ***REASONING*** The efficiency e of a Carnot engine is given by Equation 15.15, $e = 1 - (T_C / T_H)$, where, according to Equation 15.14, $|Q_C|/|Q_H| = T_C / T_H$. Since the efficiency is given along with T_C and $|Q_C|$, Equation 15.15 can be used to calculate T_H. Once T_H is known, the ratio T_C / T_H is thus known, and Equation 15.14 can be used to calculate $|Q_H|$.

SOLUTION

a. Solving Equation 15.15 for T_H gives

$$T_H = \frac{T_C}{1-e} = \frac{378\text{ K}}{1-0.700} = \boxed{1260\text{ K}}$$

b. Solving Equation 15.14 for $|Q_H|$ gives

$$|Q_H| = |Q_C|\left(\frac{T_H}{T_C}\right) = (5230\text{ J})\left(\frac{1260\text{ K}}{378\text{ K}}\right) = \boxed{1.74\times10^4\text{ J}}$$

---

53. ***REASONING*** The maximum efficiency e at which the power plant can operate is given by Equation 15.15, $e = 1-(T_C/T_H)$. The power output is given; it can be used to find the magnitude $|W|$ of the work output for a 24 hour period. With the efficiency and $|W|$ known, Equation 15.11, $e = |W|/|Q_H|$, can be used to find $|Q_H|$, the magnitude of the input heat. The magnitude $|Q_C|$ of the exhaust heat can then be found from Equation 15.12, $|Q_H| = |W| + |Q_C|$.

SOLUTION

a. The maximum efficiency is

$$e = 1-\frac{T_C}{T_H} = 1-\frac{323\text{ K}}{505\text{ K}} = \boxed{0.360}$$

b. Since the power output of the power plant is P = 84 000 kW, the required heat input $|Q_H|$ for a 24 hour period is

$$|Q_H| = \frac{|W|}{e} = \frac{Pt}{e} = \frac{(8.4\times10^7\text{ J/s})(24\text{ h})}{0.360}\left(\frac{3600\text{ s}}{1\text{ h}}\right) = 2.02\times10^{13}\text{ J}$$

Therefore, solving Equation 15.12 for $|Q_C|$, we have

$$|Q_C| = |Q_H| - |W| = 2.02\times10^{13}\text{ J} - 7.3\times10^{12}\text{ J} = \boxed{1.3\times10^{13}\text{ J}}$$

---

55. ***REASONING AND SOLUTION*** The efficiency e of the power plant is three-fourths its Carnot efficiency so, according to Equation 15.15,

$$e = 0.75\left(1 - \frac{T_C}{T_H}\right) = 0.75\left(1 - \frac{40\text{ K} + 273\text{ K}}{285\text{ K} + 273\text{ K}}\right) = 0.33$$

The power output of the plant is 1.2×10^9 watts. According to Equation 15.11, $e = |W|/|Q_H| = (\text{Power} \cdot t)/|Q_H|$. Therefore, at 33% efficiency, the magnitude of the heat input per unit time is

$$\frac{|Q_H|}{t} = \frac{\text{Power}}{e} = \frac{1.2 \times 10^9\text{ W}}{0.33} = 3.6 \times 10^9\text{ J/s}$$

From the principle of conservation of energy, the heat output per unit time must be

$$\frac{|Q_C|}{t} = \frac{|Q_H|}{t} - \text{Power} = 2.4 \times 10^9\text{ J/s}$$

The rejected heat is carried away by the flowing water and, according to Equation 12.4, $|Q_C| = cm\Delta T$. Therefore,

$$\frac{|Q_C|}{t} = \frac{cm\Delta T}{t} \qquad \text{or} \qquad \frac{|Q_C|}{t} = c\left(\frac{m}{t}\right)\Delta T$$

Solving the last equation for Δt, we have

$$\Delta T = \frac{|Q_C|}{tc(m/t)} = \frac{|Q_C|/t}{c(m/t)} = \frac{2.4 \times 10^9\text{ J/s}}{\left[4186\text{ J/(kg} \cdot \text{C°)}\right]\left(1.0 \times 10^5\text{ kg/s}\right)} = \boxed{5.7\text{ C°}}$$

---

59. ***REASONING AND SOLUTION*** Equation 15.14 holds for a Carnot air conditioner as well as a Carnot engine. Therefore, solving Equation 15.14 for $|Q_C|$, we have

$$|Q_C| = |Q_H|\left(\frac{T_C}{T_H}\right) = \left(6.12 \times 10^5\text{ J}\right)\left(\frac{299\text{ K}}{312\text{ K}}\right) = \boxed{5.86 \times 10^5\text{ J}}$$

---

65. ***REASONING*** Let the coefficient of performance be represented by the symbol COP. Then according to Equation 15.16, $\text{COP} = |Q_C|/|W|$. From the statement of energy conservation for a Carnot refrigerator (Equation 15.12), $|W| = |Q_H| - |Q_C|$. Combining Equations 15.16 and 15.12 leads to

$$\text{COP} = \frac{|Q_C|}{|Q_H| - |Q_C|} = \frac{|Q_C|/|Q_C|}{(|Q_H| - |Q_C|)/|Q_C|} = \frac{1}{(|Q_H|/|Q_C|) - 1}$$

Replacing the ratio of the heats with the ratio of the Kelvin temperatures, according to Equation 15.14, leads to

$$\text{COP} = \frac{1}{T_H/T_C - 1} \tag{1}$$

The heat $|Q_C|$ that must be removed from the refrigerator when the water is cooled can be calculated using Equation 12.4, $|Q_C| = cm\Delta T$; therefore,

$$|W| = \frac{|Q_C|}{\text{COP}} = \frac{cm\,\Delta T}{\text{COP}} \tag{2}$$

SOLUTION

a. Substituting values into Equation (1) gives

$$\text{COP} = \frac{1}{\frac{T_H}{T_C} - 1} = \frac{1}{\frac{(20.0 + 273.15)\text{ K}}{(6.0 + 273.15)\text{ K}} - 1} = \boxed{2.0 \times 10^1}$$

b. Substituting values into Equation (2) gives

$$|W| = \frac{cm\,\Delta T}{\text{COP}} = \frac{[4186\text{ J/(kg}\cdot\text{C°)}](5.00\text{ kg})(20.0\text{ °C} - 6.0\text{ °C})}{2.0 \times 10^1} = \boxed{1.5 \times 10^4\text{ J}}$$

---

67. ***REASONING*** The efficiency of the Carnot engine is, according to Equation 15.15,

$$e = 1 - \frac{T_C}{T_H} = 1 - \frac{842\text{ K}}{1684\text{ K}} = \frac{1}{2}$$

Therefore, the magnitude of the work delivered by the engine is, according to Equation 15.11,

$$|W| = e|Q_H| = \tfrac{1}{2}|Q_H|$$

The heat pump removes an amount of heat $|Q_H|$ from the cold reservoir. Thus, the amount of heat $|Q'|$ delivered to the hot reservoir of the heat pump is

$$|Q'| = |Q_H| + |W| = |Q_H| + \tfrac{1}{2}|Q_H| = \tfrac{3}{2}|Q_H|$$

Therefore, $|Q'|/|Q_H| = 3/2$. According to Equation 15.14, $|Q'|/|Q_H| = T'/T_C$, so $T'/T_C = 3/2$.

SOLUTION Solving for T' gives

$$T' = \tfrac{3}{2}T_C = \tfrac{3}{2}(842\text{ K}) = \boxed{1.26\times10^3\text{ K}}$$

---

71. ***REASONING AND SOLUTION***

a. Since the energy that becomes unavailable for doing work is zero for the process, we have from Equation 15.19, $W_{\text{unavailable}} = T_0\Delta S_{\text{universe}} = 0$. Therefore, $\Delta S_{\text{universe}} = 0$ and according to the discussion in Section 15.11, the process is $\boxed{\text{reversible}}$.

b. Since the process is reversible, we have (see Section 15.11)

$$\Delta S_{\text{universe}} = \Delta S_{\text{system}} + \Delta S_{\text{surroundings}} = 0$$

Therefore,

$$\Delta S_{\text{surroundings}} = -\Delta S_{\text{system}} = \boxed{-125\text{ J/K}}$$

---

75. ***REASONING AND SOLUTION***

a. Since the temperature of the gas is kept constant at all times, the process is isothermal; therefore, the internal energy of an ideal gas does not change and $\boxed{\Delta U = 0}$.

b. From the first law of thermodynamics (Equation 15.1), $\Delta U = Q - W$. But $\Delta U = 0$, so that $Q = W$. Since work is done on the gas, the work is negative, and $\boxed{Q = -6.1\times10^3\text{ J}}$.

c. The work done in an isothermal compression is given by Equation 15.3:

$$W = nRT\ln\left(\frac{V_f}{V_i}\right)$$

Therefore, the temperature of the gas is

$$T = \frac{W}{nR\ln\left(V_f/V_i\right)} = \frac{-6.1\times10^3\ \text{J}}{(3.0\ \text{mol})[8.31\ \text{J/(mol}\cdot\text{K)}]\ \ln\left[(2.5\times10^{-2}\ \text{m}^3)/(5.5\times10^{-2}\ \text{m}^3)\right]} = \boxed{310\ \text{K}}$$

---

79. ***REASONING AND SOLUTION*** The change in entropy ΔS of a system for a process in which heat Q enters or leaves the system reversibly at a constant temperature T is given by Equation 15.18, $\Delta S = (Q/T)_R$. For a phase change, $Q = mL$, where L is the latent heat (see Section 12.8).

a. If we imagine a reversible process in which 3.00 kg of ice melts into water at 273 K, the change in entropy of the water molecules is

$$\Delta S = \left(\frac{Q}{T}\right)_R = \left(\frac{mL_f}{T}\right)_R = \frac{(3.00\ \text{kg})\left(3.35\times10^5\ \text{J/kg}\right)}{273\ \text{K}} = \boxed{3.68\times10^3\ \text{J/K}}$$

b. Similarly, if we imagine a reversible process in which 3.00 kg of water changes into steam at 373 K, the change in entropy of the water molecules is

$$\Delta S = \left(\frac{Q}{T}\right)_R = \left(\frac{mL_v}{T}\right)_R = \frac{(3.00\ \text{kg})\left(2.26\times10^6\ \text{J/kg}\right)}{373\ \text{K}} = \boxed{1.82\times10^4\ \text{J/K}}$$

c. Since the change in entropy is greater for the vaporization process than for the fusion process, the $\boxed{\text{vaporization process creates more disorder}}$ in the collection of water molecules.

---

83. ***REASONING*** The work done in the process is equal to the "area" under the curved line between A and B in the drawing. From the graph, we find that there are about 78 "squares" under the curve. Each square has an "area" of

$$\left(2.0\times10^4\ \text{Pa}\right)\left(2.0\times10^{-3}\ \text{m}^3\right) = 4.0\times10^1\ \text{J}$$

SOLUTION

a. The work done in the process has a magnitude of

$$W = (78)\left(4.0\times10^1\ \text{J}\right) = \boxed{3100\ \text{J}}$$

b. The final volume is smaller than the initial volume, so the gas is compressed. Therefore, work is done on the gas so the work is $\boxed{\text{negative}}$.

---

85. ***REASONING*** According to Equations 15.6 and 15.7, the heat supplied to a monatomic ideal gas at constant pressure is $Q = C_P n\,\Delta T$, with $C_P = \frac{5}{2}R$. Thus, $Q = \frac{5}{2}nR\,\Delta T$. The percentage of this heat used to increase the internal energy by an amount ΔU is

$$\text{Percentage} = \left(\frac{\Delta U}{Q}\right)\times 100\ \% = \left(\frac{\Delta U}{\frac{5}{2}nR\,\Delta T}\right)\times 100\ \% \qquad (1)$$

But according to the first law of thermodynamics, $\Delta U = Q - W$. The work W is $W = P\,\Delta V$, and for an ideal gas $P\,\Delta V = nR\,\Delta T$. Therefore, the work W becomes $W = P\,\Delta V = nR\,\Delta T$ and the change in the internal energy is $\Delta U = Q - W = \frac{5}{2}nR\,\Delta T - nR\Delta T = \frac{3}{2}nR\Delta T$. Combining this expression for ΔU with Equation (1) above yields a numerical value for the percentage of heat being supplied to the gas that is used to increase its internal energy.

SOLUTION

a. The percentage is

$$\text{Percentage} = \left(\frac{\Delta U}{\frac{5}{2}nR\,\Delta T}\right)\times 100\ \% = \left(\frac{\frac{3}{2}nR\,\Delta T}{\frac{5}{2}nR\,\Delta T}\right)\times 100\ \% = \left(\frac{3}{5}\right)\times 100\ \% = \boxed{60.0\ \%}$$

b. The remainder of the heat, or $\boxed{40.0\ \%}$, is used for the work of expansion.

---

91. ***REASONING AND SOLUTION*** We wish to find an expression for the overall efficiency e in terms of the efficiencies e_1 and e_2. From the problem statement, the overall efficiency of the two-engine device is

$$e = \frac{|W_1| + |W_2|}{|Q_H|} \qquad (1)$$

where $|Q_H|$ is the input heat to engine 1. The efficiency of a heat engine is defined by Equation 15.11, $e = |W|/|Q_H|$, so we can write

$$|W_1| = e_1 |Q_H| \qquad (2)$$

and

$$|W_2| = e_2 |Q_{H2}|$$

Since the heat rejected by engine 1 is used as input heat for the second engine, $|Q_{H2}| = |Q_{C1}|$, and the expression above for $|W_2|$ can be written as

$$|W_2| = e_2 |Q_{C1}| \tag{3}$$

According to Equation 15.12, we have $|Q_{C1}| = |Q_H| - |W_1|$, so that Equation (3) becomes

$$|W_2| = e_2 \left(|Q_H| - |W_1| \right) \tag{4}$$

Substituting Equations (2) and (4) into Equation (1) gives

$$e = \frac{e_1 |Q_H| + e_2 \left(|Q_H| - |W_1| \right)}{|Q_H|} = \frac{e_1 |Q_H| + e_2 \left(|Q_H| - e_1 |Q_H| \right)}{|Q_H|}$$

Algebraically canceling the $|Q_H|$'s in the right hand side of the last expression gives the desired result:

$$\boxed{e = e_1 + e_2 - e_1 e_2}$$

CHAPTER 16 | *WAVES AND SOUND*

CONCEPTUAL QUESTIONS

---

4. ***REASONING AND SOLUTION*** A wave moves on a string with constant velocity. It is *not* correct to conclude that the particles of the string always have zero acceleration. As Conceptual Example 3 discusses, it is important to distinguish between the speed of the waves on the string, v_{wave}, and the speed of the particles in the string, v_{particle}. The wave speed v_{wave} is determined by the properties of the string; namely, the tension in the string and the linear mass density of the string. These properties determine the speed with which the disturbance travels along the string. The wave speed will remain constant as long as these properties remain unchanged.

   The particles in the string oscillate transversely in simple harmonic motion with the same amplitude and frequency as the source of the disturbance. Like all particles in simple harmonic motion, the acceleration of the particles continually changes. It is zero when the particles pass through their equilibrium positions and is a maximum when the particles are at their maximum displacements from their equilibrium positions.

---

11. ***REASONING AND SOLUTION*** A loudspeaker produces a sound wave. The sound wave travels from air into water. As indicated in Table 16.1, the speed of sound in water is approximately four times greater than it is in air. We are told in the hint that the frequency of the sound wave does not change as the sound enters the water. The relationship between the frequency f, the wavelength λ, and the speed v of a wave is given by Equation 16.1: $v = f\lambda$. Since the wave speed increases and the frequency remains the same as the sound enters the water, the wavelength of the sound must increase.

---

19. ***REASONING AND SOLUTION*** When a car is at rest, its horn emits a frequency of 600 Hz. A person standing in the middle of the street hears the horn with a frequency of 580 Hz. As shown in Figure 16.28, when a vehicle is moving, a stationary observer in front of the vehicle will hear sound of a shorter wavelength, while an observer behind the vehicle will hear sound of a longer wavelength. We can deduce, therefore, from Equation 16.1, $v = f\lambda$, that the person standing in front of the vehicle will hear sound of a higher frequency than when the car is at rest, while a person standing behind the vehicle will hear sound of a lower frequency than when the car is at rest. Since the person standing in the middle of the street hears a frequency of 580 Hz, which is less than the stationary frequency of 600 Hz, the person is behind the car. Therefore, the person need not jump out of the way.

CHAPTER 16 | *WAVES AND SOUND*

PROBLEMS

---

1. ***REASONING*** Since light behaves as a wave, its speed v, frequency f, and wavelength λ are related to according to $v = f\lambda$ (Equation 16.1). We can solve this equation for the frequency in terms of the speed and the wavelength.

SOLUTION Solving Equation 16.1 for the frequency, we find that

$$f = \frac{v}{\lambda} = \frac{3.00\times10^{8}\ \text{m/s}}{5.45\times10^{-7}\ \text{m}} = \boxed{5.50\times10^{14}\ \text{Hz}}$$

---

5. ***REASONING*** When the end of the Slinky is moved up and down continuously, a transverse wave is produced. The distance between two adjacent crests on the wave, is, by definition, one wavelength. The wavelength λ is related to the speed and frequency of a periodic wave by Equation 16.1, $\lambda = v/f$. In order to use Equation 16.1, we must first determine the frequency of the wave. The wave on the Slinky will have the same frequency as the simple harmonic motion of the hand. According to the data given in the problem statement, the frequency is $f = (2.00\ \text{cycles})/(1\ \text{s}) = 2.00\ \text{Hz}$.

SOLUTION Substituting the values for λ and f, we find that the distance between crests is

$$\lambda = \frac{v}{f} = \frac{0.50\ \text{m/s}}{2.00\ \text{Hz}} = \boxed{0.25\ \text{m}}$$

---

13. ***REASONING*** According to Equation 16.2, the linear density of the string is given by $(m/L) = F/v^2$, where the speed v of waves on the middle C string is given by Equation 16.1, $v = \lambda f = \lambda/T$.

SOLUTION Combining Equations 16.2 and 16.1 and using the given data, we obtain

$$m/L = \frac{F}{v^2} = \frac{FT^2}{\lambda^2} = \frac{(944\ \text{N})(3.82\times10^{-3}\ \text{s})^2}{(1.26\ \text{m})^2} = \boxed{8.68\times10^{-3}\ \text{kg/m}}$$

---

19. ***REASONING*** Newton's second law can be used to analyze the motion of the blocks using the methods developed in Chapter 4. We can thus determine an expression that relates the magnitude P of the pulling force to the magnitude F of the tension in the wire. Equation 16.2 [$v = \sqrt{F/(m/L)}$] can then be used to find the tension in the wire.

SOLUTION The following figures show a schematic of the situation described in the problem and the free-body diagrams for each block, where $m_1 = 42.0$ kg and $m_2 = 19.0$ kg. The pulling force is **P**, and the tension in the wire gives rise to the forces **F** and –**F**, which act on m_1 and m_2, respectively.

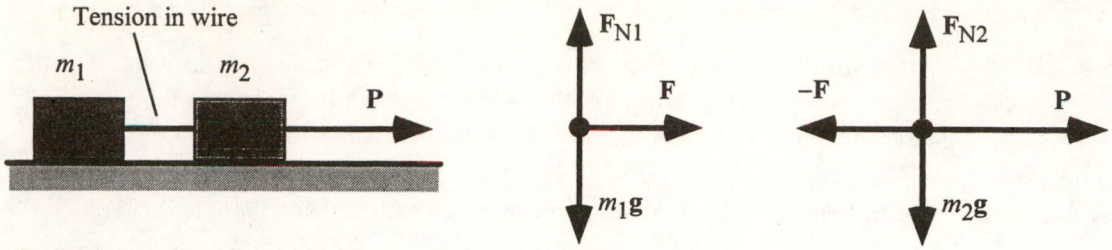

Newton's second law for block 1 is, taking forces that point to the right as positive, $F = m_1 a$, or $a = F/m_1$. For block 2, we obtain $P - F = m_2 a$. Using the expression for a obtained from the equation for block 1, we have

$$P - F = F\left(\frac{m_2}{m_1}\right) \qquad \text{or} \qquad P = F\left(\frac{m_2}{m_1}\right) + F = F\left(\frac{m_2}{m_1} + 1\right)$$

According to Equation 16.2, $F = v^2(m/L)$, where m/L is the mass per unit length of the wire. Combining this expression for F with the expression for P, we have

$$P = v^2(m/L)\left(\frac{m_2}{m_1} + 1\right) = (352 \text{ m/s})^2(8.50\times10^{-4} \text{ kg/m})\left(\frac{19.0 \text{ kg}}{42.0 \text{ kg}} + 1\right) = \boxed{153 \text{ N}}$$

---

23. ***REASONING*** Since the wave is traveling in the $+x$ direction, its form is given by Equation 16.3 as

$$y = A\sin\left(2\pi ft - \frac{2\pi x}{\lambda}\right)$$

We are given that the amplitude is $A = 0.35$ m. However, we need to evaluate $2\pi f$ and $\frac{2\pi}{\lambda}$. Although the wavelength λ is not stated directly, it can be obtained from the values for the speed v and the frequency f, since we know that $v = f\lambda$ (Equation 16.1).

SOLUTION Since the frequency is $f = 14$ Hz, we have

$$2\pi f = 2\pi(14 \text{ Hz}) = 88 \text{ rad/s}$$

It follows from Equation 16.1 that

$$\frac{2\pi}{\lambda} = \frac{2\pi f}{v} = \frac{2\pi(14 \text{ Hz})}{5.2 \text{ m/s}} = 17 \text{ m}^{-1}$$

Using these values for $2\pi f$ and $\frac{2\pi}{\lambda}$ in Equation 16.3, we have

$$y = A\sin\left(2\pi ft - \frac{2\pi x}{\lambda}\right)$$

$$\boxed{y = (0.35\text{ m})\sin\left[(88\text{ rad/s})t - \left(17\text{ m}^{-1}\right)x\right]}$$

---

27. ***REASONING*** According to Equation 16.2, the tension F in the string is given by $F = v^2(m/L)$. Since $v = \lambda f$ from Equation 16.1, the expression for F can be written

$$F = (\lambda f)^2\left(\frac{m}{L}\right) \qquad (1)$$

where the quantity m/L is the linear density of the string. In order to use Equation (1), we must first obtain values for f and λ; these quantities can be found by analyzing the expression for the displacement of a string particle.

SOLUTION The displacement is given by $y = (0.021\text{ m})\sin(25t - 2.0x)$. Inspection of this equation and comparison with Equation 16.3, $y = A\sin\left(2\pi ft - \frac{2\pi x}{\lambda}\right)$, gives

$$2\pi f = 25\text{ rad/s} \quad \text{or} \quad f = \frac{25}{2\pi}\text{ Hz}$$

and

$$\frac{2\pi}{\lambda} = 2.0\text{ m}^{-1} \quad \text{or} \quad \lambda = \frac{2\pi}{2.0}\text{ m}$$

Substituting these values f and λ into Equation (1) gives

$$F = (\lambda f)^2\left(\frac{m}{L}\right) = \left[\left(\frac{2\pi}{2.0}\text{ m}\right)\left(\frac{25\text{ Hz}}{2\pi}\right)\right]^2(1.6\times10^{-2}\text{ kg/m}) = \boxed{2.5\text{ N}}$$

---

29. ***REASONING AND SOLUTION*** The speed of sound in an ideal gas is given by Equation 16.5, $v = \sqrt{\gamma kT/m}$. The ratio of the speed of sound v_2 in the container (after the temperature change) to the speed v_1 (before the temperature change) is

$$\frac{v_2}{v_1} = \sqrt{\frac{T_2}{T_1}}$$

Thus, the new speed is

$$v_2 = v_1\sqrt{\frac{T_2}{T_1}} = (1220\text{ m/s})\sqrt{\frac{405\text{ K}}{201\text{ K}}} = \boxed{1730\text{ m/s}}$$

---

33. ***REASONING*** If we treat the sample of argon atoms like an ideal monatomic gas ($\gamma = 1.67$) at 298 K, Equation 14.6 $\left(\frac{1}{2}mv_{\text{rms}}^2 = \frac{3}{2}kT\right)$ can be solved for the root-mean-square speed v_{rms} of the argon atoms. The speed of sound in argon can be found from Equation 16.5: $v = \sqrt{\gamma kT/m}$.

SOLUTION We first find the mass of an argon atom. Since the molecular mass of argon is 39.9 u, argon has a mass per mole of 39.9×10^{-3} kg/mol. Thus, the mass of a single argon atom is

$$m = \frac{39.9\times10^{-3}\text{ kg/mol}}{6.022\times10^{23}\text{ mol}^{-1}} = 6.63\times10^{-26}\text{ kg}$$

a. Solving Equation 14.6 for v_{rms} and substituting the data given in the problem statement, we find

$$v_{\text{rms}} = \sqrt{\frac{3kT}{m}} = \sqrt{\frac{3\left(1.38\times10^{-23}\text{ J/K}\right)(298\text{ K})}{6.63\times10^{-26}\text{ kg}}} = \boxed{431\text{ m/s}}$$

b. The speed of sound in argon is, according to Equation 16.5,

$$v = \sqrt{\frac{\gamma kT}{m}} = \sqrt{\frac{(1.67)\left(1.38\times10^{-23}\text{ J/K}\right)(298\text{ K})}{6.63\times10^{-26}\text{ kg}}} = \boxed{322\text{ m/s}}$$

---

39. ***REASONING*** The sound will spread out uniformly in all directions. For the purposes of counting the echoes, we will consider only the sound that travels in a straight line parallel to the ground and reflects from the vertical walls of the cliff.

Let the distance between the hunter and the closer cliff be x_1 and the distance from the hunter to the further cliff be x_2.

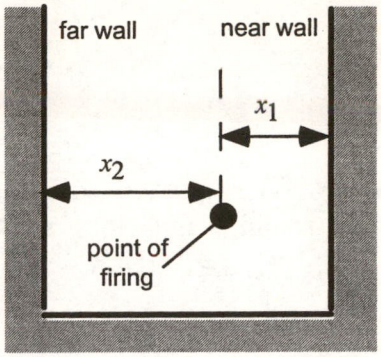

The first echo arrives at the location of the hunter after traveling a total distance $2x_1$ in a time t_1, so that, if v_s is the speed of sound, $t_1 = 2x_1 / v_s$. Similarly, the second echo arrives after reflection from the far wall and in an amount of time t_2 after the firing of the gun. The quantity t_2 is related to the distance x_2 and the speed of sound v_s according to $t_2 = 2x_2 / v_s$. The time difference between the first and second echo is, therefore

$$\Delta t = t_2 - t_1 = \frac{2}{v_s}(x_2 - x_1) \tag{1}$$

The third echo arrives in a time t_3 after the second echo. It arises from the sound of the second echo that is reflected from the closer cliff wall. Thus, $t_3 = 2x_1 / v_s$, or, solving for x_1, we have

$$x_1 = \frac{v_s t_3}{2} \tag{2}$$

Combining Equations (1) and (2), we obtain

$$\Delta t = t_2 - t_1 = \frac{2}{v_s}\left(x_2 - \frac{v_s t_3}{2}\right)$$

Solving for x_2, we have

$$x_2 = \frac{v_s}{2}(\Delta t + t_3) \tag{3}$$

The distance between the cliffs can be found from $d = x_1 + x_2$, where x_1 and x_2 can be determined from Equations (2) and (3), respectively.

SOLUTION According to Equation (2), the distance x_1 is

$$x_1 = \frac{(343\text{ m/s})(1.1\text{ s})}{2} = 190\text{ m}$$

According to Equation (3), the distance x_2 is

$$x_2 = \frac{(343\text{ m/s})}{2}(1.6\text{ s} + 1.1\text{ s}) = 460\text{ m}$$

Therefore, the distance between the cliffs is

$$d = x_1 + x_2 = 190 \text{ m} + 460 \text{ m} = \boxed{650 \text{ m}}$$

---

41. ***REASONING*** Equation 16.7 relates the Young's modulus Y, the mass density ρ, and the speed of sound v in a long slender solid bar. According to Equation 16.7, the Young's modulus is given by $Y = \rho v^2$ .The data given in the problem can be used to compute values for both ρ and v.

SOLUTION Using the values of the data given in the problem statement, we find that the speed of sound in the bar is

$$v = \frac{L}{t} = \frac{0.83 \text{ m}}{1.9\times10^{-4} \text{ s}} = 4.4\times10^3 \text{ m/s}$$

where L is the length of the rod and t is the time required for the wave to travel the length of the rod. The mass density of the bar is, from Equation 11.1, $\rho = m/V = m/(LA)$, where m and A are, respectively, the mass and the cross-sectional area of the rod. The density of the rod is, therefore,

$$\rho = \frac{m}{LA} = \frac{2.1 \text{ kg}}{(0.83 \text{ m})(1.3\times10^{-4} \text{ m}^2)} = 1.9\times10^4 \text{ kg/m}^3$$

Using these values, we find that the bulk modulus for the rod is

$$Y = \rho v^2 = \left(1.9\times10^4 \text{ kg/m}^3\right)\left(4.4\times10^3 \text{ m/s}\right)^2 = 3.7\times10^{11} \text{ N/m}^2$$

Comparing this value to those given in Table 10.1, we conclude that the bar is most likely made of $\boxed{\text{tungsten}}$.

---

47. ***REASONING*** We must determine the time t for the warning to travel the vertical distance $h = 10.0$ m from the prankster to the ears of the man when he is just under the window. The desired distance above the man's ears is the distance that the balloon would travel in this time and can be found with the aid of the equations of kinematics.

SOLUTION Since sound travels with constant speed v_s, the distance h and the time t are related by $h = v_s t$. Therefore, the time t required for the warning to reach the ground is

$$t = \frac{h}{v_s} = \frac{10.0 \text{ m}}{343 \text{ m/s}} = 0.0292 \text{ s}$$

We now proceed to find the distance that the balloon travels in this time. To this end, we must find the balloon's final speed v_y after falling from rest for 10.0 m. Since the balloon is dropped from rest, we use Equation 3.6*b* ($v_y^2 = v_{0y}^2 + 2a_y y$) with $v_{0y} = 0$ m/s:

$$v_y = \sqrt{v_{0y}^2 + 2a_y y} = \sqrt{(0\text{ m/s})^2 + 2(9.80\text{ m/s}^2)(10.0\text{ m})} = 14.0\text{ m/s}$$

Using this result, we can find the balloon's speed 0.0292 seconds before it hits the man by solving Equation 3.3b ($v_y = v_{0y} + a_y t$) for v_{0y}:

$$v_{0y} = v_y - a_y t = \left[(14.0\text{ m/s}) - (9.80\text{ m/s}^2)(0.0292\text{ s})\right] = 13.7\text{ m/s}$$

Finally, we can find the desired distance y above the man's head from Equation 3.5b:

$$y = v_{0y}t + \tfrac{1}{2}a_y t^2 = (13.7\text{ m/s})(0.0292\text{ s}) + \tfrac{1}{2}(9.80\text{ m/s}^2)(0.0292\text{ s})^2 = \boxed{0.404\text{ m}}$$

---

49. ***REASONING AND SOLUTION*** Since the sound radiates uniformly in all directions, at a distance r from the source, the energy of the sound wave is distributed over the area of a sphere of radius r. Therefore, according to Equation 16.9 [$I = P/(4\pi r^2)$] with $r = 3.8$ m, the power radiated from the source is

$$P = 4\pi I r^2 = 4\pi(3.6\times10^{-2}\text{ W/m}^2)(3.8\text{ m})^2 = \boxed{6.5\text{ W}}$$

---

53. ***REASONING AND SOLUTION*** According to Equation 16.8, the power radiated by the speaker is $P = IA = I\pi r^2$, where r is the radius of the circular opening. Thus, the radiated power is

$$P = (17.5\text{ W/m}^2)(\pi)(0.0950\text{ m})^2 = 0.496\text{ W}$$

As a percentage of the electrical power, this is

$$\frac{0.496\text{ W}}{25.0\text{ W}} \times 100\ \% = \boxed{1.98\ \%}$$

---

57. ***REASONING*** Since the sound is emitted from the rocket uniformly in all directions, the energy carried by the sound wave spreads out uniformly over concentric spheres of increasing radii r_1, r_2, r_3, . . . as the wave propagates. Let r_1 represent the radius when the measured intensity at the ground is I (position 1 for the rocket) and r_2 represent the radius when the measured intensity at the ground is $\frac{1}{3}I$ (position 2 of the rocket). The time

required for the energy to spread out over the sphere of radius r_1 is $t_1 = r_1 / v_s$, where v_s is the speed of sound. Similarly, the time required for the energy to spread out over a sphere of radius r_2 is $t_2 = r_2 / v_s$. As the sound wave emitted at position 1 spreads out uniformly, the rocket continues to accelerate upward to position 2 with acceleration a_y for a time t_{12}. Therefore, the time that elapses between the two intensity measurements is

$$t = t_2 - t_1 + t_{12} \tag{1}$$

From Equation 3.3*b*, the time t_{12} is

$$t_{12} = \frac{(v_2 - v_1)}{a_y} \tag{2}$$

where v_1 and v_2 are the speeds of the rocket at positions 1 and 2, respectively. Combining Equations (1) and (2), we obtain

$$t = t_2 - t_1 + \frac{(v_2 - v_1)}{a_y} \tag{3}$$

The respective values of v_1 and v_2 can be found from Equation 3.6b ($v_y^2 = v_{0y}^2 + 2a_y y$) with $v_{0y} = 0$ m/s, and $y = r$. Once values for v_1 and v_2 are known, Equation (3) can be solved directly.

SOLUTION From the data given in the problem statement, $r_1 = 562$ m. We can find r_2 using the following reasoning: from the definition of intensity, $I_1 = P/(4\pi r_1^2)$ and $I_2 = P/(4\pi r_2^2)$. Since $I_1 = 3I_2$, we have

$$P/(4\pi r_1^2) = 3P/(4\pi r_2^2) \qquad \text{or} \qquad r_2 = r_1\sqrt{3} = 973 \text{ m}$$

The times t_1 and t_2 are

$$t_1 = \frac{r_1}{v_s} = \frac{562 \text{ m}}{343 \text{ m/s}} = 1.64 \text{ s}$$

and

$$t_2 = \frac{r_2}{v_s} = \frac{973 \text{ m}}{343 \text{ m/s}} = 2.84 \text{ s}$$

Then, taking up as the positive direction, we have from Equation 3.6b,

$$v_1 = \sqrt{2a_y r_1} = \sqrt{2(58.0\text{ m/s}^2)(562\text{ m})} = 255\text{ m/s}$$

and

$$v_2 = \sqrt{2a_y r_2} = \sqrt{2(58.0\text{ m/s}^2)(973\text{ m})} = 336\text{ m/s}$$

Substituting these values into Equation (3), we have

$$t = 2.84\text{ s} - 1.64\text{ s} + \frac{(336\text{ m/s} - 255\text{ m/s})}{58.0\text{ m/s}^2} = \boxed{2.6\text{ s}}$$

---

61. ***REASONING AND SOLUTION*** The intensity level β in decibels (dB) is related to the sound intensity I according to Equation 16.10:

$$\beta = (10\text{ dB})\log\left(\frac{I}{I_0}\right)$$

where the quantity I_0 is the reference intensity. Therefore, we have

$$\beta_2 - \beta_1 = (10\text{ dB})\log\left(\frac{I_2}{I_0}\right) - (10\text{ dB})\log\left(\frac{I_1}{I_0}\right) = (10\text{ dB})\log\left(\frac{I_2/I_0}{I_1/I_0}\right) = (10\text{ dB})\log\left(\frac{I_2}{I_1}\right)$$

Solving for the ratio I_2/I_1, we find

$$30.0\text{ dB} = (10\text{ dB})\log\left(\frac{I_2}{I_1}\right) \qquad \text{or} \qquad \frac{I_2}{I_1} = 10^{3.0} = 1000$$

Thus, we conclude that the sound intensity $\boxed{\text{increases by a factor of 1000}}$.

---

64. ***REASONING*** We must first find the intensities that correspond to the given sound intensity levels (in decibels). The total intensity is the sum of the two intensities. Once the total intensity is known, Equation 16.10 can be used to find the total sound intensity level in decibels.

SOLUTION Since, according to Equation 16.10, $\beta = (10\text{ dB})\log(I/I_0)$, where I_0 is the reference intensity corresponding to the threshold of hearing $\left(I_0 = 1.00\times10^{-12}\text{ W/m}^2\right)$, it follows that $I = I_0\ 10^{\beta/(10\text{ dB})}$. Therefore, if $\beta_1 = 75.0$ dB and $\beta_2 = 72.0$ dB at the point in question, the corresponding intensities are

$$I_1 = I_0\ 10^{\beta_1/(10\text{ dB})} = (1.00\times10^{-12}\text{ W/m}^2)\ 10^{(75.0\text{ dB})/(10\text{ dB})} = 3.16\times10^{-5}\text{ W/m}^2$$

$$I_2 = I_0\ 10^{\beta_2/(10\text{ dB})} = (1.00\times10^{-12}\text{ W/m}^2)\ 10^{(72.0\text{ dB})/(10\text{ dB})} = 1.58\times10^{-5}\text{ W/m}^2$$

Therefore, the total intensity I_{total} at the point in question is

$$I_{\text{total}} = I_1 + I_2 = \left(3.16\times10^{-5}\text{ W/m}^2\right) + \left(1.58\times10^{-5}\text{ W/m}^2\right) = 4.74\times10^{-5}\text{ W/m}^2$$

and the corresponding intensity level β_{total} is

$$\beta_{\text{total}} = (10\text{ dB})\log\left(\frac{I_{\text{total}}}{I_0}\right) = (10\text{ dB})\log\left(\frac{4.74\times10^{-5}\text{ W/m}^2}{1.00\times10^{-12}\text{ W/m}^2}\right) = \boxed{76.8\text{ dB}}$$

---

69. ***REASONING AND SOLUTION*** The sound intensity level β in decibels (dB) is related to the sound intensity I according to Equation 16.10, $\beta = (10\text{ dB})\log(I/I_0)$, where the quantity I_0 is the reference intensity. According to the problem statement, when the sound intensity level triples, the sound intensity also triples; therefore,

$$3\beta = (10\text{ dB})\log\left(\frac{3I}{I_0}\right)$$

Then,

$$3\beta - \beta = (10\text{ dB})\log\left(\frac{3I}{I_0}\right) - (10\text{ dB})\log\left(\frac{I}{I_0}\right) = (10\text{ dB})\log\left(\frac{3I/I_0}{I/I_0}\right)$$

Thus, $2\beta = (10\text{ dB})\log 3$ and

$$\beta = (5\text{ dB})\log 3 = \boxed{2.39\text{ dB}}$$

---

73. ***REASONING*** Since you detect a frequency that is smaller than that emitted by the car when the car is stationary, the car must be moving away from you. Therefore, according to Equation 16.12, the frequency f_o heard by a stationary observer from a source moving away from the observer is given by

$$f_o = f_s\left(\frac{1}{1+\frac{v_s}{v}}\right)$$

where f_s is the frequency emitted from the source when it is stationary with respect to the observer, v is the speed of sound, and v_s is the speed of the moving source. This expression can be solved for v_s.

SOLUTION We proceed to solve for v_s and substitute the data given in the problem statement. Rearrangement gives

$$\frac{v_s}{v} = \frac{f_s}{f_o} - 1$$

Solving for v_s and noting that $f_o / f_s = 0.86$ yields

$$v_s = v\left(\frac{f_s}{f_o} - 1\right) = (343 \text{ m/s})\left(\frac{1}{0.86} - 1\right) = \boxed{56 \text{ m/s}}$$

---

75. ***REASONING AND SOLUTION*** In this situation, both the source and the observer are moving through the air. Therefore, the frequency of sound heard by the crew is given by Equation 16.15 using the plus signs in both the numerator and the denominator.

$$f_o = f_s\left(\frac{1 + \frac{v_o}{v}}{1 + \frac{v_s}{v}}\right) = (1550 \text{ Hz})\left(\frac{1 + \frac{13.0 \text{ m/s}}{343 \text{ m/s}}}{1 + \frac{67.0 \text{ m/s}}{343 \text{ m/s}}}\right) = \boxed{1350 \text{ Hz}}$$

---

79. ***REASONING***

a. Since the two submarines are approaching each other head on, the frequency f_o detected by the observer (sub B) is related to the frequency f_s emitted by the source (sub A) by

$$f_o = f_s\left(\frac{1 + \frac{v_o}{v}}{1 - \frac{v_s}{v}}\right) \qquad (16.15)$$

where v_o and v_s are the speed of the observer and source, respectively, and v is the speed of the underwater sound

b. The sound reflected from submarine B has the same frequency that it detects, namely, f_o. Now sub B becomes the source of sound and sub A is the observer. We can still use Equation 16.15 to find the frequency detected by sub A.

SOLUTION

a. The frequency f_o detected by sub B is

$$f_o = f_s\left(\frac{1+\frac{v_o}{v}}{1-\frac{v_s}{v}}\right) = (1550\text{ Hz})\left(\frac{1+\frac{8\text{ m/s}}{1522\text{ m/s}}}{1-\frac{12\text{ m/s}}{1522\text{ m/s}}}\right) = \boxed{1570\text{ Hz}}$$

b. The sound reflected from submarine B has the same frequency that it detects, namely, 1570 Hz. Now sub B is the source of sound whose frequency is f_s = 1570 Hz. The speed of sub B is v_s = 8 m/s. The frequency detected by sub A (whose speed is v_o = 12 m/s) is

$$f_o = f_s\left(\frac{1+\frac{v_o}{v}}{1-\frac{v_s}{v}}\right) = (1570\text{ Hz})\left(\frac{1+\frac{12\text{ m/s}}{1522\text{ m/s}}}{1-\frac{8\text{ m/s}}{1522\text{ m/s}}}\right) = \boxed{1590\text{ Hz}}$$

---

83. ***REASONING AND SOLUTION*** The speed of sound in a liquid is given by Equation 16.6, $v = \sqrt{B_{ad}/\rho}$, where B_{ad} is the adiabatic bulk modulus and ρ is the density of the liquid. Solving for B_{ad}, we obtain $B_{ad} = v^2\rho$. Values for the speed of sound in fresh water and in ethyl alcohol are given in Table 16.1. The ratio of the adiabatic bulk modulus of fresh water to that of ethyl alcohol at 20°C is, therefore,

$$\frac{(B_{ad})_{\text{water}}}{(B_{ad})_{\text{ethyl alcohol}}} = \frac{v^2_{\text{water}}\rho_{\text{water}}}{v^2_{\text{ethyl alcohol}}\rho_{\text{ethyl alcohol}}} = \frac{(1482\text{ m/s})^2(998\text{ kg/m}^3)}{(1162\text{ m/s})^2(789\text{ kg/m}^3)} = \boxed{2.06}$$

---

85. ***REASONING*** The tension F in the violin string can be found by solving Equation 16.2 for F to obtain $F = mv^2/L$, where v is the speed of waves on the string and can be found from Equation 16.1 as $v = f\lambda$.

SOLUTION Combining Equations 16.2 and 16.1 and using the given data, we obtain

$$F = \frac{mv^2}{L} = (m/L)f^2\lambda^2 = \left(7.8\times10^{-4}\text{ kg/m}\right)\left(440\text{ Hz}\right)^2\left(65\times10^{-2}\text{ m}\right)^2 = \boxed{64\text{ N}}$$

---

87. ***REASONING AND SOLUTION*** The sound intensity level β in decibels (dB) is related to the sound intensity I according to Equation 16.10, $\beta = (10\text{ dB})\log(I/I_0)$, where I_0 is the reference intensity. If β_1 and β_2 represent two different sound level intensities, then,

$$\beta_2 - \beta_1 = (10\text{ dB})\log\left(\frac{I_2}{I_0}\right) - (10\text{ dB})\log\left(\frac{I_1}{I_0}\right) = (10\text{ dB})\log\left(\frac{I_2/I_0}{I_1/I_0}\right) = (10\text{ dB})\log\left(\frac{I_2}{I_1}\right)$$

Therefore,

$$\frac{I_2}{I_1} = 10^{(\beta_2-\beta_1)/(10\text{ dB})}$$

When the difference in sound intensity levels is $\beta_2 - \beta_1 = 1.0\text{ dB}$, the ratio of the sound intensities is

$$\frac{I_2}{I_1} = 10^{(1\text{ dB})/(10\text{ dB})} = \boxed{1.3}$$

---

89. ***REASONING*** As the transverse wave propagates, the colored dot moves up and down in simple harmonic motion with a frequency of 5.0 Hz. The amplitude (1.3 cm) is the magnitude of the maximum displacement of the dot from its equilibrium position.

SOLUTION The period T of the simple harmonic motion of the dot is $T = 1/f = 1/(5.0\text{ Hz}) = 0.20\text{ s}$. In one period the dot travels through one complete cycle of its motion, and covers a vertical distance of $4 \times (1.3\text{ cm}) = 5.2\text{ cm}$. Therefore, in 3.0 s the dot will have traveled a *total vertical distance* of

$$\left(\frac{3.0\text{ s}}{0.20\text{ s}}\right)(5.2\text{ cm}) = \boxed{78\text{ cm}}$$

---

91. ***REASONING AND SOLUTION: METHOD 1*** Since the police car is matching the speed of the speeder, there is no relative motion between the police car and the listener (the speeder). Therefore, the frequency that the speeder hears when the siren is turned on is the same as if the police car and the listener are stationary. Thus, the speeder hears a frequency of $\boxed{860\text{ Hz}}$.

REASONING AND SOLUTION: METHOD 2 In this situation, both the source (the siren) and the observer (the speeder) are moving through the air. Therefore, the frequency of the siren heard by the speeder is given by Equation 16.15 using the minus signs in both the numerator and the denominator.

$$f_o = f_s\left(\frac{1-\frac{v_o}{v}}{1-\frac{v_s}{v}}\right) = (860\text{ Hz})\left(\frac{1-\frac{38\text{ m/s}}{343\text{ m/s}}}{1-\frac{38\text{ m/s}}{343\text{ m/s}}}\right) = \boxed{860\text{ Hz}}$$

---

93. ***REASONING*** According to Equation 16.10, the sound intensity level β in decibels (dB) is related to the sound intensity I according to $\beta = (10\text{ dB})\log\ (I/I_0)$, where the quantity I_0 is the reference intensity. Since the sound is emitted uniformly in all directions, the intensity, or power per unit area, is given by $I = P/(4\pi r^2)$. Thus, the sound intensity at position 1 can be written as $I_1 = P/(4\pi r_1^2)$, while the sound intensity at position 2 can be written as $I_2 = P/(4\pi r_2^2)$. We can obtain the sound intensity levels from Equation 16.10 for these two positions by using these expressions for the intensities.

SOLUTION Using Equation 16.10 and the expressions for the intensities at the two positions, we can write the difference in the sound intensity levels β_{21} between the two positions as follows:

$$\beta_{21} = \beta_2 - \beta_1 = (10\text{ dB})\log\left(\frac{I_2}{I_0}\right) - (10\text{ dB})\log\left(\frac{I_1}{I_0}\right)$$

$$= (10\text{ dB})\log\left(\frac{I_2/I_0}{I_1/I_0}\right) = (10\text{ dB})\ \log\left(\frac{I_2}{I_1}\right)$$

$$\beta_{21} = (10\text{ dB})\log\left[\frac{P/(4\pi r_2^2)}{P/(4\pi r_1^2)}\right] = (10\text{ dB})\log\left(\frac{r_1^2}{r_2^2}\right) = (10\text{ dB})\log\left(\frac{r_1}{r_2}\right)^2$$

$$= (20\text{ dB})\log\left(\frac{r_1}{r_2}\right) = (20\text{ dB})\log\left(\frac{r_1}{2r_1}\right) = (20\text{ dB})\log(1/2) = \boxed{-6.0\text{ dB}}$$

The negative sign indicates that the sound intensity level decreases.

---

97. ***REASONING*** Using the procedures developed in Chapter 4 for using Newton's second law to analyze the motion of bodies and neglecting the weight of the wire relative to the tension in the wire lead to the following equations of motion for the two blocks:

$$\sum F_x = F - m_1 g\ (\sin\ 30.0°) = 0 \qquad (1)$$

$$\sum F_y = F - m_2 g = 0 \qquad (2)$$

where F is the tension in the wire. In Equation (1) we have taken the direction of the $+x$ axis for block 1 to be parallel to and up the incline. In Equation (2) we have taken the direction of the $+y$ axis to be upward for block 2. This set of equations consists of two equations in three unknowns, m_1, m_2, and F. Thus, a third equation is needed in order to solve for any of the unknowns. A useful third equation can be obtained by solving Equation 16.2 for F:

$$F = (m/L)v^2 \tag{3}$$

Combining Equation (3) with Equations (1) and (2) leads to

$$(m/L)v^2 - m_1 g \sin 30.0° = 0 \tag{4}$$

$$(m/L)v^2 - m_2 g = 0 \tag{5}$$

Equations (4) and (5) can be solved directly for the masses m_1 and m_2.

SOLUTION Substituting values into Equation (4), we obtain

$$m_1 = \frac{(m/L)v^2}{g \sin 30.0°} = \frac{(0.0250 \text{ kg/m})(75.0 \text{ m/s})^2}{(9.80 \text{ m/s}^2) \sin 30.0°} = \boxed{28.7 \text{ kg}}$$

Similarly, substituting values into Equation (5), we obtain

$$m_2 = \frac{(m/L)v^2}{g} = \frac{(0.0250 \text{ kg/m})(75.0 \text{ m/s})^2}{(9.80 \text{ m/s}^2)} = \boxed{14.3 \text{ kg}}$$

CHAPTER 17 THE PRINCIPLE OF LINEAR SUPERPOSITION AND INTERFERENCE PHENOMENA

CONCEPTUAL QUESTIONS

---

3. ***REASONING AND SOLUTION*** Consider the situation in Figure 17.3. If you walk along a line that is perpendicular to the line between the speakers and passes through the overlap point, you will always be equidistant from both speakers. Therefore the sound waves along this line always overlap exactly in phase. Hence, you will always hear the same loudness; you will not observe the loudness to change from loud to faint to loud.

   On the other hand, if you walk along a line that passes through the overlap point and is parallel to the line between the speakers, your distance from the two speakers will vary such that the difference in path lengths traveled by the two waves will vary. At certain points the path length difference between the two sound waves will be an integer number of wavelengths; constructive interference will occur at these points, and the sound intensity will be a maximum. At other points, the path length difference between the two sound waves will be an odd number of half-wavelengths [$(1/2)\lambda$, $(3/2)\lambda$, $(5/2)\lambda$, etc.]; destructive interference will occur and the sound intensity will be a minimum. In between the points of constructive and destructive interference, the waves will be out of phase by varying degrees. Therefore, as you walk along this line, you will observe the sound intensity to alternate between faint and loud.

---

7. ***REASONING AND SOLUTION*** Tuning forks vibrate at 438 Hz and 440 Hz. When sounded together, a listener hears a beat frequency of 2 Hz, the difference in frequency between the two forks. The forks are then vibrated underwater with the listener also underwater. The forks vibrate at 438 and 440 Hz, just as they do in air. The speed of sound, however, is four times faster in water than in air. From the relation $v = \lambda f$, we see that the wavelength of the sound in water is four times longer than it is in air. Since the speed of the waves is four times greater than it is in air, the regions of constructive and destructive interference move past the ears of the listener four times more rapidly than they do in air. However, the regions of condensation and rarefaction in water are separated by four times the distance than they are in air. Therefore, at a given point in space, the amplitude variation of the resultant wave occurs at the same rate that it does in air. Consequently, the listener hears a beat frequency of 2 Hz, just as he does in air.

---

14. ***REASONING AND SOLUTION*** The speed of sound v in an ideal gas is given by Equation 16.5, $v = \sqrt{\gamma kT/m}$, where $\gamma = C_P / C_V$ is the ratio of the specific heat capacities, k is Boltzmann's constant, T is the Kelvin temperature, and m is the mass of a molecule of the gas. Thus, the speed of sound in an ideal gas is directly proportional to the square root of the Kelvin temperature T.

If the instrument is open at both ends, then the frequency of sound produced by the instrument is given by Equation 17.4, $f_n = n[v/(2L)]$, where v is the speed of sound, L is the distance between the two open ends of the instrument, and $n = 1, 2, 3, 4, \ldots$ Similarly, if the instrument is open at only one end and closed at the other end, the frequency of sound produced by the instrument is given by Equation 17.5, $f_n = n[v/(4L)]$, where v is the speed of sound, L is the distance between the open and closed ends of the instrument, and $n = 1, 3, 5, 7, \ldots$

When a wind instrument is brought inside from the cold outdoors, its frequency will change because the temperature inside is higher than the temperature outside. If we can treat air as an ideal gas, then according to Equation 16.5, the speed of sound will increase because T increases. According to both Equations 17.4 and 17.5, the frequency produced by a wind instrument is directly proportional to v, the speed of sound; therefore, since the speed of sound increases with increasing temperature, the frequency of the sound produced by the instrument will also increase. Note that the length L also typically increases with increasing temperature, as discussed in Section 12.4. However, the change in length is relatively negligible compared to the change in the speed of sound, as can be verified using data from Table 12.1.

CHAPTER 17 *THE PRINCIPLE OF LINEAR SUPERPOSITION AND INTERFERENCE PHENOMENA*

PROBLEMS

1. ***REASONING AND SOLUTION*** According to the principle of linear superposition, when two or more waves are present simultaneously at the same place, the resultant wave is the sum of the individual waves. Therefore, the shape of the string at the indicated times looks like the following:

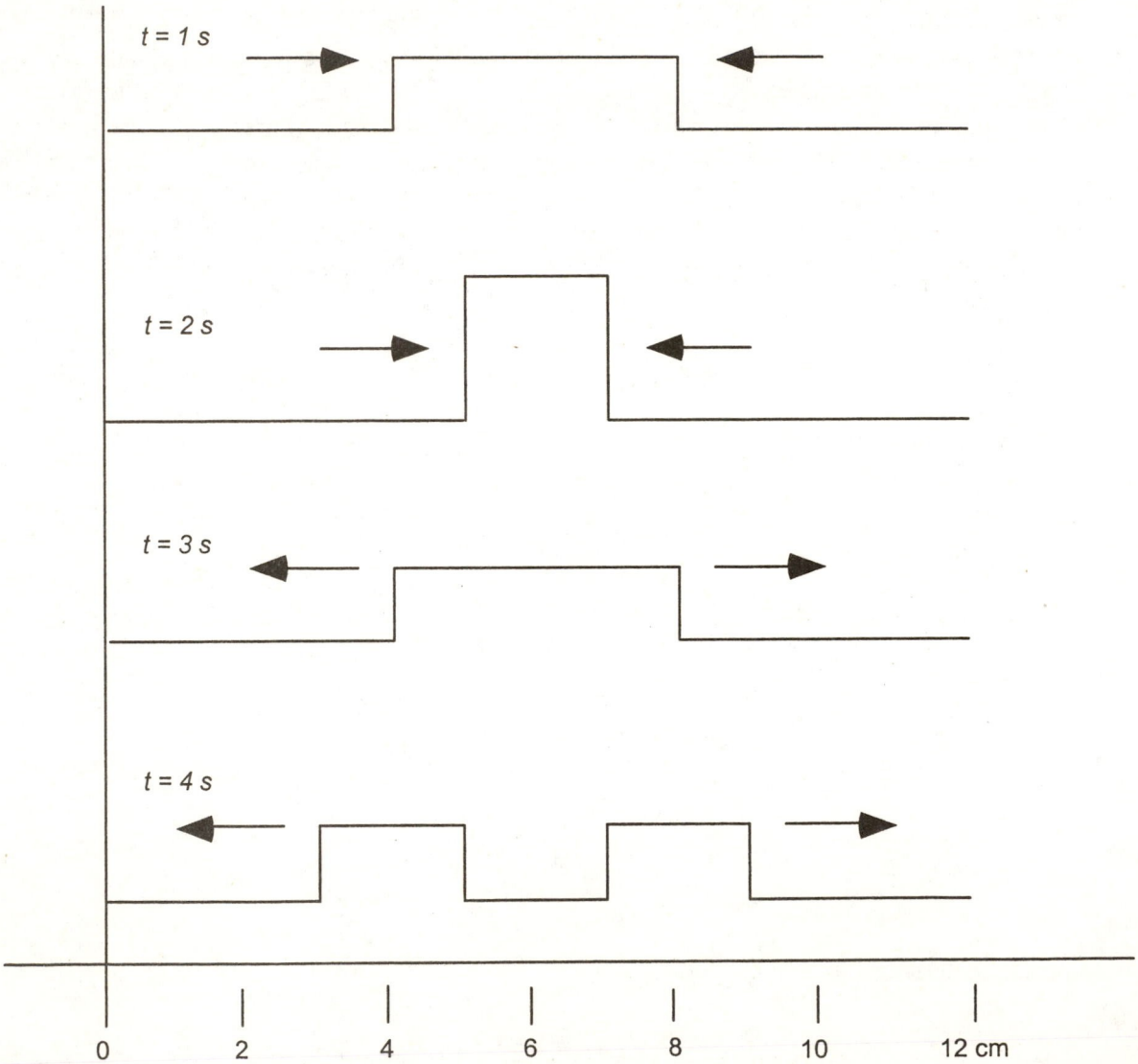

5. ***REASONING*** When constructive interference occurs again at point C, the path length difference is two wavelengths, or $\Delta s = 2\lambda = 3.20\text{ m}$. Therefore, we can write the expression for the path length difference as

$$s_{AC} - s_{BC} = \sqrt{s_{AB}^2 + s_{BC}^2} - s_{BC} = 3.20\text{ m}$$

This expression can be solved for s_{AB}.

SOLUTION Solving for s_{AB}, we find that

$$s_{AB} = \sqrt{(3.20\text{ m} + 2.40\text{ m})^2 - (2.40\text{ m})^2} = \boxed{5.06\text{ m}}$$

---

7. ***REASONING*** The geometry of the positions of the loudspeakers and the listener is shown in the following drawing.

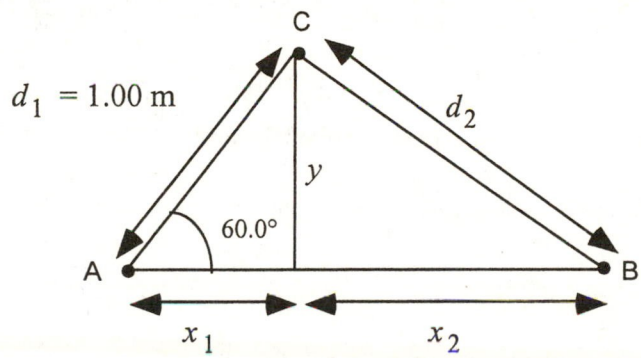

The listener at C will hear either a loud sound or no sound, depending upon whether the interference occurring at C is constructive or destructive. If the listener hears no sound, destructive interference occurs, so

$$d_2 - d_1 = \frac{n\lambda}{2} \qquad n = 1, 3, 5, \ldots \tag{1}$$

SOLUTION Since $v = \lambda f$, according to Equation 16.1, the wavelength of the tone is

$$\lambda = \frac{v}{f} = \frac{343\text{ m/s}}{68.6\text{ Hz}} = 5.00\text{ m}$$

Speaker B will be closest to Speaker A when $n = 1$ in Equation (1) above, so

$$d_2 = \frac{n\lambda}{2} + d_1 = \frac{5.00\text{ m}}{2} + 1.00\text{ m} = 3.50\text{ m}$$

From the figure above we have that,

$$x_1 = (1.00\text{ m})\cos 60.0° = 0.500\text{ m}$$

$$y = (1.00\text{ m})\sin 60.0° = 0.866\text{ m}$$

Then

$$x_2^2 + y^2 = d_2^2 = (3.50\text{ m})^2 \qquad \text{or} \qquad x_2 = \sqrt{(3.50\text{ m})^2 - (0.866\text{ m})^2} = 3.39\text{ m}$$

Therefore, the closest that speaker A can be to speaker B so that the listener hears no sound is $x_1 + x_2 = 0.500\text{ m} + 3.39\text{ m} = \boxed{3.89\text{ m}}$.

---

11. ***REASONING*** The diffraction angle for the first minimum for a circular opening is given by Equation 17.2: $\sin\theta = 1.22\lambda / D$, where D is the diameter of the opening.

SOLUTION

a. Using Equation 16.1, we must first find the wavelength of the 2.0-kHz tone:

$$\lambda = \frac{v}{f} = \frac{343\text{ m/s}}{2.0\times10^3\text{ Hz}} = 0.17\text{ m}$$

The diffraction angle for a 2.0-kHz tone is, therefore,

$$\theta = \sin^{-1}\left(1.22\times\frac{0.17\text{ m}}{0.30\text{ m}}\right) = \boxed{44°}$$

b. The wavelength of a 6.0-kHz tone is

$$\lambda = \frac{v}{f} = \frac{343\text{ m/s}}{6.0\times10^3\text{ Hz}} = 0.057\text{ m}$$

Therefore, if we wish to generate a 6.0-kHz tone whose diffraction angle is as wide as that for the 2.0-kHz tone in part (a), we will need a speaker of diameter D, where

$$D = \frac{1.22\ \lambda}{\sin\theta} = \frac{(1.22)(0.057\text{ m})}{\sin 44°} = \boxed{0.10\text{ m}}$$

---

17. ***REASONING*** The beat frequency of two sound waves is the difference between the two sound frequencies. From the graphs, we see that the period of the wave in the upper figure is 0.020 s, so its frequency is $f_1 = 1/T_1 = 1/(0.020\text{ s}) = 5.0\times10^1\text{ Hz}$. The frequency of the wave in the lower figure is $f_2 = 1/(0.024\text{ s}) = 4.2\times10^1\text{ Hz}$.

SOLUTION The beat frequency of the two sound waves is

$$f_{\text{beat}} = f_1 - f_2 = 5.0 \times 10^1 \text{ Hz} - 4.2 \times 10^1 \text{ Hz} = \boxed{8 \text{ Hz}}$$

---

19. ***REASONING AND SOLUTION*** The first case requires that the frequency be either

$$440 \text{ Hz} - 5 \text{ Hz} = 435 \text{ Hz} \quad \text{or} \quad 440 \text{ Hz} + 5 \text{ Hz} = 445 \text{ Hz}$$

The second case requires that the frequency be either

$$436 \text{ Hz} - 9 \text{ Hz} = 427 \text{ Hz} \quad \text{or} \quad 436 \text{ Hz} + 9 \text{ Hz} = 445 \text{ Hz}$$

The frequency of the tuning fork is $\boxed{445 \text{ Hz}}$.

---

23. ***REASONING*** The fundamental frequency f_1 is given by Equation 17.3 with $n = 1$: $f_1 = v/(2L)$. Since values for f_1 and L are given in the problem statement, we can use this expression to find the speed of the waves on the cello string. Once the speed is known, the tension F in the cello string can be found by using Equation 16.2, $v = \sqrt{F/(m/L)}$.

SOLUTION Combining Equations 17.3 and 16.2 yields

$$2Lf_1 = \sqrt{\frac{F}{m/L}}$$

Solving for F, we find that the tension in the cello string is

$$F = 4L^2 f_1^2 (m/L) = 4(0.800 \text{ m})^2 (65.4 \text{ Hz})^2 (1.56 \times 10^{-2} \text{ kg/m}) = \boxed{171 \text{ N}}$$

---

27. ***REASONING*** According to Equation 17.3, the fundamental (n = 1) frequency of a string fixed at both ends is related to the wave speed v by $f_1 = v/2L$, where L is the length of the string. Thus, the speed of the wave is $v = 2Lf_1$. Combining this with Equation 16.2, $v = \sqrt{F/(m/L)}$, we have, after some rearranging,

$$\frac{F}{L^2} = 4f_1^2 (m/L)$$

Since the strings have the same tension and the same lengths between their fixed ends, we have

$$f_{1\text{E}}^2 (m/L)_{\text{E}} = f_{1\text{G}}^2 (m/L)_{\text{G}}$$

where the symbols "E" and "G" represent the E and G strings on the violin. This equation can be solved for the linear density of the G string.

SOLUTION The linear density of the string is

$$(m/L)_G = \frac{f_{1E}^2}{f_{1G}^2}(m/L)_E = \left(\frac{f_{1E}}{f_{1G}}\right)^2 (m/L)_E$$

$$= \left(\frac{659.3\text{ Hz}}{196.0\text{ Hz}}\right)^2 \left(3.47\times10^{-4}\text{ kg/m}\right) = \boxed{3.93\times10^{-3}\text{ kg/m}}$$

---

31. ***REASONING*** We can find the extra length that the D-tuner adds to the E-string by calculating the length of the D-string and then subtracting from it the length of the E string. For standing waves on a string that is fixed at both ends, Equation 17.3 gives the frequencies as $f_n = n(v/2L)$. The ratio of the fundamental frequency of the D-string to that of the E-string is

$$\frac{f_D}{f_E} = \frac{v/(2L_D)}{v/(2L_E)} = \frac{L_E}{L_D}$$

This expression can be solved for the length L_D of the D-string in terms of quantities given in the problem statement.

SOLUTION The length of the D-string is

$$L_D = L_E\left(\frac{f_E}{f_D}\right) = (0.628\text{ m})\left(\frac{41.2\text{ Hz}}{36.7\text{ Hz}}\right) = 0.705\text{ m}$$

The length of the E-string is extended by the D-tuner by an amount

$$L_D - L_E = 0.705\text{ m} - 0.628\text{ m} = \boxed{0.077\text{ m}}$$

---

33. ***REASONING*** The beat frequency produced when the piano and the other instrument sound the note (three octaves higher than middle C) is $f_{beat} = f - f_0$, where f is the frequency of the piano and f_0 is the frequency of the other instrument ($f_0 = 2093$ Hz). We can find f by considering the temperature effects and the mechanical effects that occur when the temperature drops from 25.0 °C to 20.0 °C.

SOLUTION The fundamental frequency f_0 of the wire at 25.0 °C is related to the tension F_0 in the wire by

$$f_0 = \frac{v}{2L_0} = \frac{\sqrt{F_0/(m/L)}}{2L_0} \tag{1}$$

where Equations 17.3 and 16.2 have been combined.

The amount ΔL by which the piano wire attempts to contract is (see Equation 12.2) $\Delta L = \alpha L_0 \Delta T$, where α is the coefficient of linear expansion of the wire, L_0 is its length at 25.0 °C, and ΔT is the amount by which the temperature drops. Since the wire is prevented from contracting, there must be a stretching force exerted at each end of the wire. According to Equation 10.17, the magnitude of this force is

$$\Delta F = Y\left(\frac{\Delta L}{L_0}\right)A$$

where Y is the Young's modulus of the wire, and A is its cross-sectional area. Combining this relation with Equation 12.2, we have

$$\Delta F = Y\left(\frac{\alpha L_0 \Delta T}{L_0}\right)A = \alpha(\Delta T)YA$$

Thus, the frequency f at the lower temperature is

$$f = \frac{v}{2L_0} = \frac{\sqrt{(F_0 + \Delta F)/(m/L)}}{2L_0} = \frac{\sqrt{\left[F_0 + \alpha(\Delta T)YA\right]/(m/L)}}{2L_0} \tag{2}$$

Using Equations (1) and (2), we find that the frequency f is

$$f = f_0 \frac{\sqrt{\left[F_0 + \alpha(\Delta T)YA\right]/(m/L)}}{\sqrt{F_0/(m/L)}} = f_0\sqrt{\frac{F_0 + \alpha(\Delta T)YA}{F_0}}$$

$$f = (2093\text{ Hz})\sqrt{\frac{818.0\text{ N} + (12\times10^{-6}/\text{C}°)(5.0\text{ C}°)(2.0\times10^{11}\text{ N/m}^2)\,(7.85\times10^{-7}\text{ m}^2)}{818.0\text{ N}}}$$

$$= 2105\text{ Hz}$$

Therefore, the beat frequency is $2105\text{ Hz} - 2093\text{ Hz} = \boxed{12\text{ Hz}}$.

35. ***REASONING AND SOLUTION*** The distance between one node and an adjacent antinode is $\lambda/4$. Thus, we must first determine the wavelength of the standing wave. A tube open at only one end can develop standing waves only at the odd harmonic frequencies. Thus, for a tube of length L producing sound at the third harmonic ($n = 3$), $L = 3(\lambda/4)$. Therefore, the wavelength of the standing wave is

$$\lambda = \tfrac{4}{3}L = \tfrac{4}{3}(1.5\text{ m}) = 2.0\text{ m}$$

and the distance between one node and the adjacent antinode is $\lambda/4 = \boxed{0.50\text{ m}}$.

---

39. ***REASONING*** The natural frequencies of a tube open at only one end are given by Equation 17.5 as $f_n = n\left(\frac{v}{4L}\right)$, where n is any odd integer ($n = 1, 3, 5, \ldots$), v is the speed of sound, and L is the length of the tube. We can use this relation to find the value for n for the 450-Hz sound and to determine the length of the pipe.

SOLUTION

a. The frequency f_n of the 450-Hz sound is given by $450\text{ Hz} = n\left(\frac{v}{4L}\right)$. Likewise, the frequency of the next higher harmonic is $750\text{ Hz} = (n+2)\left(\frac{v}{4L}\right)$, because n is an odd integer and this means that the value of n for the next higher harmonic must be $n + 2$. Taking the ratio of these two relations gives

$$\frac{750\text{ Hz}}{450\text{ Hz}} = \frac{(n+2)\left(\frac{v}{4L}\right)}{n\left(\frac{v}{4L}\right)} = \frac{n+2}{n}$$

Solving this equation for n gives $n = \boxed{3}$.

b. Solving the equation $450\text{ Hz} = n\left(\frac{v}{4L}\right)$ for L and using $n = 3$, we find that the length of the tube is

$$L = n\left(\frac{v}{4f_n}\right) = 3\left[\frac{343\text{ m/s}}{4(450\text{ Hz})}\right] = \boxed{0.57\text{ m}}$$

---

43. ***REASONING*** According to Equation 11.4, the absolute pressure at the bottom of the mercury is $P = P_{atm} + \rho gh$, where the height h of the mercury column is the original length L_0 of the air column minus the shortened length L. Hence,

$$P = P_{\text{atm}} + \rho g(L_0 - L)$$

SOLUTION From Equation 17.5, the fundamental (n = 1) frequency f_1 of the shortened tube is $f_1 = 1(v/4L)$, where L is the length of the air column in the tube. Likewise, the frequency f_3 of the third ($n = 3$) harmonic in the original tube is $f_3 = 3(v/4L_0)$, where L_0 is the length of the air column in the original tube. Since $f_1 = f_3$, we have that

$$1\left(\frac{v}{4L}\right) = 3\left(\frac{v}{4L_0}\right) \quad \text{or} \quad L = \tfrac{1}{3}L_0$$

The pressure at the bottom of the mercury is

$$P = P_{\text{atm}} + \rho g\left(\tfrac{2}{3}L_0\right)$$

$$= 1.01\times10^5 \text{ Pa} + (13\,600 \text{ kg/m}^3)(9.80 \text{ m/s}^2)\left(\tfrac{2}{3}\times 0.75 \text{ m}\right) = \boxed{1.68\times10^5 \text{ Pa}}$$

---

47. ***REASONING*** The tones from the two speakers will produce destructive interference with the smallest frequency when the path length difference at C is one-half of a wavelength. From Figure 17.7, we see that the path length difference is $\Delta s = s_{\text{AC}} - s_{\text{BC}}$. From Example 1, we know that $s_{\text{AC}} = 4.00 \text{ m}$, and from Figure 17.7, $s_{\text{BC}} = 2.40 \text{ m}$. Therefore, the path length difference is $\Delta s = 4.00 \text{ m} - 2.40 \text{ m} = 1.60 \text{ m}$.

SOLUTION Thus, destructive interference will occur when

$$\frac{\lambda}{2} = 1.60 \text{ m} \qquad \text{or} \qquad \lambda = 3.20 \text{ m}$$

This corresponds to a frequency of

$$f = \frac{v}{\lambda} = \frac{343 \text{ m/s}}{3.20 \text{ m}} = \boxed{107 \text{ Hz}}$$

---

51. ***REASONING*** For standing waves on a string that is clamped at both ends, Equations 17.3 and 16.2 indicate that the standing wave frequencies are

$$f_n = n\left(\frac{v}{2L}\right) \qquad \text{where} \qquad v = \sqrt{\frac{F}{m/L}}$$

Combining these two expressions, we have, with $n = 1$ for the fundamental frequency,

$$f_1 = \frac{1}{2L}\sqrt{\frac{F}{m/L}}$$

This expression can be used to find the ratio of the two fundamental frequencies.

SOLUTION The ratio of the two fundamental frequencies is

$$\frac{f_{\text{old}}}{f_{\text{new}}} = \frac{\frac{1}{2L}\sqrt{\frac{F_{\text{old}}}{m/L}}}{\frac{1}{2L}\sqrt{\frac{F_{\text{new}}}{m/L}}} = \sqrt{\frac{F_{\text{old}}}{F_{\text{new}}}}$$

Since $F_{\text{new}} = 4F_{\text{old}}$, we have

$$f_{\text{new}} = f_{\text{old}}\sqrt{\frac{F_{\text{new}}}{F_{\text{old}}}} = f_{\text{old}}\sqrt{\frac{4\,F_{\text{old}}}{F_{\text{old}}}} = f_{\text{old}}\sqrt{4} = (55.0\text{ Hz})\,(2) = \boxed{1.10\times10^2\text{ Hz}}$$

---

55. ***REASONING*** The natural frequencies of the cord are, according to Equation 17.3, $f_n = nv/(2L)$, where $n = 1, 2, 3, \ldots$. The speed v of the waves on the cord is, according to Equation 16.2, $v = \sqrt{F/(m/L)}$, where F is the tension in the cord. Combining these two expressions, we have

$$f_n = \frac{nv}{2L} = \frac{n}{2L}\sqrt{\frac{F}{m/L}} \qquad \text{or} \qquad \left(\frac{f_n 2L}{n}\right)^2 = \frac{F}{m/L}$$

Applying Newton's second law of motion, $\Sigma F = ma$, to the forces that act on the block and are parallel to the incline gives

$$F - Mg\sin\theta = Ma = 0 \qquad \text{or} \qquad F = Mg\sin\theta$$

where $Mg\sin\theta$ is the component of the block's weight that is parallel to the incline. Substituting this value for the tension into the equation above gives

$$\left(\frac{f_n 2L}{n}\right)^2 = \frac{Mg\sin\theta}{m/L}$$

This expression can be solved for the angle θ and evaluated at the various harmonics. The answer can be chosen from the resulting choices.

SOLUTION Solving this result for $\sin\theta$ shows that

$$\sin\theta = \frac{(m/L)}{Mg}\left(\frac{f_n 2L}{n}\right)^2 = \frac{1.20\times10^{-2}\text{ kg/m}}{(15.0\text{ kg})(9.80\text{ m/s}^2)}\left[\frac{(165\text{ Hz})2(0.600\text{ m})}{n}\right]^2 = \frac{3.20}{n^2}$$

Thus, we have

$$\theta = \sin^{-1}\left(\frac{3.20}{n^2}\right)$$

Evaluating this for the harmonics corresponding to the range of n from $n=2$ to $n=4$, we have

$$\theta = \sin^{-1}\left(\frac{3.20}{2^2}\right) = 53.1° \text{ for } n=2$$

$$\theta = \sin^{-1}\left(\frac{3.20}{3^2}\right) = 20.8° \text{ for } n=3$$

$$\theta = \sin^{-1}\left(\frac{3.20}{4^2}\right) = 11.5° \text{ for } n=4$$

The angles between 15.0° and 90.0° are $\boxed{\theta = 20.8°}$ and $\boxed{\theta = 53.1°}$.

CHAPTER 18 | *ELECTRIC FORCES AND ELECTRIC FIELDS*

CONCEPTUAL QUESTIONS

---

8. ***REASONING AND SOLUTION*** Identical point charges are fixed to opposite corners of a square, as shown in the figure at the right.

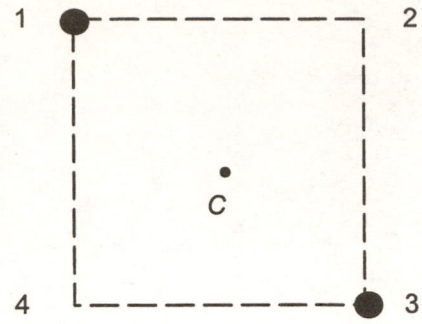

There exists an electric field at each point in space in the vicinity of this configuration. The electric field at a given point is the resultant of the electric field at that point due to the charge at corner 1 and the charge at corner 3.

The magnitude of the electric field due to a single point charge q is given by Equation 18.3: $E = k|q|/r^2$, where r is the distance from the charge to the point in question.

Since the point charges at corners 1 and 3 are identical, they each have the same value of q. Furthermore, they are equidistant from the center point C. Therefore, at the center of the square, each charge gives rise to an electric field that is equal in magnitude and opposite in direction. The resultant electric field at the center of the square is, therefore, zero. Note that this result is independent of the polarity of the charges.

The distance from either empty corner, 2 or 4, to either of the charges is the same. Since the charges are equidistant from either empty corner, each charge gives rise to an electric field that is equal in magnitude. Since the direction of the electric field due to a point charge is radial (radially inward for negative charges and radially outward for positive charges), we see that the electric fields due to each of the two point charges will be mutually perpendicular. Their resultant can be found by using the Pythagorean theorem.

The magnitude of the force experienced by a third point charge placed in this system is, from Equation 18.2: $F = |q_0|E$, where $|q_0|$ is the magnitude of the charge and E is the magnitude of the electric field at the location of the charge. Since the electric field is zero at the center of the square, a third point charge will experience no force there. Thus, a third point charge will experience the greater force at one of the empty corners of the square.

---

14. ***REASONING AND SOLUTION*** In general, electric field lines are always directed away from positive charges and toward negative charges. At a point in space, the direction of the electric field is tangent to the electric field line that passes through that point. The magnitude of the electric field is strongest in regions where the field lines are closest together. Furthermore, the text states that electric field lines created by a positive point charge are directed radially outward, while those created by a negative point charge are directed radially inward.

a. *In both drawings I and II the electric field is the same everywhere.* This statement is false. By inspection, we see that the field lines in II get closer together as we proceed from left to right in the drawing; hence, the magnitude of the electric field is getting stronger. The direction of the electric field at any point along a field line can be found by constructing a line tangent to the field line. Clearly, the direction of the electric field in II changes as we proceed from left to right in the drawing.

b. *As you move from left to right in each case, the electric field becomes stronger.* This statement is false. While this is true for the field in II by reasons that are discussed in part (a) above, we can see by inspection that the field lines in I remain equally spaced and point in the same direction as we go from left to right. Thus, the field in I remains constant in both magnitude and direction.

c. *The electric field in I is the same everywhere but becomes stronger in II as you move from left to right.* This statement is true for the reasons discussed in parts (a) and (b) above.

d. *The electric fields in both I and II could be created by negative charges located somewhere on the left and positive charges located somewhere on the right.* Since electric field lines are always directed away from positive charges and toward negative charges, this statement is false.

e. *Both I and II arise from a single point charge located somewhere on the left.* Since the electric field lines are radial in the vicinity of a single point charge, this statement is false.

---

17. ***REASONING AND SOLUTION*** Two charges $+q$ and $-q$ are inside a Gaussian surface. Since the net charge inside the Gaussian surface is zero, Gauss' law states that the electric flux Φ_E through the surface is also zero. However, the fact that Φ_E is zero does not imply that **E** is zero at any point on the Gaussian surface.

The flux through a Gaussian surface of any shape that surrounds the two point charges is given by text Equation 18.6: $\Phi_E = \sum(E \cos \phi)\Delta A$, where E is the magnitude of the electric field at a point on the surface, ϕ is the angle that the electric field makes with the normal to the surface at that point, ΔA is a tiny area that surrounds the point, and the summation is carried out over the entire Gaussian surface. The individual terms $(E \cos \phi)\Delta A$ can be positive or negative, depending on the sign of the factor $(\cos \phi)$.

We can see from Figure 18.26 that the electric field in the vicinity of the two point charges changes from point to point in both magnitude and direction. Furthermore, as Figure 18.26 indicates, the electric field lines are directed away from the positive charge and toward the negative charge. In general, the value of E is non-zero at every point on any arbitrary Gaussian surface that surrounds the charges. The magnitude and direction of **E** at each point is such that when the summation in Equation 18.6 is carried out over the entire Gaussian surface, the electric flux through the entire surface is zero.

CHAPTER 18 | *ELECTRIC FORCES AND ELECTRIC FIELDS*

PROBLEMS

---

1. ***REASONING AND SOLUTION*** The charge on a single electron is -1.60×10^{-19} C. In order to give a neutral silver dollar a charge of +2.4 μC, we must remove an amount of negative charge equal to –2.4 μC. This corresponds to

$$\left(-2.4\times10^{-6}\text{ C}\right)\left(\frac{1\text{ electron}}{-1.60\times10^{-19}\text{ C}}\right)=\boxed{1.5\times10^{13}\text{ electrons}}$$

---

5. ***REASONING*** Identical conducting spheres equalize their charge upon touching. When spheres A and B touch, an amount of charge $+q$, flows from A and instantaneously neutralizes the $-q$ charge on B leaving B momentarily neutral. Then, the remaining amount of charge, equal to $+4q$, is equally split between A and B, leaving A and B each with equal amounts of charge $+2q$. Sphere C is initially neutral, so when A and C touch, the $+2q$ on A splits equally to give $+q$ on A and $+q$ on C. When B and C touch, the $+2q$ on B and the $+q$ on C combine to give a total charge of $+3q$, which is then equally divided between the spheres B and C; thus, B and C are each left with an amount of charge $+1.5q$.

SOLUTION Taking note of the initial values given in the problem statement, and summarizing the final results determined in the *Reasoning* above, we conclude the following:

a. Sphere C ends up with an amount of charge equal to $\boxed{+1.5q}$.

b. The charges on the three spheres before they were touched, are, according to the problem statement, $+5q$ on sphere A, $-q$ on sphere B, and zero charge on sphere C. Thus, the total charge on the spheres is $+5q-q+0=\boxed{+4q}$.

c. The charges on the spheres after they are touched are $+q$ on sphere A, $+1.5q$ on sphere B, and $+1.5q$ on sphere C. Thus, the total charge on the spheres is $+q+1.5q+1.5q=\boxed{+4q}$.

---

7. ***REASONING AND SOLUTION*** The magnitude of the force of attraction between the charges is given by Coulomb's law (Equation 18.1): $F=k|q_1||q_2|/r^2$, where $|q_1|$ and $|q_2|$ are the magnitudes of the charges and r is the separation of the charges. Let F_A and F_B represent the magnitudes of the forces between the charges when the separations are r_A and $r_B=r_A/9$, respectively. Then

$$\frac{F_B}{F_A} = \frac{k|q_1||q_2|/r_B^2}{k|q_1||q_2|/r_A^2} = \left(\frac{r_A}{r_B}\right)^2 = \left(\frac{r_A}{r_A/9}\right)^2 = (9)^2 = 81$$

Therefore, we can conclude that $F_B = 81F_A = (81)(1.5\text{ N}) = \boxed{120\text{ N}}$.

---

9. ***REASONING*** Initially, the two spheres are neutral. Since negative charge is removed from the sphere which loses electrons, it then carries a net positive charge. Furthermore, the neutral sphere to which the electrons are added is then negatively charged. Once the charge is transferred, there exists an electrostatic force on each of the two spheres, the magnitude of which is given by Coulomb's law (Equation 18.1), $F = k|q_1||q_2|/r^2$.

SOLUTION

a. Since each electron carries a charge of -1.60×10^{-19} C, the amount of negative charge removed from the first sphere is

$$\left(3.0\times10^{13}\text{ electrons}\right)\left(\frac{1.60\times10^{-19}\text{ C}}{1\text{ electron}}\right) = 4.8\times10^{-6}\text{ C}$$

Thus, the first sphere carries a charge $+4.8 \times 10^{-6}$ C, while the second sphere carries a charge -4.8×10^{-6} C. The magnitude of the electrostatic force that acts on each sphere is, therefore,

$$F = \frac{k|q_1||q_2|}{r^2} = \frac{\left(8.99\times10^9\text{ N}\cdot\text{m}^2/\text{C}^2\right)\left(4.8\times10^{-6}\text{ C}\right)^2}{(0.50\text{ m})^2} = \boxed{0.83\text{ N}}$$

b. Since the spheres carry charges of opposite sign, the force is $\boxed{\text{attractive}}$.

---

11. ***REASONING AND SOLUTION*** The net electrostatic force on charge 3 at $x = +3.0$ m is the vector sum of the forces on charge 3 due to the other two charges, 1 and 2. According to Coulomb's law (Equation 18.1), the magnitude of the force on charge 3 due to charge 1 is

$$F_{13} = \frac{k|q_1||q_3|}{r_{13}^2}$$

where the distance between charges 1 and 3 is r_{13}.

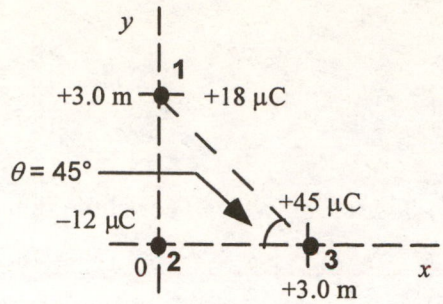

Figure 1

According to the Pythagorean theorem, $r_{13}^2 = x^2 + y^2$. Therefore,

$$F_{13} = \frac{\left(8.99\times10^9\ \text{N}\cdot\text{m}^2/\text{C}^2\right)\left(18\times10^{-6}\ \text{C}\right)\left(45\times10^{-6}\ \text{C}\right)}{(3.0\ \text{m})^2 + (3.0\ \text{m})^2} = 0.405\ \text{N}$$

Charges 1 and 3 are equidistant from the origin, so that $\theta = 45°$ (see Figure 1). Since charges 1 and 3 are both positive, the force on charge 3 due to charge 1 is repulsive and along the line that connects them, as shown in Figure 2. The components of F_{13} are:

$$F_{13x} = F_{13}\cos 45° = 0.286\ \text{N} \qquad \text{and} \qquad F_{13y} = -F_{13}\sin 45° = -0.286\ \text{N}$$

The second force on charge 3 is the attractive force (opposite signs) due to its interaction with charge 2 located at the origin. The magnitude of the force on charge 3 due to charge 2 is, according to Coulomb's law ,

$$F_{23} = \frac{k|q_2||q_3|}{r_{23}^2} = \frac{k|q_2||q_3|}{x^2}$$

$$= \frac{\left(8.99\times10^9\ \text{N}\cdot\text{m}^2/\text{C}^2\right)\left(12\times10^{-6}\ \text{C}\right)\left(45\times10^{-6}\ \text{C}\right)}{(3.0\ \text{m})^2}$$

$$= 0.539\ \text{N}$$

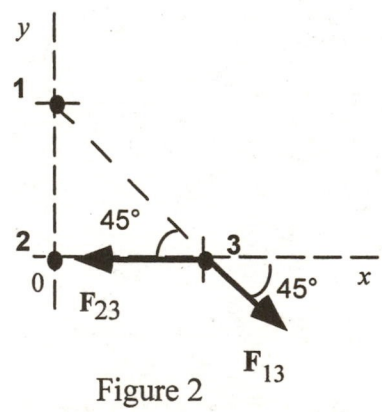

Figure 2

Since charges 2 and 3 have opposite signs, they attract each other, and charge 3 experiences a force to the left as shown in Figure 2. Taking up and to the right as the positive directions, we have

$$F_{3x} = F_{13x} + F_{23x} = +0.286\text{ N} - 0.539\text{ N} = -0.253\text{ N}$$

$$F_{3y} = F_{13y} = -0.286\text{ N}$$

Using the Pythagorean theorem, we find the magnitude of F_3 to be

$$F_3 = \sqrt{F_{3x}^2 + F_{3y}^2} = \sqrt{(-0.253\text{ N})^2 + (-0.286\text{ N})^2} = \boxed{0.38\text{ N}}$$

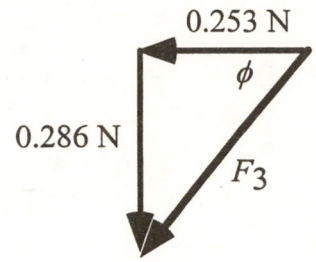

Figure 3

The direction of F_3 relative to the $-x$ axis is specified by the angle ϕ, where

$$\phi = \tan^{-1}\left(\frac{0.286\text{ N}}{0.253\text{ N}}\right) = \boxed{49^\circ \text{ below the } -x \text{ axis}}$$

---

13. ***REASONING*** Each particle will experience an electrostatic force due to the presence of the other charge. According to Coulomb's law (Equation 18.1), the magnitude of the force felt by each particle can be calculated from $F = k|q_1||q_2|/r^2$, where $|q_1|$ and $|q_2|$ are the respective charges on particles 1 and 2 and r is the distance between them. According to Newton's second law, the magnitude of the force experienced by each particle is given by $F = ma$, where a is the acceleration of the particle and we have assumed that the electrostatic force is the only force acting.

SOLUTION

a. Since the two particles have identical positive charges, $|q_1| = |q_2| = |q|$, and we have, using the data for particle 1,

$$\frac{k|q|^2}{r^2} = m_1 a_1$$

Solving for $|q|$, we find that

$$|q| = \sqrt{\frac{m_1 a_1 r^2}{k}} = \sqrt{\frac{(6.00\times10^{-6}\text{ kg})\,(4.60\times10^{3}\text{ m/s}^2)\,(2.60\times10^{-2}\text{ m})^2}{8.99\times10^{9}\text{ N}\cdot\text{m}^2/\text{C}^2}} = \boxed{4.56\times10^{-8}\text{ C}}$$

b. Since each particle experiences a force of the same magnitude (From Newton's third law), we can write $F_1 = F_2$, or $m_1 a_1 = m_2 a_2$. Solving this expression for the mass m_2 of particle 2, we have

$$m_2 = \frac{m_1 a_1}{a_2} = \frac{(6.00\times10^{-6}\ \text{kg})(4.60\times10^{3}\ \text{m/s}^2)}{8.50\times10^{3}\ \text{m/s}^2} = \boxed{3.25\times10^{-6}\ \text{kg}}$$

---

19. ***REASONING AND SOLUTION*** Before the spheres have been charged, they exert no forces on each other. After the spheres are charged, each sphere experiences a repulsive force F due to the charge on the other sphere, according to Coulomb's law (Equation 18.1). Therefore, since each sphere has the same charge, the magnitude F of this force is

$$F = \frac{k|q_1||q_2|}{r^2} = \frac{(8.99\times10^{9}\ \text{N}\cdot\text{m}^2/\text{C}^2)(1.60\times10^{-6}\ \text{C})^2}{(0.100\ \text{m})^2} = 2.30\ \text{N}$$

The repulsive force on each sphere compresses the spring to which it is attached. The magnitude of this repulsive force is related to the amount of compression by Equation 10.1: $F_x^{\text{Applied}} = kx$. Setting $F_x^{\text{Appled}} = F$ and solving for k, we find that

$$k = \frac{F}{x} = \frac{2.30\ \text{N}}{0.0250\ \text{m}} = \boxed{92.0\ \text{N/m}}$$

---

23. ***REASONING*** The charged insulator experiences an electric force due to the presence of the charged sphere shown in the drawing in the text. The forces acting on the insulator are the downward force of gravity (i.e., its weight, $W = mg$), the electrostatic force $F = k|q_1||q_2|/r^2$ (see Coulomb's law, Equation 18.1) pulling to the right, and the tension T in the wire pulling up and to the left at an angle θ with respect to the vertical as shown in the drawing in the problem statement. We can analyze the forces to determine the desired quantities θ and T.

SOLUTION.

a. We can see from the diagram given with the problem statement that

$$T_x = F \qquad \text{which gives} \qquad T\sin\theta = k|q_1||q_2|/r^2$$

and

$$T_y = W \qquad \text{which gives} \qquad T\cos\theta = mg$$

Dividing the first equation by the second yields

$$\frac{T\sin\theta}{T\cos\theta} = \tan\theta = \frac{k|q_1||q_2|/r^2}{mg}$$

Solving for θ, we find that

$$\theta = \tan^{-1}\left(\frac{k|q_1||q_2|}{mgr^2}\right)$$

$$= \tan^{-1}\left[\frac{(8.99\times10^9\ \text{N}\cdot\text{m}^2/\text{C}^2)(0.600\times10^{-6}\ \text{C})(0.900\times10^{-6}\ \text{C})}{(8.00\times10^{-2}\ \text{kg})(9.80\ \text{m/s}^2)(0.150\ \text{m})^2}\right] = \boxed{15.4°}$$

b. Since $T\cos\theta = mg$, the tension can be obtained as follows:

$$T = \frac{mg}{\cos\theta} = \frac{(8.00\times10^{-2}\ \text{kg})\ (9.80\ \text{m/s}^2)}{\cos 15.4°} = \boxed{0.813\ \text{N}}$$

---

27. ***REASONING*** Two forces act on the charged ball (charge q); they are the downward force of gravity $m\mathbf{g}$ and the electric force $\mathbf{F}$ due to the presence of the charge q in the electric field $\mathbf{E}$. In order for the ball to float, these two forces must be equal in magnitude and opposite in direction, so that the net force on the ball is zero (Newton's second law). Therefore, $\mathbf{F}$ must point upward, which we will take as the positive direction. According to Equation 18.2, $\mathbf{F} = q\mathbf{E}$. Since the charge q is negative, the electric field $\mathbf{E}$ must point downward, as the product $q\mathbf{E}$ in the expression $\mathbf{F} = q\mathbf{E}$ must be positive, since the force $\mathbf{F}$ points upward. The magnitudes of the two forces must be equal, so that $mg = |q|E$. This expression can be solved for E.

SOLUTION The magnitude of the electric field $\mathbf{E}$ is

$$E = \frac{mg}{|q|} = \frac{(0.012\ \text{kg})(9.80\ \text{m/s}^2)}{18\times10^{-6}\ \text{C}} = \boxed{6.5\times10^3\ \text{N/C}}$$

As discussed in the reasoning, this electric field points $\boxed{\text{downward}}$.

---

31. ***REASONING***

a. The drawing shows the two point charges q_1 and q_2. Point A is located at $x = 0$ cm, and point B is at $x = +6.0$ cm.

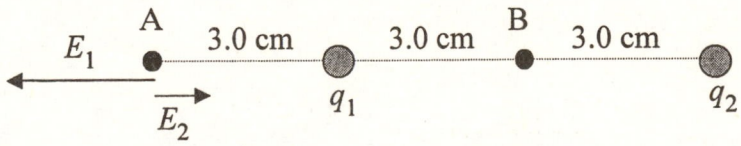

Since q_1 is positive, the electric field points away from it. At point A, the electric field E_1 points to the left, in the $-x$ direction. Since q_2 is negative, the electric field points toward it.

At point A, the electric field E_2 points to the right, in the $+x$ direction. The net electric field is $E = -E_1 + E_2$. We can use Equation 18.3, $E = k|q|/r^2$, to find the magnitude of the electric field due to each point charge.

b. The drawing shows the electric field produced by the charges q_1 and q_2 at point B, which is located at $x = +6.0$ cm.

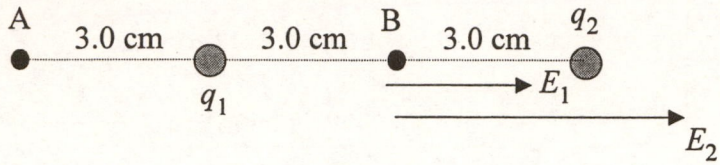

Since q_1 is positive, the electric field points away from it. At point B, the electric field points to the right, in the $+x$ direction. Since q_2 is negative, the electric field points toward it. At point B, the electric field points to the right, in the $+x$ direction. The net electric field is $E = +E_1 + E_2$.

SOLUTION

a. The net electric field at the origin (point A) is $E = -E_1 + E_2$:

$$E = -E_1 + E_2 = \frac{-k|q_1|}{r_1^2} + \frac{k|q_2|}{r_2^2}$$

$$= \frac{-\left(8.99\times10^9\ \mathrm{N\cdot m^2/C^2}\right)\left(8.5\times10^{-6}\ \mathrm{C}\right)}{\left(3.0\times10^{-2}\ \mathrm{m}\right)^2} + \frac{\left(8.99\times10^9\ \mathrm{N\cdot m^2/C^2}\right)\left(21\times10^{-6}\ \mathrm{C}\right)}{\left(9.0\times10^{-2}\ \mathrm{m}\right)^2}$$

$$= \boxed{-6.2\times10^7\ \mathrm{N/C}}$$

The minus sign tells us that the net electric field points along the $-x$ axis.

b. The net electric field at $x = +6.0$ cm (point B) is $E = E_1 + E_2$:

$$E = E_1 + E_2 = \frac{k|q_1|}{r_1^2} + \frac{k|q_2|}{r_2^2}$$

$$= \frac{\left(8.99\times10^9\ \text{N}\cdot\text{m}^2/\text{C}^2\right)\left(8.5\times10^{-6}\ \text{C}\right)}{\left(3.0\times10^{-2}\ \text{m}\right)^2} + \frac{\left(8.99\times10^9\ \text{N}\cdot\text{m}^2/\text{C}^2\right)\left(21\times10^{-6}\ \text{C}\right)}{\left(3.0\times10^{-2}\ \text{m}\right)^2}$$

$$= \boxed{+2.9\times10^8\ \text{N/C}}$$

The plus sign tells us that the net electric field points along the $+x$ axis.

---

33. ***REASONING*** Since the charged droplet (charge $= q$) is suspended motionless in the electric field **E**, the net force on the droplet must be zero. There are two forces that act on the droplet, the force of gravity $\mathbf{W} = m\mathbf{g}$, and the electric force $\mathbf{F} = q\mathbf{E}$ due to the electric field. Since the net force on the droplet is zero, we conclude that $mg = |q|E$. We can use this reasoning to determine the sign and the magnitude of the charge on the droplet.

SOLUTION

a. Since the net force on the droplet is zero, and the weight of magnitude W points downward, the electric force of magnitude $F = |q|E$ must point upward. Since the electric field points upward, the excess charge on the droplet must be $\boxed{\text{positive}}$ in order for the force **F** to point upward.

b. Using the expression $mg = |q|E$, we find that the magnitude of the excess charge on the droplet is

$$|q| = \frac{mg}{E} = \frac{(3.50\times10^{-9}\ \text{kg})(9.80\ \text{m/s}^2)}{8480\ \text{N/C}} = 4.04\times10^{-12}\ \text{C}$$

The charge on a proton is 1.60×10^{-19} C, so the excess number of protons is

$$\left(4.04\times10^{-12}\ \text{C}\right)\left(\frac{1\ \text{proton}}{1.60\times10^{-19}\ \text{C}}\right) = \boxed{2.53\times10^7\ \text{protons}}$$

---

39. ***REASONING*** The drawing shows the arrangement of the three charges. Let $\mathbf{E}_q$ represent the electric field at the empty corner due to the $-q$ charge. Furthermore, let $\mathbf{E}_1$ and $\mathbf{E}_2$ be the electric fields at the empty corner due to charges $+q_1$ and $+q_2$, respectively.

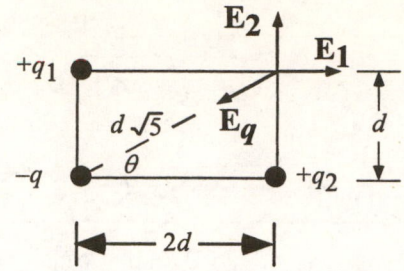

According to the Pythagorean theorem, the distance from the charge $-q$ to the empty corner along the diagonal is given by $\sqrt{(2d)^2 + d^2} = \sqrt{5d^2} = d\sqrt{5}$. The magnitude of each electric field is given by Equation 18.3, $E = k|q|/r^2$. Thus, the magnitudes of each of the electric fields at the empty corner are given as follows:

$$E_q = \frac{k|q|}{r^2} = \frac{k|q|}{\left(d\sqrt{5}\right)^2} = \frac{k|q|}{5d^2}$$

$$E_1 = \frac{k|q_1|}{(2d)^2} = \frac{k|q_1|}{4d^2} \quad \text{and} \quad E_2 = \frac{k|q_2|}{d^2}$$

The angle θ that the diagonal makes with the horizontal is $\theta = \tan^{-1}(d/2d) = 26.57°$. Since the net electric field E_{net} at the empty corner is zero, the horizontal component of the net field must be zero, and we have

$$E_1 - E_q \cos 26.57° = 0 \qquad \text{or} \qquad \frac{k|q_1|}{4d^2} - \frac{k|q|\cos 26.57°}{5d^2} = 0$$

Similarly, the vertical component of the net field must be zero, and we have

$$E_2 - E_q \sin 26.57° = 0 \qquad \text{or} \qquad \frac{k|q_2|}{d^2} - \frac{k|q|\sin 26.57°}{5d^2} = 0$$

These last two expressions can be solved for the charge magnitudes $|q_1|$ and $|q_2|$.

SOLUTION Solving the last two expressions for $|q_1|$ and $|q_2|$, we find that

$$|q_1| = \tfrac{4}{5}q \cos 26.57° = \boxed{0.716\,q}$$

$$|q_2| = \tfrac{1}{5}q \sin 26.57° = \boxed{0.0895\,q}$$

45. ***REASONING AND SOLUTION*** The net electric field at point P in Figure 1 is the vector sum of the fields $\mathbf{E}_+$ and $\mathbf{E}_-$, which are due, respectively, to the charges $+q$ and $-q$. These fields are shown in Figure 2.

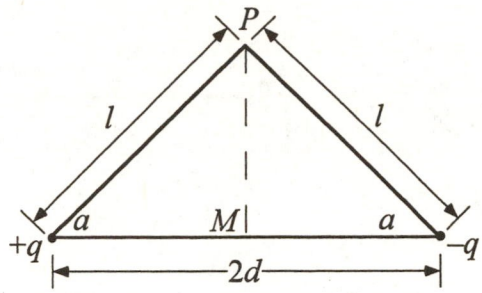

Figure 1

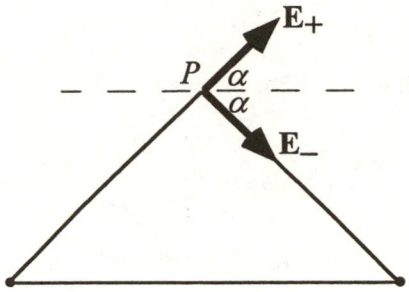

Figure 2

According to Equation 18.3, the magnitudes of the fields E_+ and E_- are the same, since the triangle is an isosceles triangle with equal sides of length ℓ. Therefore, $E_+ = E_- = k|q|/\ell^2$. The vertical components of these two fields cancel, while the horizontal components reinforce, leading to a total field at point P that is horizontal and has a magnitude of

$$E_P = E_+\cos\alpha + E_-\cos\alpha = 2\left(\frac{k|q|}{\ell^2}\right)\cos\alpha$$

At point M in Figure 1, both $\mathbf{E}_+$ and $\mathbf{E}_-$ are horizontal and point to the right. Again using Equation 18.3, we find

$$E_M = E_+ + E_- = \frac{k|q|}{d^2} + \frac{k|q|}{d^2} = \frac{2k|q|}{d^2}$$

Since $E_M/E_P = 9.0$, we have

$$\frac{E_M}{E_P} = \frac{2k|q|/d^2}{2k|q|(\cos\alpha)/\ell^2} = \frac{1}{(\cos\alpha)d^2/\ell^2} = 9.0$$

But from Figure 1, we can see that $d/\ell = \cos\alpha$. Thus, it follows that

$$\frac{1}{\cos^3\alpha} = 9.0 \qquad \text{or} \qquad \cos\alpha = \sqrt[3]{1/9.0} = 0.48$$

The value for α is, then, $\alpha = \cos^{-1}(0.48) = \boxed{61°}$.

47. ***REASONING*** As discussed in Section 18.9, the electric flux Φ_E through a surface is equal to the component of the electric field that is normal to the surface multiplied by the area of the surface, $\Phi_E = E_\perp A$, where $E_\perp$ is the component of **E** that is normal to the surface of area A. We can use this expression and the figure in the text to determine the flux through the two surfaces.

SOLUTION

a. The flux through surface 1 is

$$(\Phi_E)_1 = (E\cos 35°)A_1 = (250\text{ N/C})(\cos 35°)(1.7\text{ m}^2) = \boxed{350\text{ N}\cdot\text{m}^2/\text{C}}$$

b. Similarly, the flux through surface 2 is

$$(\Phi_E)_2 = (E\cos 55°)A_2 = (250\text{ N/C})(\cos 55°)(3.2\text{ m}^2) = \boxed{460\text{ N}\cdot\text{m}^2/\text{C}}$$

---

53. ***REASONING*** The electric flux through each face of the cube is given by $\Phi_E = (E\cos\phi)A$ (see Section 18.9) where E is the magnitude of the electric field at the face, A is the area of the face, and ϕ is the angle between the electric field and the outward normal of that face. We can use this expression to calculate the electric flux Φ_E through each of the six faces of the cube.

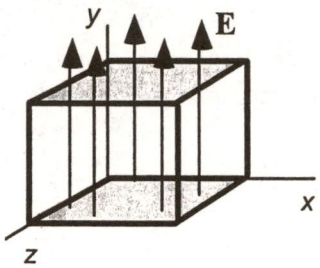

SOLUTION

a. On the bottom face of the cube, the outward normal points parallel to the $-y$ axis, in the opposite direction to the electric field, and $\phi = 180°$. Therefore,

$$(\Phi_E)_{\text{bottom}} = (1500\text{ N/C})(\cos 180°)(0.20\text{ m})^2 = \boxed{-6.0\times 10^1\text{ N}\cdot\text{m}^2/\text{C}}$$

On the top face of the cube, the outward normal points parallel to the $+y$ axis, and $\phi = 0.0°$. The electric flux is, therefore,

$$(\Phi_E)_{\text{top}} = (1500\text{ N/C})(\cos 0.0°)(0.20\text{ m})^2 = \boxed{+6.0\times 10^1\text{N}\cdot\text{m}^2/\text{C}}$$

On each of the other four faces, the outward normals are perpendicular to the direction of the electric field, so $\phi = 90°$. So for each of the four side faces,

$$(\Phi_E)_{\text{sides}} = (1500\text{ N/C})(\cos 90°)(0.20\text{ m})^2 = \boxed{0\text{ N}\cdot\text{m}^2/\text{C}}$$

b. The total flux through the cube is

$$(\Phi_E)_{total} = (\Phi_E)_{top} + (\Phi_E)_{bottom} + (\Phi_E)_{side\ 1} + (\Phi_E)_{side\ 2} + (\Phi_E)_{side\ 3} + (\Phi_E)_{side\ 4}$$

Therefore,

$$(\Phi_E)_{total} = (+6.0\times10^1 \text{N}\cdot\text{m}^2/\text{C}) + (-6.0\times10^1 \text{N}\cdot\text{m}^2/\text{C}) + 0 + 0 + 0 + 0 = \boxed{0\ \text{N}\cdot\text{m}^2/\text{C}}$$

---

59. ***REASONING*** Before the 3.0-μC point charge q is introduced into the region, the region contains a uniform electric field $\mathbf{E}$ of magnitude 1.6×10^4 N/C. After the 3.0-μC charge is introduced into the region, the net electric field changes. In addition to the uniform electric field $\mathbf{E}$, the region will also contain the electric field $\mathbf{E}_q$ due to the point charge q. The field at any point in the region is the vector sum of $\mathbf{E}$ and $\mathbf{E}_q$. The field $\mathbf{E}_q$ is radial as discussed in the text, and its magnitude at any distance r from the charge q is given by Equation 18.3, $E_q = k|q|/r^2$. There will be one point P in the region where the net electric field $\mathbf{E}_{net}$ is zero. This point is located where the field $\mathbf{E}$ has the same magnitude and points in the direction opposite to the field $\mathbf{E}_q$. We will use this reasoning to find the distance r_0 from the charge q to the point P.

SOLUTION Let us assume that the field $\mathbf{E}$ points to the right and that the charge q is *negative* (the problem is done the same way if q is positive, although then the relative positions of P and q will be reversed). Since q is negative, its electric field is radially *inward* (i.e., toward q); therefore, in order for the field $\mathbf{E}_q$ to point in the opposite direction to $\mathbf{E}$, the charge q will have to be to the left of the point P where $\mathbf{E}_{net}$ is zero, as shown in the drawing at the right. Using Equation 18.3, $E_q = k|q|/r_0^2$, and solving for the distance r_0, we find $r_0 = \sqrt{k|q|/E_q}$. Since the magnitude E_q must be equal to the magnitude of E at the point P, we have

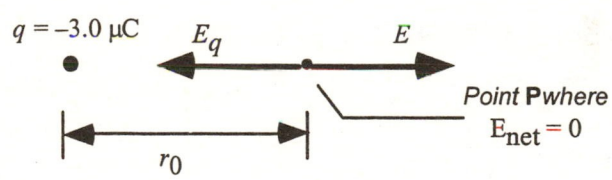

$$r_0 = \sqrt{\frac{k|q|}{E}} = \sqrt{\frac{(8.99\times10^9\ \text{Nm}^2/\text{C}^2)\ (3.0\times10^{-6}\ \text{C})}{1.6\times10^4\ \text{N/C}}} = \boxed{1.3\ \text{m}}$$

---

61. ***REASONING AND SOLUTION***

a. Since the gravitational force between the spheres is one of attraction and the electrostatic force must balance it, the electric force must be one of repulsion. Therefore, the charges must have $\boxed{\text{the same algebraic signs, both positive or both negative}}$.

b. There are two forces that act on each sphere; they are the gravitational attraction F_G of one sphere for the other, and the repulsive electric force F_E of one sphere on the other. From the problem statement, we know that these two forces balance each other, so that $F_G = F_E$. The magnitude of F_G is given by Newton's law of gravitation (Equation 4.3: $F_G = Gm_1m_2/r^2$), while the magnitude of F_E is given by Coulomb's law (Equation 18.1: $F_E = k|q_1||q_2|/r^2$). Therefore, we have

$$\frac{Gm_1m_2}{r^2} = \frac{k|q_1||q_2|}{r^2} \quad \text{or} \quad Gm^2 = k|q|^2$$

since the spheres have the same mass m and carry charges of the same magnitude $|q|$. Solving for $|q|$, we find

$$|q| = m\sqrt{\frac{G}{k}} = (2.0 \times 10^{-6}\ \text{kg})\sqrt{\frac{6.67\times10^{-11}\ \text{N}\cdot\text{m}^2/\text{kg}^2}{8.99\times10^{9}\ \text{N}\cdot\text{m}^2/\text{C}^2}} = \boxed{1.7\times10^{-16}\ \text{C}}$$

---

67. ***REASONING*** Consider the drawing at the right. It is given that the charges q_A, q_1, and q_2 are each positive. Therefore, the charges q_1 and q_2 each exert a repulsive force on the charge q_A. As the drawing shows, these forces have magnitudes F_{A1} (vertically downward) and F_{A2} (horizontally to the left). The unknown charge placed at the empty corner of the rectangle is q_U, and it exerts a force on q_A that has a magnitude F_{AU}. In order that the net force acting on q_A point in the vertical direction, the horizontal component of $\mathbf{F_{AU}}$ must cancel out the horizontal force $\mathbf{F_{A2}}$. Therefore, $\mathbf{F_{AU}}$ must point as shown in the drawing, which means that it is an attractive force and q_U must be negative, since q_A is positive.

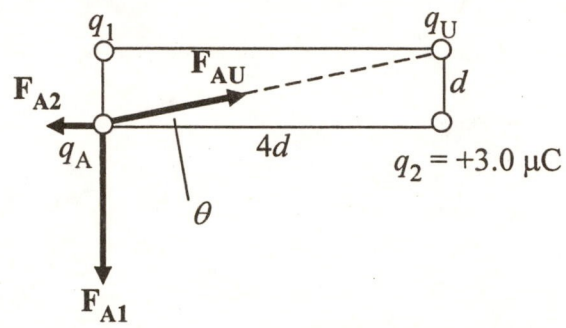

SOLUTION The basis for our solution is the fact that the horizontal component of $\mathbf{F_{AU}}$ must cancel out the horizontal force $\mathbf{F_{A2}}$. The magnitudes of these forces can be expressed using Coulomb's law $F = k|q||q'|/r^2$, where r is the distance between the charges q and q'. Thus, we have

$$F_{AU} = \frac{k|q_A||q_U|}{(4d)^2 + d^2} \quad \text{and} \quad F_{A2} = \frac{k|q_A||q_2|}{(4d)^2}$$

where we have used the fact that the distance between the charges q_A and q_U is the diagonal of the rectangle, which is $\sqrt{(4d)^2 + d^2}$ according to the Pythagorean theorem, and the fact that the distance between the charges q_A and q_2 is $4d$. The horizontal component of $\mathbf{F}_{AU}$ is $F_{AU}\cos\theta$, which must be equal to F_{A2}, so that we have

$$\frac{k|q_A||q_U|}{(4d)^2 + d^2}\cos\theta = \frac{k|q_A||q_2|}{(4d)^2} \quad \text{or} \quad \frac{|q_U|}{17}\cos\theta = \frac{|q_2|}{16}$$

The drawing in the ***REASONING***, reveals that $\cos\theta = (4\text{d})/\sqrt{(4d)^2 + d^2} = 4/\sqrt{17}$. Therefore, we find that

$$\frac{|q_U|}{17}\left(\frac{4}{\sqrt{17}}\right) = \frac{|q_2|}{16} \quad \text{or} \quad |q_U| = \frac{17\sqrt{17}}{64}|q_2| = \frac{17\sqrt{17}}{64}\left(3.0\times10^{-6}\ \text{C}\right) = \boxed{3.3\times10^{-6}\ \text{C}}$$

As discussed in the ***REASONING***, the algebraic sign of the charge q_U is $\boxed{\text{negative}}$.

CHAPTER 19 | *ELECTRIC POTENTIAL ENERGY AND THE ELECTRIC POTENTIAL*

CONCEPTUAL QUESTIONS

---

4. ***REASONING AND SOLUTION*** Four point charges of equal magnitude are placed at the corners of a square as shown in the figure at the right.

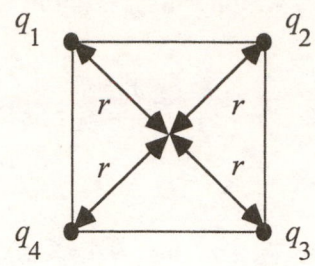

The electric field at the center of the square is the *vector sum* of the electric field at the center due to each of the charges individually. The potential at the center of the square is equal to the *algebraic sum* of the potentials at the center due to each of the charges individually.

All four charges are equidistant from the center (a distance r in the figure). If two diagonal charges have the same magnitude and sign, then the electric field at the center due to these two charges have equal magnitude and opposite directions. Their resultant is, therefore, zero. Thus, the electric field at the center will be zero if each diagonal pair of charges has the same magnitude and sign.

If a diagonal pair of charges has the same magnitude and sign, they will give rise to a non-zero potential at the center. Thus, the potential due to one diagonal pair of charges must cancel the potential due to the other diagonal pair of charges. This will be the case if the two pairs of diagonal charges have opposite signs.

Thus, both the electric field and the electric potential will be zero at the center of the square if all four charges have the same magnitude, q_1 and q_3 have the same sign, and q_2 and q_4 have the same sign, which is opposite to the signs of q_1 and q_3.

---

7. ***REASONING AND SOLUTION*** An electric potential energy exists when two protons are separated by a certain distance. It is equal to the work that must be done by an external agent to assemble the configuration. Suppose that we imagine assembling the system, one particle at a time. If there are no other charges in the region, there are no existing electric fields; therefore, no work is required to put the first proton in place. That proton, however, gives rise to an electric field that fills the region. Its magnitude at a distance r from the proton is given by Equation 18.3, $E = ke/r^2$, where $+e$ is the magnitude of the charge on the proton. Since the region contains an electric field due to the first proton, there also exists an electric potential, and the external agent must do work to place the second proton at a distance d from the first proton. The electric potential energy of the final configuration is equal to the work that must be done to bring the second proton from infinity and place it at a distance d from the first proton. The electric potential at a distance d from the first proton is $V_{\text{proton}} = +ke/d$ (Equation 19.6). According to Equation 19.3, the electric potential energy of the final configuration is therefore

$$\text{EPE} = V_{\text{proton}}(+e) = +\frac{ke^2}{d}$$

a. If both protons are replaced by electrons, similar arguments apply. However, since the electron carries a negative charge ($-e$), the electric potential at a distance d from the first electron is $V_{\text{electron}} = -ke/d$. The electric potential energy of the final configuration is now given by

$$\text{EPE} = V_{\text{electron}}(-e) = -\frac{ke}{d}(-e) = +\frac{ke^2}{d}$$

Therefore, if both protons are replaced by electrons, the electric potential energy *remains the same.*

b. When only one of the protons is replaced by an electron, we find that

$$\text{EPE} = V_{\text{proton}}(-e) = \left(+\frac{ke}{d}\right)(-e) = -\frac{ke^2}{d}$$

Thus, when only one of the protons is replaced by an electron, the electric potential energy *decreases* from $+ke^2/d$ to $-ke^2/d$.

---

16. ***REASONING AND SOLUTION*** Since both particles are released from rest, their initial kinetic energies are zero. They both have electric potential energy by virtue of their respective positions in the electric field between the plates. Since the particles are oppositely charged, they move in opposite directions toward opposite plates of the capacitor. As they move toward the plates, the particles gain kinetic energy and lose potential energy. Using $(\text{EPE})_0$ and $(\text{EPE})_\text{f}$ to denote the initial and final electric potential energies of the particle, respectively, we find from energy conservation that

$$(\text{EPE})_0 = \tfrac{1}{2}m_{\text{particle}}v_\text{f}^2 + (\text{EPE})_\text{f}$$

The final speed of each particle is given by

$$v_\text{f} = \sqrt{\frac{2\left[(\text{EPE})_0 - (\text{EPE})_\text{f}\right]}{m_{\text{particle}}}}$$

Since both particles travel through the same distance between the plates of the capacitor, the change in the electric potential energy is the same for both particles. Since the mass of the electron is smaller than the mass of the proton, the final speed of the electron will be greater than that of the proton. Therefore, the electron travels faster than the proton as the particles move toward the respective plates. The electron, therefore, strikes the capacitor plate first.

CHAPTER 19 | *ELECTRIC POTENTIAL ENERGY AND THE ELECTRIC POTENTIAL*

PROBLEMS

1. ***REASONING AND SOLUTION*** Combining Equations 19.1 and 19.3, we have

$$W_{AB} = \text{EPE}_A - \text{EPE}_B = q_0(V_A - V_B) = (+1.6\times10^{-19}\,\text{C})(0.070\,\text{V}) = \boxed{1.1\times10^{-20}\,\text{J}}$$

7. ***REASONING*** The only force acting on the moving electron is the conservative electric force. Therefore, the total energy of the electron (the sum of the kinetic energy KE and the electric potential energy EPE) remains constant throughout the trajectory of the electron. Let the subscripts A and B refer to the initial and final positions, respectively, of the electron. Then,

$$\tfrac{1}{2}mv_A^2 + \text{EPE}_A = \tfrac{1}{2}mv_B^2 + \text{EPE}_B$$

Solving for v_B gives

$$v_B = \sqrt{v_A^2 - \frac{2}{m}(\text{EPE}_B - \text{EPE}_A)}$$

Since the electron starts from rest, $v_A = 0\,\text{m/s}$. The difference in potential energies is related to the difference in potentials by Equation 19.4, $\text{EPE}_B - \text{EPE}_A = q(V_B - V_A)$.

SOLUTION The speed v_B of the electron just before it reaches the screen is

$$v_B = \sqrt{-\frac{2q}{m}(V_B - V_A)} = \sqrt{-\frac{2(-1.6\times10^{-19}\,\text{C})}{9.11\times10^{-31}\,\text{kg}}(25\,000\,\text{V})} = \boxed{9.4\times10^{7}\,\text{m/s}}$$

9. ***REASONING*** The only force acting on the moving charge is the conservative electric force. Therefore, the total energy of the charge remains constant. Applying the principle of conservation of energy between locations A and B, we obtain

$$\tfrac{1}{2}mv_A^2 + \text{EPE}_A = \tfrac{1}{2}mv_B^2 + \text{EPE}_B$$

Since the charged particle starts from rest, $v_A = 0$. The difference in potential energies is related to the difference in potentials by Equation 19.4, $\text{EPE}_B - \text{EPE}_A = q(V_B - V_A)$. Thus, we have

$$q(V_A - V_B) = \tfrac{1}{2}mv_B^2 \qquad (1)$$

Similarly, applying the conservation of energy between locations C and B gives

$$q(V_C - V_B) = \tfrac{1}{2}m(2v_B)^2 \qquad (2)$$

Dividing Equation (1) by Equation (2) yields

$$\frac{V_A - V_B}{V_C - V_B} = \frac{1}{4}$$

This expression can be solved for V_B.

SOLUTION Solving for V_B, we find that

$$V_B = \frac{4V_A - V_C}{3} = \frac{4(452\text{ V}) - 791\text{ V}}{3} = \boxed{339\text{ V}}$$

---

11. ***REASONING*** The potential of each charge q at a distance r away is given by Equation 19.6 as $V = kq/r$. By applying this expression to each charge, we will be able to find the desired ratio, because the distances are given for each charge.

SOLUTION According to Equation 19.6, the potentials of each charge are

$$V_A = \frac{kq_A}{r_A} \quad \text{and} \quad V_B = \frac{kq_B}{r_B}$$

Since we know that $V_A = V_B$, it follows that

$$\frac{kq_A}{r_A} = \frac{kq_B}{r_B} \quad \text{or} \quad \frac{q_B}{q_A} = \frac{r_B}{r_A} = \frac{0.43\text{ m}}{0.18\text{ m}} = \boxed{2.4}$$

---

15. ***REASONING*** Initially, suppose that one charge is at C and the other charge is held fixed at B. The charge at C is then moved to position A. According to Equation 19.4, the work W_{CA} done by the electric force as the charge moves from C to A is $W_{CA} = q(V_C - V_A)$, where, from Equation 19.6, $V_C = kq/d$ and $V_A = kq/r$. From the figure at the right we see that $d = \sqrt{r^2 + r^2} = \sqrt{2}r$. Therefore, we find that

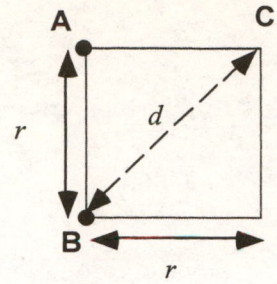

$$W_{CA} = q\left(\frac{kq}{\sqrt{2}r} - \frac{kq}{r}\right) = \frac{kq^2}{r}\left(\frac{1}{\sqrt{2}} - 1\right)$$

SOLUTION Substituting values, we obtain

$$W_{CA} = \frac{(8.99\times10^9\ \text{N}\cdot\text{m}^2/\text{C}^2)(3.0\times10^{-6}\ \text{C})^2}{0.500\ \text{m}}\left(\frac{1}{\sqrt{2}} - 1\right) = \boxed{-4.7\times10^{-2}\ \text{J}}$$

---

19. ***REASONING*** The only force acting on the moving charge is the conservative electric force. Therefore, the sum of the kinetic energy KE and the electric potential energy EPE is the same at points A and B:

$$\tfrac{1}{2}mv_A^2 + \text{EPE}_A = \tfrac{1}{2}mv_B^2 + \text{EPE}_B$$

Since the particle comes to rest at B, $v_B = 0$. Combining Equations 19.3 and 19.6, we have

$$\text{EPE}_A = qV_A = q\left(\frac{kq_1}{d}\right)$$

and

$$\text{EPE}_B = qV_B = q\left(\frac{kq_1}{r}\right)$$

where d is the initial distance between the fixed charge and the moving charged particle, and r is the distance between the charged particles after the moving charge has stopped. Therefore, the expression for the conservation of energy becomes

$$\tfrac{1}{2}mv_A^2 + \frac{kqq_1}{d} = \frac{kqq_1}{r}$$

This expression can be solved for r. Once r is known, the distance that the charged particle moves can be determined.

SOLUTION Solving the expression above for r gives

$$r = \frac{kqq_1}{\frac{1}{2}mv_A^2 + \frac{kqq_1}{d}}$$

$$= \frac{(8.99\times10^9\,\text{N}\cdot\text{m}^2/\text{C}^2)(-8.00\times10^{-6}\,\text{C})(-3.00\times10^{-6}\,\text{C})}{\frac{1}{2}(7.20\times10^{-3}\,\text{kg})(65.0\text{ m/s})^2 + \frac{(8.99\times10^9\,\text{N}\cdot\text{m}^2/\text{C}^2)(-8.00\times10^{-6}\,\text{C})(-3.00\times10^{-6}\,\text{C})}{0.0450\text{ m}}}$$

$$= 0.0108\text{ m}$$

Therefore, the charge moves a distance of $0.0450\text{ m} - 0.0108\text{ m} = \boxed{0.0342\text{ m}}$.

---

23. ***REASONING*** Initially, the three charges are infinitely far apart. We will proceed as in Example 8 by adding charges to the triangle, one at a time, and determining the electric potential energy at each step. According to Equation 19.3, the electric potential energy EPE is the product of the charge q and the electric potential V at the spot where the charge is placed, EPE = qV. The total electric potential energy of the group is the sum of the energies of each step in assembling the group.

SOLUTION Let the corners of the triangle be numbered clockwise as 1, 2 and 3, starting with the top corner. When the first charge (q_1 = 8.00 μC) is placed at a corner 1, the charge has no electric potential energy, $\text{EPE}_1 = 0$. This is because the electric potential V_1 produced by the other two charges at corner 1 is zero, since they are infinitely far away.

Once the 8.00-μC charge is in place, the electric potential V_2 that it creates at corner 2 is

$$V_2 = \frac{kq_1}{r_{21}}$$

where r_{21} = 5.00 m is the distance between corners 1 and 2, and q_1 = 8.00 μC. When the 20.0-μC charge is placed at corner 2, its electric potential energy EPE_2 is

$$\text{EPE}_2 = q_2V_2 = q_2\left(\frac{kq_1}{r_{21}}\right)$$

$$= \left(20.0\times10^{-6}\,\text{C}\right)\left[\frac{\left(8.99\times10^9\,\text{N}\cdot\text{m}^2/\text{C}^2\right)\left(8.00\times10^{-6}\,\text{C}\right)}{5.00\text{ m}}\right] = 0.288\text{ J}$$

The electric potential V_3 at the remaining empty corner is the sum of the potentials due to the two charges that are already in place on corners 1 and 2:

$$V_3 = \frac{kq_1}{r_{31}} + \frac{kq_2}{r_{32}}$$

where $q_1 = 8.00\ \mu\text{C}$, $r_{31} = 3.00$ m, $q_2 = 20.0\ \mu\text{C}$, and $r_{32} = 4.00$ m. When the third charge ($q_3 = -15.0\ \mu\text{C}$) is placed at corner 3, its electric potential energy EPE_3 is

$$\text{EPE}_3 = q_3V_3 = q_3\left(\frac{kq_1}{r_{31}} + \frac{kq_2}{r_{32}}\right) = q_3k\left(\frac{q_1}{r_{31}} + \frac{q_2}{r_{32}}\right)$$

$$= \left(-15.0\times10^{-6}\ \text{C}\right)\left(8.99\times10^{9}\ \text{N}\cdot\text{m}^2/\text{C}^2\right)\left(\frac{8.00\times10^{-6}\ \text{C}}{3.00\ \text{m}} + \frac{20.0\times10^{-6}\ \text{C}}{4.00\ \text{m}}\right) = -1.034\ \text{J}$$

The electric potential energy of the entire array is given by

$$\text{EPE} = \text{EPE}_1 + \text{EPE}_2 + \text{EPE}_3 = 0 + 0.288\ \text{J} + (-1.034\ \text{J}) = \boxed{-0.746\ \text{J}}$$

---

27. ***REASONING*** The magnitude E of the electric field is given by Equation 19.7a (without the minus sign) as $E = \dfrac{\Delta V}{\Delta s}$, where ΔV is the potential difference between the two metal conductors of the spark plug, and Δs is the distance between the two conductors. We can use this relation to find ΔV.

SOLUTION The potential difference between the conductors is

$$\Delta V = E\Delta s = \left(4.7\times10^{7}\ \text{V/m}\right)\left(0.75\times10^{-3}\ \text{m}\right) = \boxed{3.5\times10^{4}\ \text{V}}$$

---

31. ***REASONING AND SOLUTION*** As described in the problem statement, the charges jump between your hand and a doorknob. If we assume that the electric field is uniform, Equation 19.7a applies, and we have

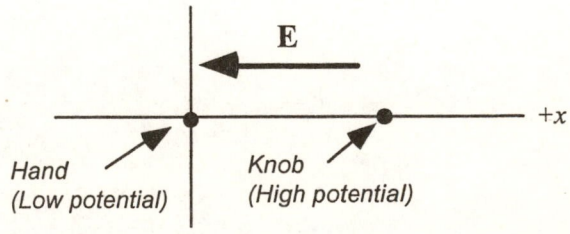

$$E = -\frac{\Delta V}{\Delta s} = -\frac{V_{\text{knob}} - V_{\text{hand}}}{\Delta s}$$

Therefore, solving for the potential difference between your hand and the doorknob, we have

$$V_{\text{knob}} - V_{\text{hand}} = -E\Delta s = -(-3.0\times10^{6}\,\text{N/C})(3.0\times10^{-3}\,\text{m}) = \boxed{+9.0\times10^{3}\,\text{V}}$$

---

35. ***REASONING*** The total energy of the particle on the equipotential surface A is $E_A = \frac{1}{2}mv_A^2 + qV_A$. Similarly, the total energy of the particle when it reaches equipotential surface B is $E_B = \frac{1}{2}mv_B^2 + qV_B$. According to Equation 6.8, the work W done by the outside force is equal to the final total energy E_B minus the initial total energy E_A; $W = E_B - E_A$.

SOLUTION The work done by the outside force in moving the particle from A to B is,

$$W = E_B - E_A = \left(\tfrac{1}{2}mv_B^2 + qV_B\right) - \left(\tfrac{1}{2}mv_A^2 + qV_A\right) = \tfrac{1}{2}m\left(v_B^2 - v_A^2\right) + q(V_B - V_A)$$

$$= \tfrac{1}{2}(5.00\times10^{-2}\,\text{kg})\left[(3.00\,\text{m/s})^2 - (2.00\,\text{m/s})^2\right]$$

$$+ (4.00\times10^{-5}\,\text{C})\left[7850\,\text{V} - 5650\,\text{V}\right] = \boxed{0.213\,\text{J}}$$

---

37. ***REASONING*** The charge that resides on the outer surface of the cell membrane is $q = CV$, according to Equation 19.8. Before we can use this expression, however, we must first determine the capacitance of the membrane. If we assume that the cell membrane behaves like a parallel plate capacitor filled with a dielectric, Equation 19.10 ($C = \kappa\varepsilon_0 A/d$) applies as well.

SOLUTION The capacitance of the cell membrane is

$$C = \frac{\kappa\varepsilon_0 A}{d} = \frac{(5.0)(8.85\times10^{-12}\,\text{F/m})(5.0\times10^{-9}\,\text{m}^2)}{1.0\times10^{-8}\,\text{m}} = 2.2\times10^{-11}\,\text{F}$$

a. The charge on the outer surface of the membrane is, therefore,

$$q = CV = (2.2\times10^{-11}\,\text{F})(60.0\times10^{-3}\,\text{V}) = \boxed{1.3\times10^{-12}\,\text{C}}$$

b. If the charge in part (a) is due to K^+ ions with charge $+e$ ($e = 1.6\times10^{-19}$ C), the number of ions present on the outer surface of the membrane is

$$\text{Number of } K^+ \text{ ions} = \frac{1.3\times10^{-12}\,\text{C}}{1.6\times10^{-19}\,\text{C}} = \boxed{8.1\times10^{6}}$$

---

41. ***REASONING AND SOLUTION*** Equation 19.10 gives the capacitance for a parallel plate capacitor filled with a dielectric of constant κ: $C = \kappa\varepsilon_0 A/d$. Solving for κ, we have

$$\kappa = \frac{Cd}{\varepsilon_0 A} = \frac{(7.0\times10^{-6}\,\text{F})(1.0\times10^{-5}\,\text{m})}{(8.85\times10^{-12}\,\text{F/m})(1.5\,\text{m}^2)} = \boxed{5.3}$$

---

47. ***REASONING*** According to Equation 19.10, the capacitance of a parallel plate capacitor filled with a dielectric is $C = \kappa\varepsilon_0 A/d$, where κ is the dielectric constant, A is the area of one plate, and d is the distance between the plates.

From the definition of capacitance (Equation 19.8), $q = CV$. Thus, using Equation 19.10, we see that the charge q on a parallel plate capacitor that contains a dielectric is given by $q = \left(\kappa\varepsilon_0 A/d\right)V$. Since each dielectric occupies one-half of the volume between the plates, the area of each plate in contact with each material is $A/2$. Thus,

$$q_1 = \frac{\kappa_1\varepsilon_0(A/2)}{d}V = \frac{\kappa_1\varepsilon_0 A}{2d}V \qquad \text{and} \qquad q_2 = \frac{\kappa_2\varepsilon_0(A/2)}{d}V = \frac{\kappa_2\varepsilon_0 A}{2d}V$$

According to the problem statement, the total charge stored by the capacitor is

$$q_1 + q_2 = CV \tag{1}$$

where q_1 and q_2 are the charges on the plates in contact with dielectrics 1 and 2, respectively.
Using the expressions for q_1 and q_2 above, Equation (1) becomes

$$CV = \frac{\kappa_1\varepsilon_0 A}{2d}V + \frac{\kappa_2\varepsilon_0 A}{2d}V = \frac{\kappa_1\varepsilon_0 A + \kappa_2\varepsilon_0 A}{2d}V = \frac{(\kappa_1+\kappa_2)\varepsilon_0 A}{2d}V$$

This expression can be solved for C.

SOLUTION Solving for C, we obtain $\boxed{C = \dfrac{\varepsilon_0 A(\kappa_1+\kappa_2)}{2d}}$

---

51. ***REASONING*** According to Equation 19.11b, the energy stored in a capacitor with capacitance C and potential V across its plates is Energy $= \frac{1}{2}CV^2$.

SOLUTION Therefore, solving Equation 19.11b for V, we have

$$V = \sqrt{\frac{2(\text{Energy})}{C}} = \sqrt{\frac{2(73\text{ J})}{120\times10^{-6}\text{F}}} = \boxed{1.1\times10^{3}\text{ V}}$$

---

55. ***REASONING AND SOLUTION*** Equation 19.7a gives the result directly:

$$E = -\frac{\Delta V}{\Delta s} = -\frac{V_B - V_A}{\Delta s} = -\frac{28\text{ V}-95\text{ V}}{0.016\text{ m}} = \boxed{4.2\times10^{3}\text{ V/m}}$$

The electric field points from high potential to low potential. Thus, it points from $\boxed{A \text{ to } B}$.

---

59. ***REASONING*** If we assume that the motion of the proton and the electron is horizontal in the $+x$ direction, the motion of the proton is determined by Equation 2.8, $x = v_0 t + \frac{1}{2}a_p t^2$, where x is the distance traveled by the proton, v_0 is its initial speed, and a_p is its acceleration. If the distance between the capacitor places is d, then this relation becomes $\frac{1}{2}d = v_0 t + \frac{1}{2}a_p t^2$, or

$$d = 2v_0 t + a_p t^2 \tag{1}$$

We can solve Equation (1) for the initial speed v_0 of the proton, but, first, we must determine the time t and the acceleration a_p of the proton . Since the proton strikes the negative plate at the same instant the electron strikes the positive plate, we can use the motion of the electron to determine the time t.

For the electron, $\frac{1}{2}d = \frac{1}{2}a_e t^2$, where we have taken into account the fact that the electron is released from rest. Solving this expression for t we have $t = \sqrt{d/a_e}$. Substituting this expression into Equation (1), we have

$$d = 2v_0\sqrt{\frac{d}{a_e}} + \left(\frac{a_p}{a_e}\right)d \tag{2}$$

The accelerations can be found by noting that the magnitudes of the forces on the electron and proton are equal, since these particles have the same magnitude of charge. The force on the electron is $F = eE = eV/d$, and the acceleration of the electron is, therefore,

$$a_e = \frac{F}{m_e} = \frac{eV}{m_e d} \tag{3}$$

Newton's second law requires that $m_e a_e = m_p a_p$, so that

$$\frac{a_p}{a_e} = \frac{m_e}{m_p} \tag{4}$$

Combining Equations (2), (3) and (4) leads to the following expression for v_0, the initial speed of the proton:

$$v_0 = \frac{1}{2}\left(1 - \frac{m_e}{m_p}\right)\sqrt{\frac{eV}{m_e}}$$

SOLUTION Substituting values into the expression above, we find

$$v_0 = \frac{1}{2}\left(1 - \frac{9.11\times10^{-31}\text{kg}}{1.67\times10^{-27}\text{kg}}\right)\sqrt{\frac{(1.60\times10^{-19}\text{C})(175\text{ V})}{9.11\times10^{-31}\text{kg}}} = \boxed{2.77\times10^{6}\text{m/s}}$$

CHAPTER 20 | *ELECTRIC CIRCUITS*

CONCEPTUAL QUESTIONS

---

3. ***REASONING AND SOLUTION*** Two materials have different resistivities. Two wires of the same length are made, one from each of the materials. The resistance of each wire is given by Equation 20.3: $R = \rho L / A$, where ρ is the resistivity of the wire material, and L and A are, respectively, the length and cross-sectional area of the wire. Even when the wires have the same length, they may have the same resistance, if the cross-sectional areas of the wires are chosen so that the ratio $\rho L / A$ is the same for each.

---

10. ***REASONING AND SOLUTION*** A number of light bulbs are to be connected to a single electrical outlet. The bulbs will provide more brightness when they are connected in such a way that their power output is greatest. Since the voltage at the outlet is constant, the power delivered by the bulbs is given by Equation 20.6c: $P = V^2 / R$, where R is the equivalent resistance of the combination of light bulbs. When the light bulbs are connected in series, their equivalent resistance is given by Equation 20.16: $R_S = R_1 + R_2 + R_3 + \ldots$ When the light bulbs are connected in parallel, their equivalent resistance is given by Equation 20.17: $\frac{1}{R_P} = \frac{1}{R_1} + \frac{1}{R_2} + \frac{1}{R_3} + \ldots$ Clearly, the equivalent resistance will be less when the bulbs are connected in parallel. From Equation 20.6c, we can conclude that the power output will be greatest, and the light bulbs will provide more brightness, when they are connected in parallel.

---

13. ***REASONING AND SOLUTION*** One way is to make two combinations that consist of two resistors in parallel. The resistance of each combination is

$$R_p = [(1/R) + (1/R)]^{-1} = R/2$$

These two combinations can then be placed in series, as shown at the right. The equivalent resistance of the series combination is

$$R_s = (R/2) + (R/2) = R$$

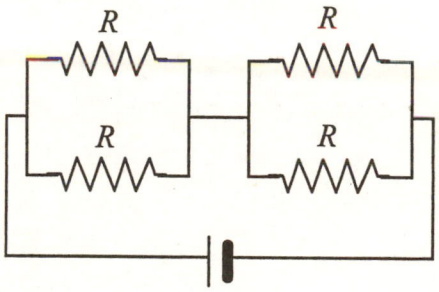

A second way is to make two combinations that consist of two resistors in series. The resistance of each combination is

$$R_s = R + R = 2R$$

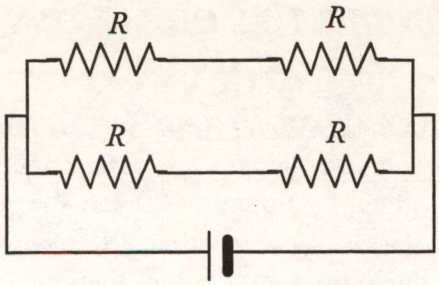

These two combinations can then be placed in parallel, as shown at the right. The equivalent resistance of the parallel combination is

$$R_p = [1/(2R) + 1/(2R)]^{-1} = R$$

CHAPTER 20 | *ELECTRIC CIRCUITS*

PROBLEMS

---

5. ***REASONING AND SOLUTION*** Ohm's law (Equation 20.2, $V = IR$) gives the result directly

$$I = \frac{V}{R} = \frac{240\text{ V}}{11\ \Omega} = \boxed{22\text{ A}}$$

---

9. ***REASONING*** The number N of protons that strike the target is equal to the amount of electric charge Δq striking the target divided by the charge e of a proton, $N = (\Delta q)/e$. From Equation 20.1, the amount of charge is equal to the product of the current I and the time Δt. We can combine these two relations to find the number of protons that strike the target in 15 seconds.

The heat Q that must be supplied to change the temperature of the aluminum sample of mass m by an amount ΔT is given by Equation 12.4 as $Q = cm\Delta T$, where c is the specific heat capacity of aluminum. The heat is provided by the kinetic energy of the protons and is equal to the number of protons that strike the target times the kinetic energy per proton. Using this reasoning, we can find the change in temperature of the block for the 15 second-time interval.

SOLUTION

a. The number N of protons that strike the target is

$$N = \frac{\Delta q}{e} = \frac{I\,\Delta t}{e} = \frac{(0.50\times10^{-6}\text{ A})(15\text{ s})}{1.6\times10^{-19}\text{ C}} = \boxed{4.7\times10^{13}}$$

b. The amount of heat Q provided by the kinetic energy of the protons is

$$Q = (4.7\times10^{13}\text{ protons})(4.9\times10^{-12}\text{ J/proton}) = 230\text{ J}$$

Since $Q = cm\Delta T$ and since Table 12.2 gives the specific heat of aluminum as $c = 9.00\times10^{2}$ J/(kg·C°), the change in temperature of the block is

$$\Delta T = \frac{Q}{cm} = \frac{230\text{ J}}{\left(9.00\times10^{2}\ \dfrac{\text{J}}{\text{kg}\cdot\text{C}^\circ}\right)(15\times10^{-3}\text{ kg})} = \boxed{17\text{ C}^\circ}$$

---

13. ***REASONING*** The resistance of a metal wire of length L, cross-sectional area A and resistivity ρ is given by Equation 20.3: $R = \rho L / A$. Solving for A, we have $A = \rho L / R$. We can use this expression to find the ratio of the cross-sectional area of the aluminum wire to that of the copper wire.

SOLUTION Forming the ratio of the areas and using resistivity values from Table 20.1, we have

$$\frac{A_{\text{aluminum}}}{A_{\text{copper}}} = \frac{\rho_{\text{aluminum}} L/R}{\rho_{\text{copper}} L/R} = \frac{\rho_{\text{aluminum}}}{\rho_{\text{copper}}} = \frac{2.82\times10^{-8}\ \Omega\cdot\text{m}}{1.72\times10^{-8}\ \Omega\cdot\text{m}} = \boxed{1.64}$$

---

19. ***REASONING*** We will ignore any changes in length due to thermal expansion. Although the resistance of each section changes with temperature, the total resistance of the composite does not change with temperature. Therefore,

$$\underbrace{\left(R_{\text{tungsten}}\right)_0 + \left(R_{\text{carbon}}\right)_0}_{\text{At room temperature}} = \underbrace{R_{\text{tungsten}} + R_{\text{carbon}}}_{\text{At temperature } T}$$

From Equation 20.5, we know that the temperature dependence of the resistance for a wire of resistance R_0 at temperature T_0 is given by $R = R_0[1 + \alpha(T - T_0)]$, where α is the temperature coefficient of resistivity. Thus,

$$\left(R_{\text{tungsten}}\right)_0 + \left(R_{\text{carbon}}\right)_0 = \left(R_{\text{tungsten}}\right)_0 (1 + \alpha_{\text{tungsten}}\Delta T) + \left(R_{\text{carbon}}\right)_0 (1 + \alpha_{\text{carbon}}\Delta T)$$

Since ΔT is the same for each wire, this simplifies to

$$\left(R_{\text{tungsten}}\right)_0 \alpha_{\text{tungsten}} = -\left(R_{\text{carbon}}\right)_0 \alpha_{\text{carbon}} \qquad (1)$$

This expression can be used to find the ratio of the resistances. Once this ratio is known, we can find the ratio of the lengths of the sections with the aid of Equation 20.3 ($L = RA/\rho$).

SOLUTION From Equation (1), the ratio of the resistances of the two sections of the wire is

$$\frac{\left(R_{\text{tungsten}}\right)_0}{\left(R_{\text{carbon}}\right)_0} = -\frac{\alpha_{\text{carbon}}}{\alpha_{\text{tungsten}}} = -\frac{-0.0005\ [(\text{C}^\circ)^{-1}]}{0.0045\ [(\text{C}^\circ)^{-1}]} = \frac{1}{9}$$

Thus, using Equation 20.3, we find the ratio of the tungsten and carbon lengths to be

$$\frac{L_{\text{tungsten}}}{L_{\text{carbon}}} = \frac{(R_0 A/\rho)_{\text{tungsten}}}{(R_0 A/\rho)_{\text{carbon}}} = \frac{(R_{\text{tungsten}})_0}{(R_{\text{carbon}})_0}\left(\frac{\rho_{\text{carbon}}}{\rho_{\text{tungsten}}}\right) = \left(\frac{1}{9}\right)\left(\frac{3.5\times10^{-5}\,\Omega\cdot\text{m}}{5.6\times10^{-8}\,\Omega\cdot\text{m}}\right) = \boxed{70}$$

where we have used resistivity values from Table 20.1 and the fact that the two sections have the same cross-sectional areas.

---

25. ***REASONING*** According to Equation 6.10b, the energy used is Energy = Pt, where P is the power and t is the time. According to Equation 20.6a, the power is $P = IV$, where I is the current and V is the voltage. Thus, Energy = IVt, and we apply this result first to the dryer and then to the computer.

SOLUTION The energy used by the dryer is

$$\text{Energy} = Pt = IVt = (16\text{ A})(240\text{ V})(45\text{ min})\underbrace{\left(\frac{60\text{ s}}{1.00\text{ min}}\right)}_{\substack{\text{Converts minutes}\\ \text{to seconds}}} = 1.04\times10^7\text{ J}$$

For the computer, we have

$$\text{Energy} = 1.04\times10^7\text{ J} = IVt = (2.7\text{ A})(120\text{ V})t$$

Solving for t we find

$$t = \frac{1.04\times10^7\text{ J}}{(2.7\text{ A})(120\text{ V})} = 3.21\times10^4\text{ s} = \left(3.21\times10^4\text{ s}\right)\left(\frac{1.00\text{ h}}{3600\text{ s}}\right) = \boxed{8.9\text{ h}}$$

---

27. ***REASONING AND SOLUTION*** As a function of temperature, the resistance of the wire is given by Equation 20.5: $R = R_0\left[1+\alpha(T-T_0)\right]$, where α is the temperature coefficient of resistivity. From Equation 20.6c, we have $P = V^2/R$. Combining these two equations, we have

$$P = \frac{V^2}{R_0\left[1+\alpha\left(T-T_0\right)\right]} = \frac{P_0}{1+\alpha\left(T-T_0\right)}$$

where $P_0 = V^2/R_0$, since the voltage is constant. But $P = \frac{1}{2}P_0$, so we find

$$\frac{P_0}{2} = \frac{P_0}{1+\alpha\left(T-T_0\right)} \qquad \text{or} \qquad 2 = 1+\alpha\left(T-T_0\right)$$

Solving for T, we find

$$T = \frac{1}{\alpha} + T_0 = \frac{1}{0.0045\ (\text{C}°)^{-1}} + 28° = \boxed{250\ °\text{C}}$$

---

33. ***REASONING*** The average power is given by Equation 20.15c as $\overline{P} = V_{\text{rms}}^2 / R$. In this expression the rms voltage V_{rms} appears. However, we seek the peak voltage V_0. The relation between the two types of voltage is given by Equation 20.13 as $V_{\text{rms}} = V_0 / \sqrt{2}$, so we can obtain the peak voltage by using Equation 20.13 to substitute into Equation 20.15c.

SOLUTION Substituting V_{rms} from Equation 20.13 into Equation 20.15c gives

$$\overline{P} = \frac{V_{\text{rms}}^2}{R} = \frac{\left(V_0 / \sqrt{2}\right)^2}{R} = \frac{V_0^2}{2R}$$

Solving for the peak voltage V_0 gives

$$V_0 = \sqrt{2R\overline{P}} = \sqrt{2(4.0\ \Omega)(55\ \text{W})} = \boxed{21\ \text{V}}$$

---

37. ***REASONING*** According to Equation 6.10b, the energy supplied to the water in the form of heat is $Q = Pt$, where P is the power and t is the time. According to Equation 20.6c, the power is $P = V^2/R$, so that $Q = (V^2/R)t$, where V is the voltage and R is the resistance. But for a mass m of water with a specific heat capacity c, the heat Q is related to the change in temperature ΔT by $Q = mc\Delta T$ according to Equation 12.4. Our solution uses these two expressions for Q.

SOLUTION Combining the two expressions for Q, we have $Q = mc\Delta T = V^2 t/R$. Solving for t, we have

$$t = \frac{Rcm\Delta T}{V^2} \tag{1}$$

The mass of the water can be determined from its volume V_{water}. From Equation 11.1, we have $m = \rho_{\text{water}} V_{\text{water}}$, where $\rho_{\text{water}} = 1.000 \times 10^3\ \text{kg/m}^3$ is the mass density of water (see Table 11.1). Therefore, Equation (1) becomes

$$t = \frac{Rc\rho_{\text{water}} V_{\text{water}} \Delta T}{V^2} \tag{2}$$

The value (see Table 12.2) for the specific heat of water is 4186 J/(kg•C°). The volume of water is

$$V_{\text{water}} = (52\text{ gal})\left(\frac{3.79\times10^{-3}\text{ m}^3}{1.00\text{ gal}}\right) = 0.197\text{ m}^3$$

Substituting values into Equation (2), we obtain

$$t = \frac{(3.0\ \Omega)[4186\text{ J/(kg}\cdot\text{C°)}](1.000\times10^3\text{ kg/m}^3)(0.197\text{ m}^3)(53\text{ °C}-11\text{ °C})}{(120\text{ V})^2} = 7220\text{ s}$$

$$= (7220\text{ s})\left(\frac{1.00\text{ h}}{3600\text{ s}}\right) = \boxed{2.0\text{ h}}$$

---

39. ***REASONING*** Using Ohm's law (Equation 20.2) we can write an expression for the voltage across the original circuit as $V = I_0R_0$. When the additional resistor R is inserted in series, assuming that the battery remains the same, the voltage across the new combination is given by $V = I(R+R_0)$. Since V is the same in both cases, we can write $I_0R_0 = I(R+R_0)$. This expression can be solved for R_0.

SOLUTION Solving for R_0, we have

$$I_0R_0 - IR_0 = IR \quad \text{or} \quad R_0(I_0 - I) = IR$$

Therefore, we find that

$$R_0 = \frac{IR}{I_0 - I} = \frac{(12.0\text{ A})(8.00\ \Omega)}{15.0\text{ A}-12.0\text{ A}} = \boxed{32\ \Omega}$$

---

43. ***REASONING*** The equivalent series resistance R_s is the sum of the resistances of the three resistors. The potential difference V can be determined from Ohm's law as $V = IR_s$.

SOLUTION

a. The equivalent resistance is

$$R_s = 25\ \Omega + 45\ \Omega + 75\ \Omega = \boxed{145\ \Omega}$$

b. The potential difference across the three resistors is

$$V = IR_s = (0.51\text{ A})(145\ \Omega) = \boxed{74\text{ V}}$$

---

45. ***REASONING*** The resistance of one of the wires in the extension cord is given by Equation 20.3: $R = \rho L / A$, where the resistivity of copper is $\rho = 1.72 \times 10^{-8}\ \Omega \cdot \text{m}$, according to Table 20.1. Since the two wires in the cord are in series with each other, their total resistance is $R_{\text{cord}} = R_{\text{wire 1}} + R_{\text{wire 2}} = 2\rho L / A$. Once we find the equivalent resistance of the entire circuit (extension cord + trimmer), Ohm's law can be used to find the voltage applied to the trimmer.

SOLUTION

a. The resistance of the extension cord is

$$R_{\text{cord}} = \frac{2\rho L}{A} = \frac{2(1.72\times10^{-8}\ \Omega\cdot\text{m})(46\ \text{m})}{1.3\times10^{-6}\ \text{m}^2} = \boxed{1.2\ \Omega}$$

b. The total resistance of the circuit (cord + trimmer) is, since the two are in series,

$$R_s = 1.2\ \Omega + 15.0\ \Omega = 16.2\ \Omega$$

Therefore from Ohm's law (Equation 20.2: $V = IR$), the current in the circuit is

$$I = \frac{V}{R_s} = \frac{120\ \text{V}}{16.2\ \Omega} = 7.4\ \text{A}$$

Finally, the voltage applied to the trimmer alone is (again using Ohm's law),

$$V_{\text{trimmer}} = (7.4\ \text{A})(15.0\ \Omega) = \boxed{110\ \text{V}}$$

---

49. ***REASONING AND SOLUTION*** Since the circuit elements are in parallel, the equivalent resistance can be obtained directly from Equation 20.17:

$$\frac{1}{R_p} = \frac{1}{R_1} + \frac{1}{R_2} = \frac{1}{16\ \Omega} + \frac{1}{8.0\ \Omega} \qquad \text{or} \qquad \boxed{R_p = 5.3\ \Omega}$$

---

53. ***REASONING*** Since the resistors are connected in parallel, the voltage across each one is the same and can be calculated from Ohm's Law (Equation 20.2: $V = IR$). Once the voltage across each resistor is known, Ohm's law can again be used to find the current in the second resistor. The total power consumed by the parallel combination can be found calculating the power consumed by each resistor from Equation 20.6b: $P = I^2R$. Then, the total power consumed is the sum of the power consumed by each resistor.

SOLUTION Using data for the second resistor, the voltage across the resistors is equal to

$$V = IR = (3.00\text{ A})(64.0\ \Omega) = 192\text{ V}$$

a. The current through the 42.0-Ω resistor is

$$I = \frac{V}{R} = \frac{192\text{ V}}{42.0\ \Omega} = \boxed{4.57\text{ A}}$$

b. The power consumed by the 42.0-Ω resistor is

$$P = I^2 R = (4.57\text{ A})^2 (42.0\ \Omega) = 877\text{ W}$$

while the power consumed by the 64.0-Ω resistor is

$$P = I^2 R = (3.00\text{ A})^2 (64.0\ \Omega) = 576\text{ W}$$

Therefore the total power consumed by the two resistors is 877 W + 576 W = $\boxed{1450\text{ W}}$.

---

55. ***REASONING*** The equivalent resistance of the three devices in parallel is R_p, and we can find the value of R_p by using our knowledge of the total power consumption of the circuit; the value of R_p can be found from Equation 20.6c, $P = V^2 / R_p$. Ohm's law (Equation 20.2, $V = IR$) can then be used to find the current through the circuit.

SOLUTION
a. The total power used by the circuit is $P = 1650\text{ W} + 1090\text{ W} + 1250\text{ W} = 3990\text{ W}$. The equivalent resistance of the circuit is

$$R_p = \frac{V^2}{P} = \frac{(120\text{ V})^2}{3990\text{ W}} = \boxed{3.6\ \Omega}$$

b. The total current through the circuit is

$$I = \frac{V}{R_p} = \frac{120\text{ V}}{3.6\ \Omega} = \boxed{33\text{ A}}$$

This current is larger than the rating of the circuit breaker; therefore, the $\boxed{\text{breaker will open}}$.

---

59. ***REASONING*** To find the current, we will use Ohm's law, together with the proper equivalent resistance. The coffee maker and frying pan are in series, so their equivalent resistance is given by Equation 20.16 as $R_{\text{coffee}} + R_{\text{pan}}$. This total resistance is in parallel with the resistance of the bread maker, so the equivalent resistance of the parallel combination can be obtained from Equation 20.17 as $R_{\text{p}}^{-1} = (R_{\text{coffee}} + R_{\text{pan}})^{-1} + R_{\text{bread}}^{-1}$.

SOLUTION Using Ohm's law and the expression developed above for R_{p}^{-1}, we find

$$I = \frac{V}{R_{\text{p}}} = V\left(\frac{1}{R_{\text{coffee}} + R_{\text{pan}}} + \frac{1}{R_{\text{bread}}}\right) = (120\text{ V})\left(\frac{1}{14\ \Omega + 16\ \Omega} + \frac{1}{23\ \Omega}\right) = \boxed{9.2\text{ A}}$$

---

61. ***REASONING*** When two or more resistors are in series, the equivalent resistance is given by Equation 20.16: $R_{\text{s}} = R_1 + R_2 + R_3 + \ldots$. Likewise, when resistors are in parallel, the expression to be solved to find the equivalent resistance is given by Equation 20.17: $\frac{1}{R_{\text{p}}} = \frac{1}{R_1} + \frac{1}{R_2} + \frac{1}{R_3} + \ldots$. We will successively apply these to the individual resistors in the figure in the text beginning with the resistors on the right side of the figure.

SOLUTION Since the 4.0-Ω and the 6.0-Ω resistors are in series, the equivalent resistance of the combination of those two resistors is 10.0 Ω. The 9.0-Ω and 8.0-Ω resistors are in parallel; their equivalent resistance is 4.24 Ω. The equivalent resistances of the parallel combination (9.0 Ω and 8.0 Ω) and the series combination (4.0 Ω and the 6.0 Ω) are in parallel; therefore, their equivalent resistance is 2.98 Ω. The 2.98-Ω combination is in series with the 3.0-Ω resistor, so that equivalent resistance is 5.98 Ω. Finally, the 5.98-Ω combination and the 20.0-Ω resistor are in parallel, so the equivalent resistance between the points A and B is $\boxed{4.6\ \Omega}$.

---

65. ***REASONING*** Since we know that the current in the 8.00-Ω resistor is 0.500 A, we can use Ohm's law ($V = IR$) to find the voltage across the 8.00-Ω resistor. The 8.00-Ω resistor and the 16.0-Ω resistor are in parallel; therefore, the voltages across them are equal. Thus, we can also use Ohm's law to find the current through the 16.0-Ω resistor. The currents that flow through the 8.00-Ω and the 16.0-Ω resistors combine to give the total current that flows through the 20.0-Ω resistor. Similar reasoning can be used to find the current through the 9.00-Ω resistor.

SOLUTION

a. The voltage across the 8.00-Ω resistor is $V_8 = (0.500\text{ A})(8.00\ \Omega) = 4.00\text{ V}$. Since this is also the voltage that is across the 16.0-Ω resistor, we find that the current through the

16.0-Ω resistor is $I_{16} = (4.00\text{ V})/(16.0\,\Omega) = 0.250\text{ A}$. Therefore, the total current that flows through the 20.0-Ω resistor is

$$I_{20} = 0.500\text{ A} + 0.250\text{ A} = \boxed{0.750\text{ A}}$$

b. The 8.00-Ω and the 16.0-Ω resistors are in parallel, so their equivalent resistance can be obtained from Equation 20.17, $\frac{1}{R_p} = \frac{1}{R_1} + \frac{1}{R_2} + \frac{1}{R_3} + \ldots$, and is equal to 5.33 Ω. Therefore, the equivalent resistance of the upper branch of the circuit is $R_{\text{upper}} = 5.33\,\Omega + 20.0\,\Omega = 25.3\,\Omega$, since the 5.33-Ω resistance is in series with the 20.0-Ω resistance. Using Ohm's law, we find that the voltage across the upper branch must be $V = (0.750\text{ A})(25.3\,\Omega) = 19.0\text{ V}$. Since the lower branch is in parallel with the upper branch, the voltage across both branches must be the same. Therefore, the current through the 9.00-Ω resistor is, from Ohm's law,

$$I_9 = \frac{V_{\text{lower}}}{R_9} = \frac{19.0\text{ V}}{9.00\,\Omega} = \boxed{2.11\text{ A}}$$

---

67. ***REASONING*** The terminal voltage of the battery is given by $V_{\text{terminal}} = \text{Emf} - Ir$, where r is the internal resistance of the battery. Since the terminal voltage is observed to be one-half of the emf of the battery, we have $V_{\text{terminal}} = \text{Emf}/2$ and $I = \text{Emf}/(2r)$. From Ohm's law, the equivalent resistance of the circuit is $R = \text{emf}/I = 2r$. We can also find the equivalent resistance of the circuit by considering that the identical bulbs are in parallel across the battery terminals, so that the equivalent resistance of the N bulbs is found from

$$\frac{1}{R_p} = \frac{N}{R_{\text{bulb}}} \qquad \text{or} \qquad R_p = \frac{R_{\text{bulb}}}{N}$$

This equivalent resistance is in series with the battery, so we find that the equivalent resistance of the circuit is

$$R = 2r = \frac{R_{\text{bulb}}}{N} + r$$

This expression can be solved for N.

SOLUTION Solving the above expression for N, we have

$$N = \frac{R_{\text{bulb}}}{2r - r} = \frac{R_{\text{bulb}}}{r} = \frac{15\,\Omega}{0.50\,\Omega} = \boxed{30}$$

---

75. ***REASONING*** The current I can be found by using Kirchhoff's loop rule. Once the current is known, the voltage between points A and B can be determined.

SOLUTION

a. We assume that the current is directed clockwise around the circuit. Starting at the upper-left corner and going clockwise around the circuit, we set the potential drops equal to the potential rises:

$$\underbrace{(5.0\ \Omega)I + (27\ \Omega)I + 10.0\ \text{V} + (12\ \Omega)I + (8.0\ \Omega)I}_{\text{Potential drops}} = \underbrace{30.0\ \text{V}}_{\text{Potential rises}}$$

Solving for the current gives $\boxed{I = 0.38\ \text{A}}$.

b. The voltage between points A and B is

$$V_{AB} = 30.0\ \text{V} - (0.38\ \text{A})(27\ \Omega) = \boxed{2.0 \times 10^1\ \text{V}}$$

c. $\boxed{\text{Point B}}$ is at the higher potential.

---

77. ***REASONING*** We begin by labeling the currents in the three resistors. The drawing below shows the directions chosen for these currents. The directions are arbitrary, and if any of them is incorrect, then the analysis will show that the corresponding value for the current is negative.

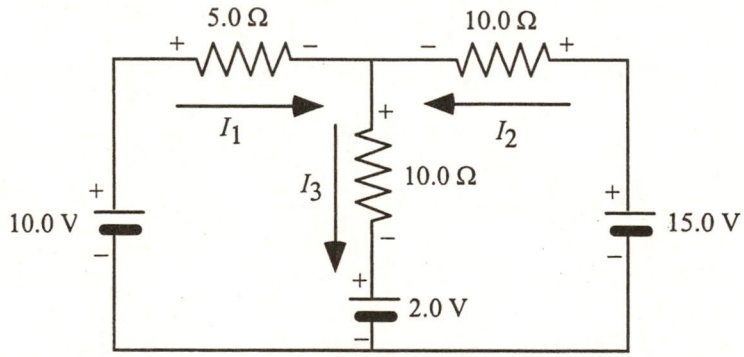

We then mark the resistors with the plus and minus signs that serve as an aid in identifying the potential drops and rises for the loop rule, recalling that conventional current is always directed from a higher potential (+) toward a lower potential (–). Thus, given the directions chosen for I_1, I_2, and I_3, the plus and minus signs *must* be those shown in the drawing. We can now use Kirchhoff's rules to find the voltage across the 5.0-Ω resistor.

SOLUTION Applying the loop rule to the left loop (and suppressing units for onvenience) gives

$$5.0I_1 + 10.0I_3 + 2.0 = 10.0 \qquad (1)$$

Similarly, for the right loop,

$$10.0I_2 + 10.0I_3 + 2.0 = 15.0 \qquad (2)$$

If we apply the junction rule to the upper junction, we obtain

$$I_1 + I_2 = I_3 \qquad (3)$$

Subtracting Equation (2) from Equation (1) gives

$$5.0I_1 - 10.0I_2 = -5.0 \qquad (4)$$

We now multiply Equation (3) by 10 and add the result to Equation (2); the result is

$$10.0I_1 + 20.0I_2 = 13.0 \qquad (5)$$

If we then multiply Equation (4) by 2 and add the result to Equation (5), we obtain $20.0I_1 = 3.0$, or solving for I_1, we obtain $I_1 = 0.15\ \text{A}$. The fact that I_1 is positive means that the current in the drawing has the correct direction. The voltage across the 5.0-Ω resistor can be found from Ohm's law:

$$V = (0.15\ \text{A})(5.0\ \Omega) = \boxed{0.75\ \text{V}}$$

Current flows from the higher potential to the lower potential, and the current through the 5.0-Ω flows from left to right, so the $\boxed{\text{left end of the resistor}}$ is at the higher potential.

---

81. ***REASONING*** Since only 0.100 mA out of the available 60.0 mA is needed to cause a full-scale deflection of the galvanometer, the shunt resistor must allow the excess current of 59.9 mA to detour around the meter coil, as the drawing at the right indicates. The value for the shunt resistance can be obtained by recognizing that the 50.0-Ω coil resistance and the shunt resistance are in parallel, both being connected between points A and B in the drawing. Thus, the voltage across each resistance is the same.

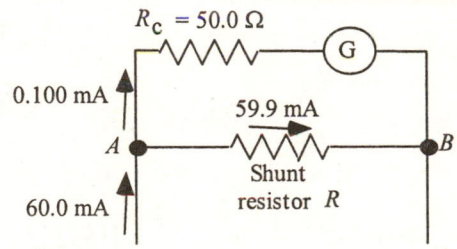

SOLUTION Expressing voltage as the product of current and resistance, we find that

$$\underbrace{\left(59.9\times10^{-3}\ \text{A}\right)(R)}_{\text{Voltage across shunt resistance}} = \underbrace{\left(0.100\times10^{-3}\ \text{A}\right)(50.0\ \Omega)}_{\text{Voltage across coil resistance}}$$

$$R = \frac{\left(0.100\times10^{-3}\ \text{A}\right)(50.0\ \Omega)}{59.9\times10^{-3}\ \text{A}} = \boxed{0.0835\ \Omega}$$

85. ***REASONING AND SOLUTION***
a. According to Ohm's law (Equation 20.2, $V = IR$) the current in the circuit is

$$I = \frac{V}{R+R} = \frac{V}{2R}$$

The voltage across either resistor is IR, so that we find

$$IR = \left(\frac{V}{2R}\right)R = \frac{V}{2} = \frac{60.0\text{ V}}{2} = \boxed{30.0\text{ V}}$$

b. The voltmeter's resistance is $R_v = V/I = (60.0\text{ V})/(5.00\times10^{-3}\text{ A}) = 12.0\times10^{3}\ \Omega$, and this resistance is in parallel with the resistance $R = 1550\ \Omega$. The equivalent resistance of this parallel combination can be obtained as follows:

$$\frac{1}{R_p} = \frac{1}{R_v} + \frac{1}{R} \qquad \text{or} \qquad R_p = \frac{RR_v}{R+R_v} = \frac{(1550\ \Omega)(12.0\times10^3\ \Omega)}{1550\ \Omega + 12.0\times10^3\ \Omega} = 1370\ \Omega$$

The voltage registered by the voltmeter is IR_p, where I is the current supplied by the battery to the series combination of the other 1550-Ω resistor and R_p. According to Ohm's law

$$I = \frac{60.0\text{ V}}{1550\ \Omega + 1370\ \Omega} = 0.0205\text{ A}$$

Thus, the voltage registered by the voltmeter is

$$IR_p = (0.0205\text{ A})(1370\ \Omega) = \boxed{28.1\text{ V}}$$

89. ***REASONING AND SOLUTION*** Let C_0 be the capacitance of an empty capacitor. Then the capacitances are as follows, according to the discussion following Equation 19.10:

$$C_1 = 3.00C_0 \quad \text{and} \quad C_2 = 5.00C_0$$

The series capacitance of the two is

$$\frac{1}{C_s} = \frac{1}{2.50C_0} + \frac{1}{4.00C_0} \qquad \text{or} \qquad C_s = 1.54C_0$$

Now

$$\kappa C_0 = 1.54 C_0 \qquad \text{or} \qquad \kappa = \boxed{1.54}$$

---

93. ***REASONING*** When two or more capacitors are in series, the equivalent capacitance of the combination can be obtained from Equation 20.19, $\frac{1}{C_s} = \frac{1}{C_1} + \frac{1}{C_2} + \frac{1}{C_3} \ldots$. Equation 20.18 gives the equivalent capacitance for two or more capacitors in parallel: $C_p = C_1 + C_2 + C_3 + \ldots$. The energy stored in a capacitor is given by $\frac{1}{2}CV^2$, according to Equation 19.11. Thus, the energy stored in the series combination is $\frac{1}{2}C_s V_s^2$, where

$$\frac{1}{C_s} = \frac{1}{7.0\ \mu\text{F}} + \frac{1}{3.0\ \mu\text{F}} = 0.476\,(\mu\text{F})^{-1} \qquad \text{or} \qquad C_s = \frac{1}{0.476\,(\mu\text{F})^{-1}} = 2.10\ \mu\text{F}$$

Similarly, the energy stored in the parallel combination is $\frac{1}{2}C_p V_p^2$ where

$$C_p = 7.0\ \mu\text{F} + 3.0\ \mu\text{F} = 10.0\ \mu\text{F}$$

The voltage required to charge the parallel combination of the two capacitors to the same total energy as the series combination can be found by equating the two energy expressions and solving for V_p.

SOLUTION Equating the two expressions for the energy, we have

$$\tfrac{1}{2}C_s V_s^2 = \tfrac{1}{2}C_p V_p^2$$

Solving for V_p, we obtain the result

$$V_p = V_s\sqrt{\frac{C_s}{C_p}} = (24\ \text{V})\sqrt{\frac{2.10\ \mu\text{F}}{10.0\ \mu\text{F}}} = \boxed{11\text{V}}$$

---

95. ***REASONING*** The charge q on a discharging capacitor in a RC circuit is given by Equation 20.22: $q = q_0 e^{-t/RC}$, where q_0 is the original charge at time $t = 0$ s. Once t (time for one pulse) and the ratio q/q_0 are known, this expression can be solved for C.

SOLUTION Since the pacemaker delivers 81 pulses per minute, the time for one pulse is

$$\frac{1\ \text{min}}{81\ \text{pulses}} \times \frac{60.0\ \text{s}}{1.00\ \text{min)}} = 0.74\ \text{s/pulse}$$

Since one pulse is delivered every time the fully-charged capacitor loses 63.2% of its original charge, the charge remaining is 36.8% of the original charge. Thus, we have $q = (0.368)q_0$, or $q/q_0 = 0.368$.

From Equation 20.22, we have

$$\frac{q}{q_0} = e^{-t/RC}$$

Taking the natural logarithm of both sides, we have,

$$\ln\left(\frac{q}{q_0}\right) = -\frac{t}{RC}$$

Solving for C, we find

$$C = \frac{-t}{R\ln(q/q_0)} = \frac{-(0.74\text{ s})}{(1.8\times10^6\ \Omega)\ln(0.368)} = \boxed{4.1\times10^{-7}\text{ F}}$$

---

101. ***REASONING AND SOLUTION*** The equivalent capacitance of the combination can be obtained immediately from Equation 20.19, $\frac{1}{C_s} = \frac{1}{C_1} + \frac{1}{C_2} + \frac{1}{C_3} \ldots$.

$$\frac{1}{C_s} = \frac{1}{3.0\ \mu\text{F}} + \frac{1}{7.0\ \mu\text{F}} + \frac{1}{9.0\ \mu\text{F}} = 0.587\,(\mu\text{F})^{-1} \quad \text{or} \quad C_s = \frac{1}{0.587\,(\mu\text{F})^{-1}} = \boxed{1.7\ \mu\text{F}}$$

---

102. ***REASONING*** Electric current is the amount of charge flowing per unit time (see Equation 20.1). Thus, the amount of charge is the current times the time. Furthermore, the potential difference is the difference in electric potential energy per unit charge (see Equation 19.4), so that, once the amount of charge has been calculated, we can determine the energy by multiplying the potential difference by the charge.

SOLUTION

a. According to Equation 20.1, the current I is the amount of charge Δq divided by the time Δt, or $I = \Delta q/\Delta t$. Therefore, the additional charge that passes through the machine in normal mode versus standby mode is

$$q_{\text{additional}} = \underbrace{I_{\text{normal}}\Delta t}_{\Delta q \text{ in normal mode}} - \underbrace{I_{\text{standby}}\Delta t}_{\Delta q \text{ in standby mode}} = \left(I_{\text{normal}} - I_{\text{standby}}\right)\Delta t$$

$$= (0.110\text{ A} - 0.067\text{ A})(60.0\text{ s}) = \boxed{2.6\text{ C}}$$

b. According to Equation 19.4, the potential difference ΔV is the difference Δ(EPE) in the electric potential energy divided by the charge $q_{\text{additional}}$, or $\Delta V = \Delta(\text{EPE})/q_{\text{additional}}$. As a result, the additional energy used in the normal mode as compared to the standby mode is

$$\Delta(\text{EPE}) = q_{\text{additional}}\Delta V = (2.6\text{ C})(120\text{ V}) = \boxed{310\text{ J}}$$

---

103. ***REASONING AND SOLUTION*** Ohm's law (Equation 20.2), $V = IR$, gives the result directly:

$$R = \frac{V}{I} = \frac{9.0\text{ V}}{0.11\text{ A}} = \boxed{82\ \Omega}$$

---

109. ***REASONING*** When two or more resistors are in series, the equivalent resistance is given by Equation 20.16: $R_s = R_1 + R_2 + R_3 + \ldots$. When resistors are in parallel, the expression to be solved to find the equivalent resistance is Equation 20.17: $\frac{1}{R_p} = \frac{1}{R_1} + \frac{1}{R_2} + \frac{1}{R_3} + \ldots$.

We will use these relations to determine the eight different values of resistance that can be obtained by connecting together the three resistors: 1.00, 2.00, and 3.00 Ω.

SOLUTION When all the resistors are connected in series, the equivalent resistance is the sum of all three resistors and the equivalent resistance is $\boxed{6.00\ \Omega}$. When all three are in parallel, we have from Equation 20.17, the equivalent resistance is $\boxed{0.545\ \Omega}$.

We can also connect two of the resistors in parallel and connect the parallel combination in series with the third resistor. When the 1.00 and 2.00-Ω resistors are connected in parallel and the 3.00-Ω resistor is connected in series with the parallel combination, the equivalent resistance is $\boxed{3.67\ \Omega}$. When the 1.00 and 3.00-Ω resistors are connected in parallel and the 2.00-Ω resistor is connected in series with the parallel combination, the equivalent resistance is $\boxed{2.75\ \Omega}$. When the 2.00 and 3.00-Ω resistors are connected in parallel and the 1.00-Ω resistor is connected in series with the parallel combination, the equivalent resistance is $\boxed{2.20\ \Omega}$.

We can also connect two of the resistors in series and put the third resistor in parallel with the series combination. When the 1.00 and 2.00-Ω resistors are connected in series and the 3.00-Ω resistor is connected in parallel with the series combination, the equivalent resistance is $\boxed{1.50\ \Omega}$. When the 1.00 and 3.00-Ω resistors are connected in series and the 2.00-Ω resistor is connected in parallel with the series combination, the equivalent resistance is $\boxed{1.33\ \Omega}$. Finally, when the 2.00 and 3.00-Ω resistors are connected in series and the 1.00-Ω resistor is connected in parallel with the series combination, the equivalent resistance is $\boxed{0.833\ \Omega}$.

---

113. ***REASONING*** The resistance of a metal wire of length L, cross-sectional area A and resistivity ρ is given by Equation 20.3: $R = \rho L / A$. The volume V_2 of the new wire will be the same as the original volume V_1 of the wire, where volume is the product of length and cross-sectional area. Thus, $V_1 = V_2$ or $A_1 L_1 = A_2 L_2$. Since the new wire is three times longer than the first wire, we can write

$$A_1 L_1 = A_2 L_2 = A_2(3L_1) \quad \text{or} \quad A_2 = A_1/3$$

We can form the ratio of the resistances, use this expression for the area A_2, and find the new resistance.

SOLUTION The resistance of the new wire is determined as follows:

$$\frac{R_2}{R_1} = \frac{\rho L_2 / A_2}{\rho L_1 / A_1} = \frac{L_2 A_1}{L_1 A_2} = \frac{(3L_1)A_1}{L_1(A_1/3)} = 9$$

Solving for R_2, we find that

$$R_2 = 9R_1 = 9(21.0\ \Omega) = \boxed{189\ \Omega}$$

---

115. ***REASONING*** The foil effectively converts the capacitor into two capacitors in series. Equation 19.10 gives the expression for the capacitance of a capacitor of plate area A and plate separation d (no dielectric): $C_0 = \varepsilon_0 A / d$. We can use this expression to determine the capacitance of the individual capacitors created by the presence of the foil. Then using the fact that the "two capacitors" are in series, we can use Equation 20.19 to find the equivalent capacitance of the system.

SOLUTION Since the foil is placed one-third of the way from one plate of the original capacitor to the other, we have $d_1 = (2/3)d$, and $d_2 = (1/3)d$. Then

$$C_1 = \frac{\varepsilon_0 A}{(2/3)d} = \frac{3\varepsilon_0 A}{2d}$$

and

$$C_2 = \frac{\varepsilon_0 A}{(1/3)d} = \frac{3\varepsilon_0 A}{d}$$

Since these two capacitors are effectively in series, it follows that

$$\frac{1}{C_s}=\frac{1}{C_1}+\frac{1}{C_2}=\frac{1}{3\varepsilon_0 A/(2d)}+\frac{1}{3\varepsilon_0 A/d}=\frac{3d}{3\varepsilon_0 A}=\frac{d}{\varepsilon_0 A}$$

But $C_0=\varepsilon_0 A/d$, so that $d/(\varepsilon_0 A)=1/C_0$, and we have

$$\frac{1}{C_s}=\frac{1}{C_0} \qquad \text{or} \qquad \boxed{C_s=C_0}$$

CHAPTER 21 | *MAGNETIC FORCES AND MAGNETIC FIELDS*

CONCEPTUAL QUESTIONS

---

3. ***REASONING AND SOLUTION*** A charged particle, passing through a certain region of space, has a velocity whose magnitude and direction remain constant.

a. If it is known that the external magnetic field is zero everywhere in the region, we can conclude that the electric field is also zero. Any charged particle placed in an electric field will experience a force given by $\mathbf{F} = q\mathbf{E}$, where q is the charge and $\mathbf{E}$ is the electric field. If the magnitude and direction of the velocity of the particle are constant, then the particle has zero acceleration. From Newton's second law, we know that the net force on the particle is zero. But there is no magnetic field and, hence, no magnetic force. Therefore, the net force is the electric force. Since the electric force is zero, the electric field must be zero.

b. If it is known that the external electric field is zero everywhere, we *cannot* conclude that the external magnetic field is also zero. In order for a moving charged particle to experience a magnetic force when it is placed in a magnetic field, the velocity of the moving charge must have a component that is perpendicular to the direction of the magnetic field. If the moving charged particle enters the region such that its velocity is parallel or antiparallel to the magnetic field, it will experience no magnetic force, even though a magnetic field is present. In the absence of an external electric field, there is no electric force either. Thus, there is no net force, and the velocity vector will not change in any way.

---

10. ***REASONING AND SOLUTION*** When the particle is launched in the x, y plane, its initial velocity will be perpendicular to the magnetic field; therefore, the particle will travel on a circular path in the x, y plane. In order for the charged particle to hit the target, the target must lie on the circular path of the moving particle. This will occur if the particle moves in a counterclockwise circle that passes through the third quadrant of the coordinate system. RHR-1 can be used to determine possible trajectories in the following way. Place the fingers of the right hand into the page (direction of the magnetic field) and orient the thumb along one of the coordinate axes (direction of the particle's initial velocity). The palm of the right hand will face the direction in which the force on the charged particle is directed. Since the particle travels on a circle, the direction of the force will point toward the center of the particle's trajectory. The figures below show the results for all four possible cases.

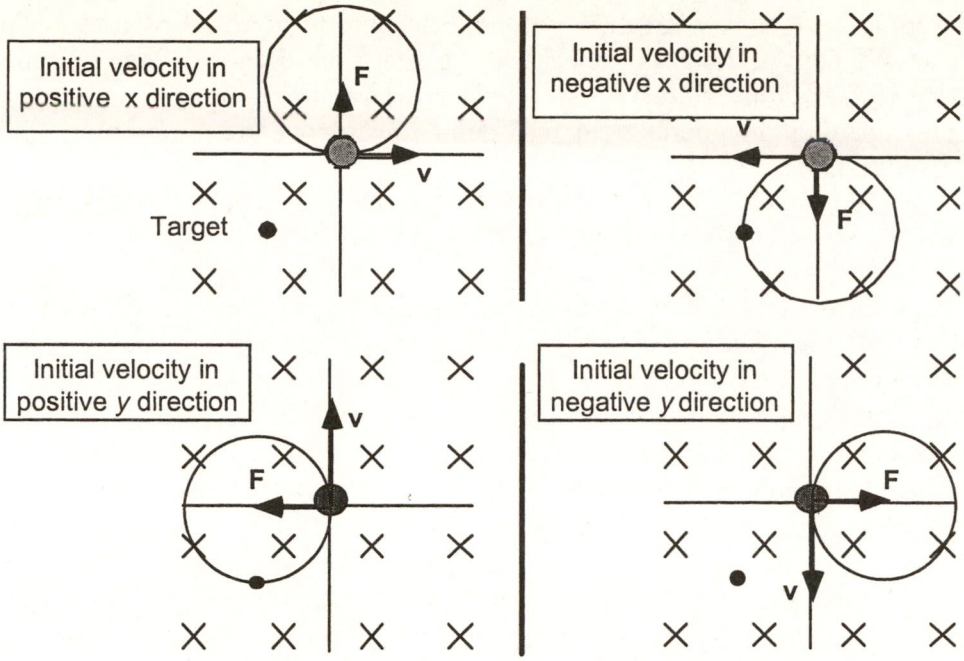

Clearly, the charged particle can hit the target only if the initial velocity of the particle points either in the negative x direction or the positive y direction.

---

18. ***REASONING AND SOLUTION*** The total magnetic field at the point P is the resultant of the magnetic field at P due to each individual wire. If the current in all four wires is directed into the page, then from RHR-2, the magnetic field at P due to the current in wire 1 must point toward wire 3, the magnetic field at P due to the current in wire 2 must point toward wire 1, the magnetic field at P due to the current in wire 3 must point toward wire 4, and the magnetic field at P due to the current in wire 4 must point toward wire 2, as shown at the right.

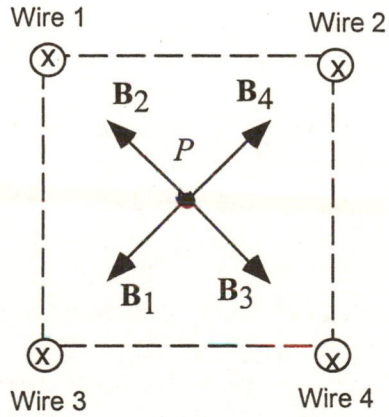

The magnetic field at a distance r from a long straight wire that carries a current I is $B = \mu_0 I/(2\pi r)$. Since the current in all four wires has the same magnitude and all four wires are equidistant from the point P, each wire gives rise to a magnetic field at P of the same magnitude. When current is flowing through all four wires, the total magnetic field at P is zero. If the current in any single wire is turned off, the total magnetic field will point toward one of the corners. For example, if the current in wire 1 is turned off, the resultant of $\mathbf{B}_2$ and $\mathbf{B}_3$ is still zero and the total magnetic field is $\mathbf{B}_4$. If the current in wire 2 is turned off, then the total magnetic field is $\mathbf{B}_3$, and so on.

We could have achieved similar results if the current in all four wires was directed out of the page. If fact, as long as opposite wires along the diagonal have currents in the same direction (for example, wires 1 and 4 with outward currents and wires 2 and 3 with inward currents) the total magnetic field will point toward one of the corners when one of the currents is turned off.

CHAPTER 21 | *MAGNETIC FORCES AND MAGNETIC FIELDS*

PROBLEMS

---

1. ***REASONING*** According to Equation 21.1, the magnitude of the magnetic force on a moving charge is $F = |q_0| vB \sin\theta$. Since the magnetic field points due north and the proton moves eastward, $\theta = 90.0°$. Furthermore, since the magnetic force on the moving proton balances its weight, we have $mg = |q_0| vB \sin\theta$, where m is the mass of the proton. This expression can be solved for the speed v.

SOLUTION Solving for the speed v, we have

$$v = \frac{mg}{|q_0| B \sin\theta} = \frac{(1.67\times10^{-27}\ \text{kg})(9.80\ \text{m/s}^2)}{(1.6\times10^{-19}\ \text{C})(2.5\times10^{-5}\ \text{T})\sin 90.0°} = \boxed{4.1\times10^{-3}\ \text{m/s}}$$

---

5. ***REASONING AND SOLUTION*** The magnitude of the force can be determined using Equation 21.1, $F = |q| vB \sin\theta$, where θ is the angle between the velocity and the magnetic field. The direction of the force is determined by using Right-Hand Rule No. 1.

a. $F = |q| vB \sin 30.0° = (8.4 \times 10^{-6}\ \text{C})(45\ \text{m/s})(0.30\ \text{T}) \sin 30.0° = \boxed{5.7 \times 10^{-5}\ \text{N}}$, directed $\boxed{\text{into the paper}}$.

b. $F = |q| vB \sin 90.0° = (8.4 \times 10^{-6}\ \text{C})(45\ \text{m/s})(0.30\ \text{T}) \sin 90.0° = \boxed{1.1 \times 10^{-4}\ \text{N}}$, directed $\boxed{\text{into the paper}}$.

c. $F = |q| vB \sin 150° = (8.4 \times 10^{-6}\ \text{C})(45\ \text{m/s})(0.30\ \text{T}) \sin 150° = \boxed{5.7 \times 10^{-5}\ \text{N}}$, directed $\boxed{\text{into the paper}}$.

---

9. ***REASONING*** The direction in which the electrons are deflected can be determined using Right-Hand Rule No. 1 and reversing the direction of the force (RHR-1 applies to positive charges, and electrons are negatively charged).

Each electron experiences an acceleration a given by Newton's second law of motion, $a = F/m$, where F is the net force and m is the mass of the electron. The only force acting on the electron is the magnetic force, $F = |q_0| vB \sin\theta$, so it is the net force. The speed v of the electron is related to its kinetic energy KE by the relation KE $= \frac{1}{2}mv^2$. Thus, we have enough information to find the acceleration.

SOLUTION

a. According to RHR-1, if you extend your right hand so that your fingers point along the direction of the magnetic field **B** and your thumb points in the direction of the velocity **v** of a *positive* charge, your palm will face in the direction of the force **F** on the positive charge.

For the electron in question, the fingers of the right hand should be oriented downward (direction of **B**) with the thumb pointing to the east (direction of **v**). The palm of the right hand points due north (the direction of **F** on a positive charge). Since the electron is negatively charged, it will be deflected $\boxed{\text{due south}}$.

b. The acceleration of an electron is given by Newton's second law, where the net force is the magnetic force. Thus,

$$a = \frac{F}{m} = \frac{|q_0| vB \sin\theta}{m}$$

Since the kinetic energy is KE $= \frac{1}{2}mv^2$, the speed of the electron is $v = \sqrt{2(\text{KE})/m}$. Thus, the acceleration of the electron is

$$a = \frac{|q_0| vB \sin\theta}{m} = \frac{|q_0| \sqrt{\frac{2(\text{KE})}{m}}\, B \sin\theta}{m}$$

$$= \frac{(1.60\times10^{-19}\ \text{C})\sqrt{\frac{2(2.40\times10^{-15}\ \text{J})}{9.11\times10^{-31}\ \text{kg}}}\,(2.00\times10^{-5}\ \text{T})\sin 90.0°}{9.11\times10^{-31}\ \text{kg}} = \boxed{2.55\times10^{14}\ \text{m/s}^2}$$

---

13. ***REASONING AND SOLUTION***

a. The speed of a proton can be found from Equation 21.2 $\left(r = \frac{mv}{|q|B}\right)$,

$$v = \frac{|q|Br}{m} = \frac{(1.6\times10^{-19}\ \text{C})(0.30\ \text{T})(0.25\ \text{m})}{1.67\times10^{-27}\ \text{kg}} = \boxed{7.2\times10^{6}\ \text{m/s}}$$

b. The magnitude F_c of the centripetal force is given by Equation 5.3,

$$F_c = \frac{mv^2}{r} = \frac{(1.67\times10^{-27}\text{ kg})(7.2\times10^{6}\text{ m/s})^2}{0.25\text{ m}} = \boxed{3.5\times10^{-13}\text{ N}}$$

---

15. ***REASONING AND SOLUTION*** The radius of curvature for a particle in a mass spectrometer is discussed in Section 21.4. According that discussion, the radius for a charged particle with a charge of $+e$ ($e = 1.60 \times 10^{-19}$ C) is $r = \sqrt{\frac{2mV}{eB^2}}$. In this problem, the charged particle has a charge of $+2e$, so that the radius becomes $r = \sqrt{\frac{2mV}{2eB^2}}$. Thus, the desired radius is

$$r = \sqrt{\frac{2mV}{2eB^2}} = \sqrt{\frac{2\left(3.27\times10^{-25}\text{ kg}\right)\left(1.00\times10^{3}\text{ V}\right)}{2\left(1.60\times10^{-19}\text{ C}\right)\left(0.500\text{ T}\right)^2}} = \boxed{0.0904\text{ m}}$$

---

19. ***REASONING AND SOLUTION*** According to Right-Hand Rule No. 1, the magnetic force on the positively charged particle is toward the bottom of the page in the drawing in the text. If the presence of the electric field is to double the magnitude of the net force on the charge, the electric field must also be $\boxed{\text{directed toward the bottom of the page}}$. Note that this results in the electric field being perpendicular to the magnetic field, even though the electric force and the magnetic force are in the same direction.

Furthermore, if the magnitude of the net force on the particle is twice the magnetic force, the electric force must be equal in magnitude to the magnetic force. In other words, combining Equations 18.2 and 21.1, we find $|q|E = |q|vB\sin\theta$, with $\sin\theta = \sin 90.0° = 1.0$. Then, solving for E

$$E = vB\sin\theta = (270\text{ m/s})(0.52\text{ T})(1.0) = \boxed{140\text{ V/m}}$$

---

23. ***REASONING*** The particle travels in a semicircular path of radius r, where r is given by Equation 21.2 $\left(r = \frac{mv}{|q|B}\right)$. The time spent by the particle in the magnetic field is given by $t = s/v$, where s is the distance traveled by the particle and v is its speed. The distance s is equal to one-half the circumference of a circle ($s = \pi r$).

SOLUTION We find that

$$t = \frac{s}{v} = \frac{\pi r}{v} = \frac{\pi}{v}\left(\frac{mv}{|q|B}\right) = \frac{\pi m}{|q|B} = \frac{\pi(6.0\times10^{-8}\text{ kg})}{(7.2\times10^{-6}\text{ C})(3.0\text{ T})} = \boxed{8.7\times10^{-3}\text{ s}}$$

27. ***REASONING AND SOLUTION*** The force on a current-carrying wire is given by Equation 21.3: $F = ILB\sin\theta$. Solving for the angle θ, we find that the angle between the wire and the magnetic field is

$$\theta = \sin^{-1}\left(\frac{F}{ILB}\right) = \sin^{-1}\left[\frac{5.46\text{ N}}{(21.0\text{ A})(0.655\text{ m})(0.470\text{ T})}\right] = \boxed{57.6°}$$

31. ***REASONING*** According to Equation 21.3, the magnetic force has a magnitude of $F = ILB\sin\theta$, where I is the current, B is the magnitude of the magnetic field, L is the length of the wire, and $\theta = 90°$ is the angle of the wire with respect to the field.

SOLUTION Using Equation 21.3, we find that

$$L = \frac{F}{IB\sin\theta} = \frac{7.1\times10^{-5}\text{ N}}{(0.66\text{ A})(4.7\times10^{-5}\text{ T})\sin 58°} = \boxed{2.7\text{ m}}$$

33. ***REASONING AND SOLUTION***

a. From Right-Hand Rule No. 1, if we extend the right hand so that the fingers point in the direction of the magnetic field, and the thumb points in the direction of the current, the palm of the hand faces the direction of the magnetic force on the current.

The springs will stretch when the magnetic force exerted on the copper rod is downward, toward the bottom of the page. Therefore, if you extend your right hand with your fingers pointing out of the page and the palm of your hand facing the bottom of the page, your thumb points left-to-right along the copper rod. Thus, the current flows $\boxed{\text{left-to-right}}$ in the copper rod.

b. The downward magnetic force exerted on the copper rod is, according to Equation 21.3

$$F = ILB\sin\theta = (12\text{ A})(0.85\text{ m})(0.16\text{ T})\sin 90.0° = 1.6\text{ N}$$

According to Equation 10.1, the force F_x^{Applied} required to stretch each spring is $F_x^{\text{Applied}} = kx$, where k is the spring constant. Since there are two springs, we know that the magnetic force F exerted on the current must equal $2F_x^{\text{Applied}}$, so that $F = 2F_x^{\text{Applied}} = 2kx$. Solving for x, we find that

$$x = \frac{F}{2k} = \frac{1.6\text{ N}}{2(75\text{ N/m})} = \boxed{1.1\times10^{-2}\text{ m}}$$

37. ***REASONING*** The magnitude τ of the torque that acts on a current-carrying coil placed in a magnetic field is specified by $\tau = NIAB \sin\phi$ (Equation 21.4), where N is the number of loops in the coil, I is the current, A is the area of one loop, B is the magnitude of the magnetic field, and ϕ is the angle between the normal to the coil and the magnetic field. All the variables in this relation are known except for the current, which can, therefore, be obtained.

SOLUTION Solving the equation $\tau = NIAB \sin\phi$ for the current I and noting that $\phi = 90.0°$ since τ is specified to be the maximum torque, we have

$$I = \frac{\tau}{NAB\sin\phi} = \frac{5.8\ \text{N}\cdot\text{m}}{(1200)\left(1.1 \times 10^{-2}\ \text{m}^2\right)(0.20\ \text{T})\sin 90.0°} = \boxed{2.2\ \text{A}}$$

---

41. ***REASONING*** The torque on the loop is given by Equation 21.4, $\tau = NIAB\sin\phi$. From the drawing in the text, we see that the angle ϕ between the normal to the plane of the loop and the magnetic field is $90° - 35° = 55°$. The area of the loop is $0.70\ \text{m} \times 0.50\ \text{m} = 0.35\ \text{m}^2$.

SOLUTION

a. The magnitude of the net torque exerted on the loop is

$$\tau = NIAB\sin\phi = (75)(4.4\ \text{A})(0.35\ \text{m}^2)(1.8\ \text{T})\ \sin 55° = \boxed{170\ \text{N}\cdot\text{m}}$$

b. As discussed in the text, when a current-carrying loop is placed in a magnetic field, the loop tends to rotate such that its normal becomes aligned with the magnetic field. The normal to the loop makes an angle of 55° with respect to the magnetic field. Since this angle decreases as the loop rotates, the $\boxed{\text{35° angle increases}}$.

---

45. ***REASONING*** The magnetic moment of the orbiting electron can be found from the expression *Magnetic moment* $= NIA$. For this situation, $N = 1$. Thus, we need to find the current and the area for the orbiting charge.

SOLUTION The current for the orbiting charge is, by definition (see Equation 20.1), $I = \Delta q / \Delta t$, where Δq is the amount of charge that passes a given point during a time interval Δt. Since the charge ($\Delta q = e$) passes by a given point once per revolution, we can find the current by dividing the total orbiting charge by the period T of revolution.

$$I = \frac{\Delta q}{T} = \frac{\Delta q}{2\pi r / v} = \frac{(1.6\times10^{-19}\ \text{C})(2.2\times10^{6}\ \text{m/s})}{2\pi(5.3\times10^{-11}\ \text{m})} = 1.06\times10^{-3}\ \text{A}$$

The area of the orbiting charge is

$$A = \pi r^2 = \pi(5.3\times10^{-11}\ \text{m})^2 = 8.82\times10^{-21}\ \text{m}^2$$

Therefore, the magnetic moment is

$$\textit{Magnetic moment} = NIA = (1)(1.06\times10^{-3}\ \text{A})(8.82\times10^{-21}\ \text{m}^2) = \boxed{9.3\times10^{-24}\ \text{A}\cdot\text{m}^2}$$

---

49. ***REASONING AND SOLUTION*** The current associated with the lightning bolt is

$$I = \frac{\Delta q}{\Delta t} = \frac{15\ \text{C}}{1.5\times10^{-3}\text{s}} = 1.0\times10^4\ \text{A}$$

The magnetic field near this current is given by

$$B = \frac{\mu_0 I}{2\pi r} = \frac{\left(4\pi\times10^{-7}\ \text{T}\cdot\text{m/A}\right)\left(1.0\times10^4\ \text{A}\right)}{2\pi(25\ \text{m})} = \boxed{8.0\times10^{-5}\ \text{T}}$$

---

51. ***REASONING*** The magnitude of the magnetic field at the center of a circular loop of current is given by Equation 21.6 as $B = N\mu_0 I/(2R)$, where N is the number of turns, μ_0 is the permeability of free space, I is the current, and R is the radius of the loop. The field is perpendicular to the plane of the loop. Magnetic fields are vectors, and here we have two fields, each perpendicular to the plane of the loop producing it. Therefore, the two field vectors are perpendicular, and we must add them as vectors to get the net field. Since they are perpendicular, we can use the Pythagorean theorem to calculate the magnitude of the net field.

SOLUTION Using Equation 21.6 and the Pythagorean theorem, we find that the magnitude of the net magnetic field at the common center of the two loops is

$$B_{\text{net}} = \sqrt{\left(\frac{N\mu_0 I}{2R}\right)^2 + \left(\frac{N\mu_0 I}{2R}\right)^2} = \sqrt{2}\left(\frac{N\mu_0 I}{2R}\right)$$

$$= \frac{\sqrt{2}(1)\left(4\pi\times10^{-7}\ \text{T}\cdot\text{m/A}\right)(1.7\ \text{A})}{2(0.040\ \text{m})} = \boxed{3.8\times10^{-5}\ \text{T}}$$

---

57. ***REASONING*** According to Equation 21.6 the magnetic field at the center of a circular, current-carrying loop of N turns and radius r is $B = N\mu_0 I/(2r)$. The number of turns N in the coil can be found by dividing the total length L of the wire by the circumference after it

has been wound into a circle. The current in the wire can be found by using Ohm's law, $I = V/R$.

SOLUTION The number of turns in the wire is

$$N = \frac{L}{2\pi r}$$

The current in the wire is

$$I = \frac{V}{R} = \frac{12.0\text{ V}}{(5.90\times10^{-3}\ \Omega/\text{m})L} = \frac{2.03\times10^{3}}{L}\text{ A}$$

Therefore, the magnetic field at the center of the coil is

$$B = N\left(\frac{\mu_0 I}{2r}\right) = \left(\frac{L}{2\pi r}\right)\left(\frac{\mu_0 I}{2r}\right) = \frac{\mu_0 L I}{4\pi r^2}$$

$$= \frac{\left(4\pi\times10^{-7}\text{ T}\cdot\text{m/A}\right)L\left(\frac{2.03\times10^{3}}{L}\text{ A}\right)}{4\pi(0.140\text{ m})^2} = \boxed{1.04\times10^{-2}\text{ T}}$$

---

61. ***REASONING AND SOLUTION*** Ampère's law in the form of Equation 21.8 indicates that $\Sigma B_{\parallel}\Delta\ell = \mu_0 I$. Since the magnetic field is everywhere perpendicular to the plane of the paper, it is everywhere perpendicular to the circular path and has no component $B_{\parallel}$ that is parallel to the circular path. Therefore, Ampère's law reduces to $\Sigma B_{\parallel}\Delta\ell = 0 = \mu_0 I$, so that $\boxed{\text{the net current passing through the circular surface is zero}}$.

---

65. ***REASONING*** The coil carries a current and experiences a torque when it is placed in an external magnetic field. Thus, when the coil is placed in the magnetic field due to the solenoid, it will experience a torque given by Equation 21.4: $\tau = NIAB\sin\phi$, where N is the number of turns in the coil, A is the area of the coil, B is the magnetic field inside the solenoid, and ϕ is the angle between the normal to the plane of the coil and the magnetic field. The magnetic field in the solenoid can be found from Equation 21.7: $B = \mu_0 nI$, where n is the number of turns per unit length of the solenoid and I is the current.

SOLUTION The magnetic field inside the solenoid is

$$B = \mu_0 nI = (4\pi\times10^{-7}\text{ T}\cdot\text{m/A})(1400\text{ turns/m})(3.5\text{ A}) = 6.2\times10^{-3}\text{ T}$$

The torque exerted on the coil is

$$\tau = NIAB\sin\phi = (50)(0.50\text{ A})(1.2\times10^{-3}\text{ m}^2)(6.2\times10^{-3}\text{ T})(\sin 90.0°) = \boxed{1.9\times10^{-4}\text{ N}\cdot\text{m}}$$

---

69. ***REASONING AND SOLUTION*** According to Equation 21.1, $B = \dfrac{F}{|q_0|(v\sin\theta)}$. Solving this for the angle θ, we find

$$\theta = \sin^{-1}\left(\frac{F}{|q_0|vB}\right) = \sin^{-1}\left[\frac{2.30\times10^{-7}\text{ N}}{(1.70\times10^{-5}\text{ C})(2.80\times10^{2}\text{ m/s})(5.00\times10^{-5}\text{ T})}\right] = \boxed{75.1° \text{ and } 105°}$$

---

73. ***REASONING*** From the discussion in Section 21.3, we know that when a charged particle moves perpendicular to a magnetic field, the trajectory of the particle is a circle. The drawing at the right shows a particle moving in the plane of the paper (the magnetic field is perpendicular to the paper). If the particle is moving initially through the coordinate origin and to the right (along the $+x$ axis), the subsequent circular path of the particle will intersect the y axis at the greatest possible value, which is equal to twice the radius r of the circle.

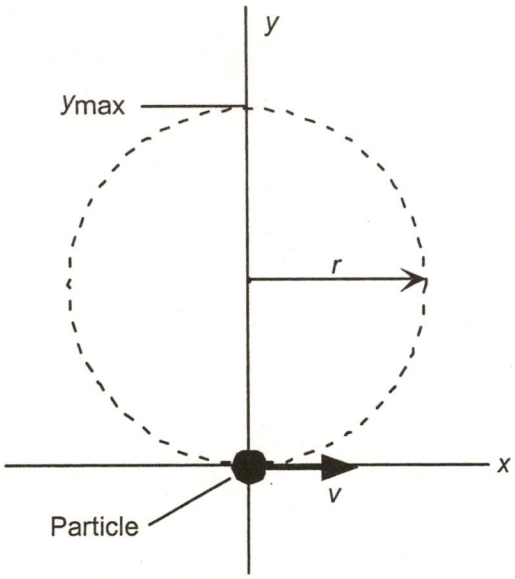

SOLUTION

a. From the drawing above, it can be seen that the largest value of y is equal to the diameter ($2r$) of the circle. When the particle passes through the coordinate origin its velocity must be parallel to the $+x$ axis. Thus, the angle is $\boxed{\theta = 0°}$.

b. The maximum value of y is twice the radius r of the circle. According to Equation 21.2, the radius of the circular path is $r = mv/(|q|B)$. The maximum value y_{max} is, therefore,

$$y_{max} = 2r = 2\left(\frac{mv}{|q|B}\right) = 2\left[\frac{(3.8\times10^{-8}\text{ kg})(44\text{ m/s})}{(7.3\times10^{-6}\text{ C})(1.6\text{ T})}\right] = \boxed{0.29\text{ m}}$$

---

77. ***REASONING*** The magnetic moment of the rotating charge can be found from the expression *Magnetic moment* $= NIA$, as discussed in Section 21.6. For this situation, $N = 1$. Thus, we need to find the current and the area for the rotating charge. This can be done by resorting to first principles.

SOLUTION The current for the rotating charge is, by definition (see Equation 20.1), $I = \Delta q / \Delta t$, where Δq is the amount of charge that passes by a given point during a time interval Δt. Since the charge passes by once per revolution, we can find the current by dividing the total rotating charge by the period T of revolution.

$$I = \frac{\Delta q}{T} = \frac{\Delta q}{2\pi / \omega} = \frac{\omega\, \Delta q}{2\pi} = \frac{(150 \text{ rad/s})(4.0 \times 10^{-6} \text{ C})}{2\pi} = 9.5 \times 10^{-5} \text{ A}$$

The area of the rotating charge is $A = \pi r^2 = \pi (0.20 \text{ m})^2 = 0.13 \text{ m}^2$

Therefore, the magnetic moment is

$$\textit{Magnetic moment} = NIA = (1)(9.5 \times 10^{-5} \text{ A})(0.13 \text{ m}^2) = \boxed{1.2 \times 10^{-5} \text{ A} \cdot \text{m}^2}$$

CHAPTER 22 | *ELECTROMAGNETIC INDUCTION*

CONCEPTUAL QUESTIONS

---

4. ***REASONING AND SOLUTION*** A magnetic field **B** is necessary if there is to be a magnetic flux Φ passing through a coil of wire. Yet, just because there is a magnetic field does not mean that a magnetic flux will pass through a coil.

The general expression for magnetic flux is given by Equation 22.2: $\Phi = (B \cos \phi)A$, where B is the magnitude of the magnetic field, A is the cross-sectional area of the coil, and ϕ is the angle between the magnetic field **B** and the normal to the surface of the coil. Equation 22.2 shows that the flux depends only on the component of the magnetic field that is perpendicular to the surface of the coil. As shown in Example 4 and in Figure 22.10, when the coil is oriented so that it is parallel to the field, $\phi = 90°$, **B** has no component normal to the surface of the coil, and the magnetic flux through the coil is zero. Therefore, the magnetic flux through a coil can be zero even though there is a magnetic field present.

---

7. ***REASONING AND SOLUTION*** A bolt of lightning contains moving charges, and hence, is a current. This current is surrounded by its magnetic field. Since the charges in a bolt of lightning move erratically, the current is a time-dependent and gives rise to a time-dependent magnetic field. Many household appliances contain coils. If the time-dependent magnetic field passes through these coils, the magnetic flux through the coils will be time-dependent. From Faraday's law, there will be an induced emf and, hence, an induced current in the coils of the appliance. This will result in an induced current in the circuit that contains the appliance. Therefore, a bolt of lightning can produce a current in the circuit of an electrical appliance, even when the lightning does not directly strike the appliance.

---

14. ***REASONING AND SOLUTION*** The current drawn by an electric motor is given by Equation 22.5: $I = (V - \xi)/R$, where V is the applied emf, ξ is the back emf developed because the coil of the motor is rotating, and R is the resistance of the wire in the coil.

An electric motor in a hair dryer is running at normal speed; therefore, the back emf is comparable to the applied emf, and the motor is drawing a relatively small current. If the shaft of the motor is prevented from turning, the back emf is suddenly reduced to zero. We see from Equation 22.5 that the current drawn by the motor will increase to the value V/R. Heat will be produced in the coil at a rate of $I^2R = (V/R)^2 R = V^2/R$, and the temperature of the coil will increase.

CHAPTER 22 | *ELECTROMAGNETIC INDUCTION*

PROBLEMS

---

3. ***REASONING AND SOLUTION*** According to Equation 22.1, the motional emf ξ is given by $\xi = vBL$ when the velocity **v**, the magnetic field **B**, and the rod are mutually perpendicular. Therefore in order for a spark to jump across the gap, the rod would have to be moving in the magnetic field with a speed that is

$$v = \frac{\xi}{BL} = \frac{940\text{ V}}{(4.8\text{ T})(1.3\text{ m})} = \boxed{150\text{ m/s}}$$

---

5. ***REASONING AND SOLUTION*** For the three rods in the drawing, we have the following:

***Rod A*:** The motional emf is $\boxed{\text{zero}}$, because the velocity of the rod is parallel to the direction of the magnetic field, and the charges do not experience a magnetic force.

***Rod B*:** The motional emf ξ is, according to Equation 22.1,

$$\xi = vBL = (2.7\text{ m/s})(0.45\text{ T})(1.3\text{ m}) = \boxed{1.6\text{ V}}$$

The positive end of Rod B is $\boxed{\text{end 2}}$.

***Rod C*:** The motional emf is $\boxed{\text{zero}}$, because the magnetic force F on each charge is directed perpendicular to the length of the rod. For the ends of the rod to become charged, the magnetic force must be directed parallel to the length of the rod.

---

7. ***REASONING*** The minimum length d of the rails is the speed v of the rod times the time t, or $d = vt$. We can obtain the speed from the expression for the motional emf given in Equation 22.1. Solving this equation for the speed gives $v = \frac{\xi}{BL}$, where ξ is the motional emf, B is the magnitude of the magnetic field, and L is the length of the rod. Thus, the length of the rails is $d = vt = \left(\frac{\xi}{BL}\right)t$. While we have no value for the motional emf, we do know that the bulb dissipates a power of $P = 60.0$ W, and has a resistance of $R = 240\ \Omega$. Power is

related to the emf and the resistance according to $P = \frac{\xi^2}{R}$ (Equation 20.6c), which can be solved to show that $\xi = \sqrt{PR}$. Substituting this expression into the equation for d gives

$$d = \left(\frac{\xi}{BL}\right)t = \left(\frac{\sqrt{PR}}{BL}\right)t$$

SOLUTION Using the above expression for the minimum necessary length of the rails, we find that

$$d = \left(\frac{\sqrt{PR}}{BL}\right)t = \left[\frac{\sqrt{(60.0\text{ W})(240\ \Omega)}}{(0.40\text{ T})(0.60\text{ m})}\right](0.50\text{ s}) = \boxed{250\text{ m}}$$

---

13. ***REASONING*** The general expression for the magnetic flux through an area A is given by Equation 22.2: $\Phi = BA\cos\phi$, where B is the magnitude of the magnetic field and ϕ is the angle of inclination of the magnetic field **B** with respect to the normal to the area.

The magnetic flux through the door is a maximum when the magnetic field lines are perpendicular to the door and $\phi_1 = 0.0°$ so that $\Phi_1 = \Phi_{max} = BA(\cos\ 0.0°) = BA$.

SOLUTION When the door rotates through an angle ϕ_2, the magnetic flux that passes through the door decreases from its maximum value to one-third of its maximum value. Therefore, $\Phi_2 = \frac{1}{3}\Phi_{max}$, and we have

$$\Phi_2 = BA\cos\phi_2 = \tfrac{1}{3}BA \qquad \text{or} \qquad \cos\phi_2 = \tfrac{1}{3} \qquad \text{or} \qquad \phi_2 = \cos^{-1}\left(\tfrac{1}{3}\right) = \boxed{70.5°}$$

---

15. ***REASONING*** The general expression for the magnetic flux through an area A is given by Equation 22.2: $\Phi = BA\cos\phi$, where B is the magnitude of the magnetic field and ϕ is the angle of inclination of the magnetic field **B** with respect to the normal to the surface.

SOLUTION Since the magnetic field **B** is parallel to the surface for the triangular ends and the bottom surface, the flux through each of these three surfaces is $\boxed{0\text{ Wb}}$.

The flux through the 1.2 m by 0.30 m face is

$$\Phi = (0.25\text{ T})(1.2\text{ m})(0.30\text{ m})\cos 0.0° = \boxed{0.090\text{ Wb}}$$

For the 1.2 m by 0.50 m side, the area makes an angle ϕ with the magnetic field **B**, where

$$\phi = 90° - \tan^{-1}\left(\frac{0.30\text{ m}}{0.40\text{ m}}\right) = 53°$$

Therefore,

$$\Phi = (0.25\text{ T})(1.2\text{ m})(0.50\text{ m})\cos 53° = \boxed{0.090\text{ Wb}}$$

---

19. ***REASONING*** According to Equation 22.3, the average emf induced in a coil of N loops is $\xi = -N\Delta\Phi/\Delta t$.

SOLUTION For the circular coil in question, the flux through a single turn changes by

$$\Delta\Phi = BA\cos 45° - BA\cos 90° = BA\cos 45°$$

during the interval of $\Delta t = 0.010$ s. Therefore, for N turns, Faraday's law gives the magnitude of the emf as

$$|\xi| = \left|-N\frac{BA\cos 45°}{\Delta t}\right|$$

Since the loops are circular, the area A of each loop is equal to πr^2. Solving for B, we have

$$B = \frac{|\xi|\Delta t}{N\pi r^2\cos 45°} = \frac{(0.065\text{ V})(0.010\text{ s})}{(950)\pi(0.060\text{ m})^2\cos 45°} = \boxed{8.6\times10^{-5}\text{ T}}$$

---

25. ***REASONING AND SOLUTION*** According to Ohm's law, the emf ξ developed by the coil is $\xi = IR$, which, when combined with the definition of electric current becomes $\xi = (\Delta q)R/(\Delta t)$. The magnitude of the emf is given by Equation 22.3 (Faraday's law):

$$|\xi| = \left|-N\frac{\Delta\Phi}{\Delta t}\right| = \left|-N\left(\frac{BA\cos 0° - 0}{\Delta t}\right)\right| = \frac{NBA}{\Delta t}$$

Combining the two expressions for the emf, we have

$$\frac{(\Delta q)R}{\Delta t} = \frac{NBA}{\Delta t} \qquad \text{or} \qquad B = \frac{(\Delta q)R}{NA} = \frac{(8.87\times10^{-3}\text{ C})(45.0\ \Omega)}{(1850)(4.70\times10^{-4}\text{ m}^2)} = \boxed{0.459\text{ T}}$$

---

27. ***REASONING*** According to Ohm's law, the average current I induced in the coil is given by $I = |\xi|/R$, where $|\xi|$ is the magnitude of the induced emf and R is the resistance of the coil. To find the induced emf, we use Faraday's law of electromagnetic induction

SOLUTION The magnitude of the induced emf can be found from Faraday's law of electromagnetic induction and is given by Equation 22.3:

$$|\xi| = \left| -N\frac{\Delta\Phi}{\Delta t} \right| = \left| -N\frac{\Delta(BA\cos 0°)}{\Delta t} \right| = NA\frac{\Delta B}{\Delta t}$$

We have used the fact that the field within a long solenoid is perpendicular to the cross-sectional area A of the solenoid and makes an angle of 0° with respect to the normal to the area. The field is given by Equation 21.7 as $B = \mu_0 n I$, so the change ΔB in the field is $\Delta B = \mu_0 n \Delta I$, where ΔI is the change in the current. The induced current is, then,

$$I = \frac{NA\dfrac{\Delta B}{\Delta t}}{R} = \frac{NA\dfrac{\mu_0 n\,\Delta I}{\Delta t}}{R}$$

$$= \frac{(10)(6.0\times10^{-4}\ \text{m}^2)\dfrac{(4\pi\times10^{-7}\ \text{T}\cdot\text{m/A})(400\ \text{turns/m})(0.40\ \text{A})}{(0.050\ \text{s})}}{1.5\ \Omega} = \boxed{1.6\times10^{-5}\ \text{A}}$$

---

29. ***REASONING AND SOLUTION*** According to Faraday's law, only the relative motion between the coil and magnet is significant.. It makes no difference if the coil is held fixed and the magnet is moved or the magnet is held fixed and the coil is moved. The results are the same. Therefore,

a. The current will flow from $\boxed{\text{left to right}}$ as shown in Figure 22.1*b*.

b. The current will flow from $\boxed{\text{right to left}}$ as shown in Figure 22.1*c*.

---

31. ***REASONING*** In solving this problem, we apply Lenz's law, which essentially says that the change in magnetic flux must be opposed by the induced magnetic field.

SOLUTION

a. The magnetic field due to the wire in the vicinity of position 1 is directed out of the paper. The coil is moving closer to the wire into a region of higher magnetic field, so the flux through the coil is increasing. Lenz's law demands that the induced field counteract this increase. The direction of the induced field, therefore, must be into the paper. The current in the coil must be $\boxed{\text{clockwise}}$.

b. At position 2 the magnetic field is directed into the paper and is decreasing as the coil moves away from the wire. The induced magnetic field, therefore, must be directed into the paper, so the current in the coil must be $\boxed{\text{clockwise}}$.

---

35. ***REASONING*** We can use the information given in the problem statement to determine the area of the coil A. Since it is square, the length of one side is $\ell = \sqrt{A}$.

SOLUTION According to Equation 22.4, the maximum emf ξ_0 induced in the coil is $\xi_0 = NAB\omega$. Therefore, the length of one side of the coil is

$$\ell = \sqrt{A} = \sqrt{\frac{\xi_0}{NB\omega}} = \sqrt{\frac{75.0\text{ V}}{(248)(0.170\text{ T})(79.1\text{ rad/s})}} = \boxed{0.150\text{ m}}$$

---

41. ***REASONING AND SOLUTION*** When the motor just begins to turn the fan blade, there is no back emf. Therefore, Ohm's law indicates that $120.0\text{ V} = I_0R$, where I_0 is the current in the motor and R is the resistance of the motor coils. Solving for I_0 gives

$$I_0 = \frac{120.0\text{ V}}{R} \qquad (1)$$

At normal operating speed, there is a back emf ξ, and Equation 22.5 gives the current in the motor as

$$I = \frac{120.0\text{ V} - \xi}{R} \qquad (2)$$

But we know that $I = 0.150I_0$. Therefore, using Equations (1) and (2), we find that

$$\frac{120.0\text{ V} - \xi}{R} = 0.150\left(\frac{120.0\text{ V}}{R}\right)$$

Eliminating R algebraically, we have

$$120.0\text{ V} - \xi = 0.150(120.0\text{ V}) \qquad \text{or} \qquad \xi = 0.850(120.0\text{ V}) = \boxed{102\text{ V}}$$

---

45. ***REASONING*** The energy density is given by Equation 22.11 as

$$\text{Energy density} = \frac{\text{Energy}}{\text{Volume}} = \frac{1}{2\mu_0}B^2$$

The energy stored is the energy density times the volume.

SOLUTION The volume is the area A times the height h. Therefore, the energy stored is

$$\text{Energy} = \frac{B^2 Ah}{2\mu_0} = \frac{(7.0\times10^{-5}\ \text{T})^2(5.0\times10^8\ \text{m}^2)(1500\ \text{m})}{2(4\pi\times10^{-7}\ \text{T}\cdot\text{m/A})} = \boxed{1.5\times10^9\ \text{J}}$$

---

49. ***REASONING*** The rate of change $\Delta I_P/\Delta t$ in the primary current is related to the emf ξ_S induced in the secondary coil according to $\xi_S = -M\Delta I_P/\Delta t$ (Equation 22.7), where M is the mutual inductance between the coils. We can use this expression to determine ΔI_P directly. However, in doing so, we will omit the minus sign, since the direction of the current is unspecified. A value for the induced emf can be obtained from the induced current I_S and the resistance R in the secondary circuit, according to $\xi_S = I_S R$, which is Ohm's law.

SOLUTION Substituting the induced emf from Ohm's law into Equation 22.7 (without the minus sign), and solving for ΔI_P gives

$$\xi_S = I_S R = M\frac{\Delta I_P}{\Delta t} \quad \text{or} \quad \Delta I_P = \frac{I_S R(\Delta t)}{M} = \frac{(6.0\times10^{-3}\ \text{A})(12\ \Omega)(72\times10^{-3}\ \text{s})}{3.2\times10^{-3}\ \text{H}} = \boxed{1.6\ \text{A}}$$

---

57. ***REASONING AND SOLUTION*** Since the secondary voltage is less than the primary voltage, we can conclude that the transformer used in the doorbell described in the problem is a $\boxed{\text{step-down transformer}}$. The turns ratio N_s/N_p is given by the transformer equation, Equation 22.12:

$$\frac{N_s}{N_p} = \frac{V_s}{V_p} = \frac{10.0\ \text{V}}{120\ \text{V}} = \frac{1}{12}$$

so the turns ratio is $\boxed{1:12}$.

---

59. ***REASONING*** The power used to heat the wires is given by Equation 20.6b: $P = I^2 R$. Before we can use this equation, however, we must determine the total resistance R of the wire and the current that flows through the wire.

SOLUTION

a. The total resistance of one of the wires is equal to the resistance per unit length multiplied by the length of the wire. Thus, we have

$$(5.0\times10^{-2}\ \Omega/\text{km})(7.0\ \text{km}) = 0.35\ \Omega$$

and the total resistance of the transmission line is twice this value or 0.70 Ω. According to Equation 20.6a ($P = IV$), the current flowing into the town is

$$I = \frac{P}{V} = \frac{1.2\times10^6\ \text{W}}{1200\ \text{V}} = 1.0\times10^3\ \text{A}$$

Thus, the power used to heat the wire is

$$P = I^2R = \left(1.0\times10^3\ \text{A}\right)^2(0.70\ \Omega) = \boxed{7.0\times10^5\ \text{W}}$$

b. According to the transformer equation (Equation 22.12), the stepped-up voltage is

$$V_s = V_p\left(\frac{N_s}{N_p}\right) = (1200\ \text{V})\left(\frac{100}{1}\right) = 1.2\times10^5\ \text{V}$$

According to Equation 20.6a ($P = IV$), the current in the wires is

$$I = \frac{P}{V} = \frac{1.2\times10^6\ \text{W}}{1.2\times10^5\ \text{V}} = 1.0\times10^1\ \text{A}$$

The power used to heat the wires is now

$$P = I^2R = \left(1.0\times10^1\ \text{A}\right)^2(0.70\ \Omega) = \boxed{7.0\times10^1\ \text{W}}$$

---

63. ***REASONING AND SOLUTION***

a. From the drawing, we see that the period of the generator (the time for one full cycle) is 0.42 s; therefore, the frequency of the generator is

$$f = \frac{1}{T} = \frac{1}{0.42\ \text{s}} = \boxed{2.4\ \text{Hz}}$$

b. The angular speed of the generator is related to its frequency by $\omega = 2\pi f$, so the angular speed is

$$\omega = 2\pi(2.4\ \text{Hz}) = \boxed{15\ \text{rad/s}}$$

c. The maximum emf ξ_0 induced in a rotating planar coil is given by $\xi_0 = NAB\omega$ (see Equation 22.4). The magnitude of the magnetic field can be found by solving this expression for B and noting from the drawing that $\xi_0 = 28$ V:

$$B = \frac{\xi_0}{NA\omega} = \frac{28\text{ V}}{(150)(0.020\text{ m}^2)(15\text{ rad/s})} = \boxed{0.62\text{ T}}$$

---

65. ***REASONING AND SOLUTION*** If the north and south poles of the magnet are interchanged, the currents in the ammeter would simply be reversed in direction. This follows from Lenz's law. As the south pole approaches the coil in Figure 22.1*b*, the field that the coil experiences is becoming stronger and points toward the south pole. To oppose the increasing flux from this field (as specified by Lenz's law) the current induced in the coil must produce a field that points away from the approaching south pole. RHR-2 indicates that to do this, the induced current must flow into the ammeter on the right and out of the ammeter on the left. Another way to look at things is to realize that, with this direction for the induced current, the coil becomes an electromagnet with a south pole located at its left side. This induced south pole opposes the motion of the approaching south pole of the moving magnet.

Therefore, in Figure 22.1*b*, the current will flow $\boxed{\text{right to left}}$ through the ammeter.

Similar reasoning (with the poles of the magnet interchanged in Figure 22.1) leads to the conclusion that in Figure 22.1*c* the induced current must flow into the ammeter on the left and out of the ammeter on the right.

Thus, in Figure 22.1*c*, the current will flow $\boxed{\text{left to right}}$ through the ammeter.

---

67. ***REASONING AND SOLUTION*** The emf ξ induced in the coil of an ac generator is given by Equation 22.4 as

$$\xi = NAB\omega \sin\omega t = (500)(1.2\times10^{-2}\text{ m}^2)(0.13\text{ T})(34\text{ rad/s})\sin 27° = \boxed{12\text{ V}}$$

---

71. ***REASONING*** According to Equation 22.3, the average emf ξ induced in a single loop ($N = 1$) is $\xi = -\Delta\Phi/\Delta t$. Since the magnitude of the magnetic field is changing, the area of the loop remains constant, and the direction of the field is parallel to the normal to the loop, the change in flux through the loop is given by $\Delta\Phi = (\Delta B)A$. Thus the magnitude $|\xi|$ of the induced emf in the loop is given by $|\xi| = |-(\Delta B)A/\Delta t|$.

Similarly, when the area of the loop is changed and the field B has a given value, we find the magnitude of the induced emf to be $|\xi| = |-B(\Delta A)/\Delta t|$.

SOLUTION

a. The magnitude of the induced emf when the field changes in magnitude is

$$|\xi| = \left| -A\left(\frac{\Delta B}{\Delta t}\right) \right| = (0.018\ \text{m}^2)(0.20\ \text{T/s}) = \boxed{3.6\times10^{-3}\ \text{V}}$$

b. At a particular value of B (when B is changing), the rate at which the area must change can be obtained from

$$|\xi| = \left| -\frac{B\Delta A}{\Delta t} \right| \quad \text{or} \quad \frac{\Delta A}{\Delta t} = \frac{|\xi|}{B} = \frac{3.6\times10^{-3}\ \text{V}}{1.8\ \text{T}} = \boxed{2.0\times10^{-3}\ \text{m}^2/\text{s}}$$

In order for the induced emf to be zero, the magnitude of the magnetic field and the area of the loop must change in such a way that the flux remains constant. Since the magnitude of the magnetic field is increasing, the area of the loop must decrease, if the flux is to remain constant. Therefore, $\boxed{\text{the area of the loop must be shrunk}}$.

CHAPTER 23 | *ALTERNATING CURRENT CIRCUITS*

CONCEPTUAL QUESTIONS

---

2. ***REASONING AND SOLUTION*** The ends of a long straight wire are connected to the terminals of an ac generator, and the current is measured. The wire is then disconnected, wound into the shape of a multiple-turn coil, and reconnected to the generator. After the wire is wound into a coil, it has a greater inductance. When the generator is turned on, the coil will develop a voltage that opposes a change in the current according to Faraday's law. Since the induced voltage opposes the rise in current, the rms current in the circuit will be less than it was before the wire was wound into a coil.

   This can also be seen by considering Equations 23.6 and 23.7. Before the wire was wound into a coil, its primary property was that of resistance, and the current through the wire was given by $I_{\text{rms}} = V_{\text{rms}} / Z$ with $Z = R$, or $I_{\text{rms}} = V_{\text{rms}} / R$. After the wire is wound into a coil, the wire possesses both resistance and inductance. Now the current in the coil is given by $I_{\text{rms}} = V_{\text{rms}} / Z$, where $Z = \sqrt{R^2 + X_{\text{L}}^{\ 2}}$. Since Z is necessarily greater than R, the rms current is less when the wire is wound into a coil.

---

7. ***REASONING AND SOLUTION*** The rms current is given by Equation 23.6: $I_{\text{rms}} = V_{\text{rms}} / Z$, where V_{rms} is the rms voltage of the generator and Z is the impedance of the circuit. Since the rms voltage of the generator is the same in both cases, the greater current is delivered to the circuit with the smaller impedance. We see from Equation 23.2 [$X_{\text{C}} = 1/(2\pi f C)$] and Equation 23.4 [$X_{\text{L}} = 2\pi f L$] that in the high frequency limit, the capacitive reactance is nearly zero and the inductive reactance is essentially infinite. In other words, in the high frequency limit, the capacitors behave as if they have been replaced by wires of zero resistance and the inductors behave as if they have been cut from the circuit leaving gaps in the wires. The figure below shows the two circuits in this limit.

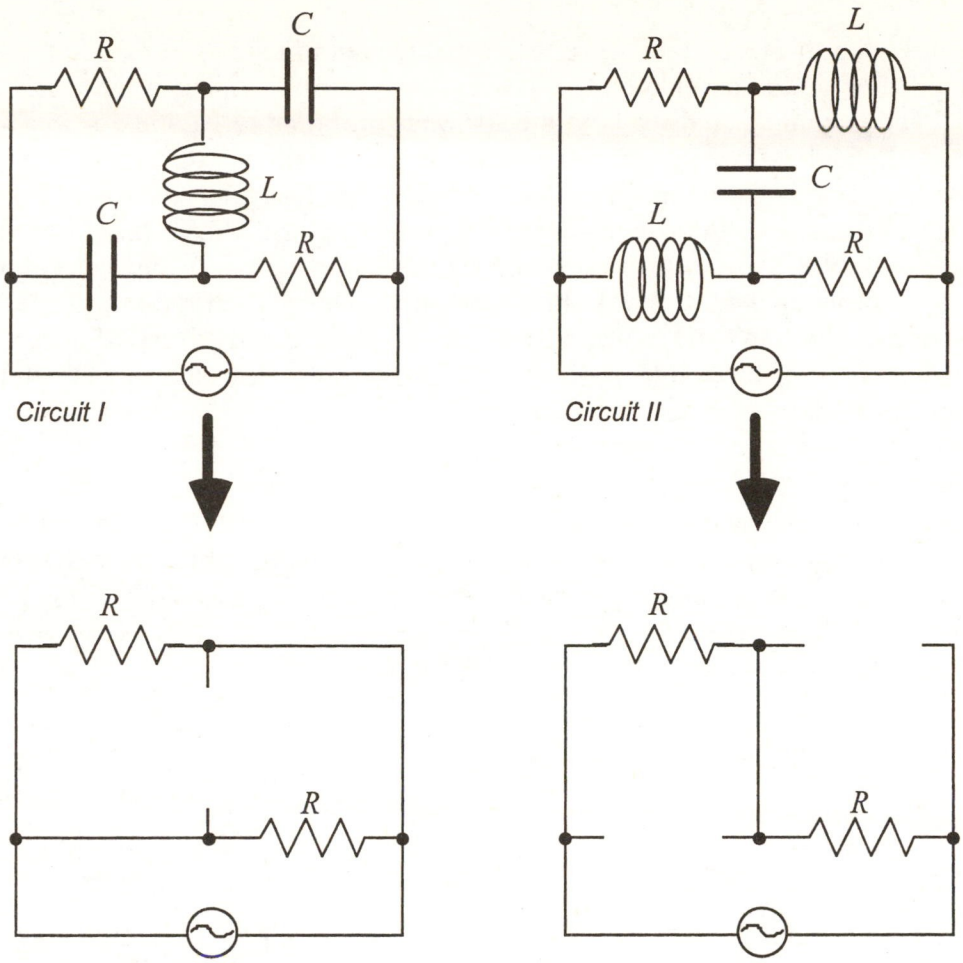

Circuit I behaves as if it consists only of two identical resistors in parallel; therefore, the impedance of circuit I is $1/Z = 1/R + 1/R$, or $Z = R/2$. Circuit II behaves as if it consists only of two identical resistors in series; therefore, the impedance of circuit II is $Z = 2R$. At a very high frequency, then, circuit I has the smaller impedance, and, therefore, its generator supplies the greater rms current.

---

11. ***REASONING AND SOLUTION*** The drawing at the right shows a full-wave rectifier circuit, in which the direction of the current through the load resistor R is the same for both positive and negative halves of the generator's voltage cycle. The points A and B are connected directly to the top and bottom of the generator, respectively. The diodes are labeled a, b, c, and d. The direction of the current through the circuit can be found by recalling the fact that charge flows through a diode only when the diode is in a forward bias condition. When the diode is in a forward bias condition, the side of the diode symbol that contains the arrowhead has a positive potential relative to the other side.

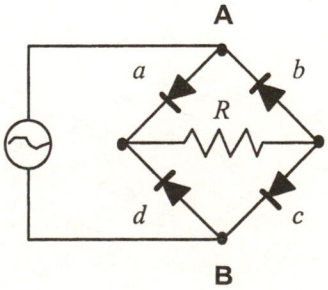

a. When the top of the generator is positive and the bottom is negative, point *A* in the figure has the positive potential and point *B* the negative potential. Since point *A* has the positive potential and is connected directly to the arrowhead for diode *a*, diode *a* is in a forward bias condition. In contrast, we see that for diode *b*, the side opposite the arrowhead is connected to point *A*, so diode *b* is in a reverse bias condition. Similarly, since point *B* is negative, and is connected directly to the side of diode *c* opposite the arrowhead, the arrowhead must be positive, and diode *c* is in a forward bias condition. In contrast, the arrowhead of diode *d* is negative, so diode *d* is in a reverse bias condition. Thus, when the top of the generator is positive and the bottom is negative, only diodes *a* and *c* are in the forward bias condition and allow charge to flow through the resistance *R*. Drawing 1 below shows the path taken by the current under these circumstances.

b. When the top of the generator is negative and the bottom is positive, reasoning similar to that given above in part (a) indicates that diodes *b* and *d* are now forward biased, while diodes *a* and *c* are in reverse bias. Furthermore, the forward bias diodes allow charge to flow from left to right through the resistance *R*. Drawing 2 below shows the path taken by the current under these circumstances.

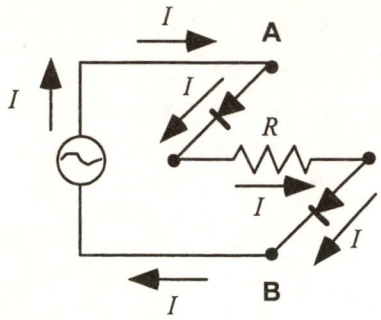

Drawing 1

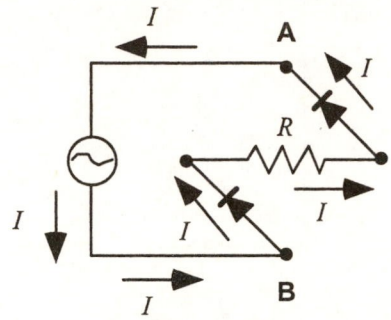

Drawing 2

CHAPTER 23 | *ALTERNATING CURRENT CIRCUITS*

PROBLEMS

---

1. ***REASONING AND SOLUTION*** The rms voltage can be calculated using $V = IX_C$, where the capacitive reactance X_C can be found using

$$X_C = \frac{1}{2\pi f C} = \frac{1}{2\pi(3.4\times10^3\ \text{Hz})(0.86\times10^{-6}\text{F})} = 54\ \Omega$$

The voltage is, therefore,

$$V = IX_C = (35\times10^{-3}\ \text{A})(54\ \Omega) = \boxed{1.9\ \text{V}}$$

---

5. ***REASONING AND SOLUTION***

a. The equivalent capacitance C_s of two capacitors in series is given by Equation 20.19 as

$$\frac{1}{C_s} = \frac{1}{C_1} + \frac{1}{C_2} = \frac{1}{3.00\times10^{-6}\ \text{F}} + \frac{1}{6.00\times10^{-6}\ \text{F}} \qquad \text{or} \qquad \boxed{C_s = 2.00\ \mu\text{F}}$$

b. The current in the circuit can be found by solving Equation 23.1, $V_{rms} = I_{rms}X_C$, for I_{rms}. However, we must first find the capacitive reactance X_C. From Equation 23.2, we have

$$X_C = \frac{1}{2\pi f C_s} = \frac{1}{2\pi(510\ \text{Hz})(2.00\times10^{-6}\ \text{F})} = 156\ \Omega$$

The current in the circuit is given by

$$I_{rms} = \frac{V_{rms}}{X_C} = \frac{120\ \text{V}}{156\ \Omega} = \boxed{0.77\ \text{A}}$$

---

9. ***REASONING*** The individual reactances are given by Equations 23.2 and 23.4, respectively,

Capacitive reactance $$X_C = \frac{1}{2\pi f C}$$

Inductive reactance $X_L = 2\pi f L$

When the reactances are equal, we have $X_C = X_L$, from which we find

$$\frac{1}{2\pi f C} = 2\pi f L \qquad \text{or} \qquad 4\pi^2 f^2 LC = 1$$

The last expression may be solved for the frequency f.

SOLUTION Solving for f with $L = 52 \times 10^{-3}$ H and $C = 76 \times 10^{-6}$ F, we obtain

$$f = \frac{1}{2\pi\sqrt{LC}} = \frac{1}{2\pi\sqrt{(52\times10^{-3}\ \text{H})(76\times10^{-6}\ \text{F})}} = \boxed{8.0\times10^{1}\ \text{Hz}}$$

---

11. ***REASONING*** The rms current can be calculated from Equation 23.3, $I_{rms} = V_{rms} / X_L$, provided that the inductive reactance is obtained first. Then the peak value of the current I_0 supplied by the generator can be calculated from the rms current I_{rms} by using Equation 20.12, $I_0 = \sqrt{2}\, I_{rms}$.

SOLUTION At the frequency of $f = 620$ Hz, we find, using Equations 23.4 and 23.3, that

$$X_L = 2\pi f L = 2\pi(620\ \text{Hz})(8.2\times10^{-3}\ \text{H}) = 32\ \Omega$$

$$I_{rms} = \frac{V_{rms}}{X_L} = \frac{10.0\ \text{V}}{32\ \Omega} = 0.31\ \text{A}$$

Therefore, from Equation 20.12, we find that the *peak value* I_0 of the current supplied by the generator must be

$$I_0 = \sqrt{2}\, I_{rms} = \sqrt{2}\,(0.31\ \text{A}) = \boxed{0.44\ \text{A}}$$

---

15. ***REASONING*** The voltage supplied by the generator can be found from Equation 23.6, $V_{rms} = I_{rms} Z$. The value of I_{rms} is given in the problem statement, so we must obtain the impedance of the circuit.

SOLUTION The impedance of the circuit is, according to Equation 23.7,

$$Z = \sqrt{R^2 + (X_L - X_C)^2} = \sqrt{(275\ \Omega)^2 + (648\ \Omega - 415\ \Omega)^2} = 3.60\times10^{2}\ \Omega$$

The rms voltage of the generator is

$$V_{\text{rms}} = I_{\text{rms}} Z = (0.233\text{ A})(3.60 \times 10^2\ \Omega) = \boxed{83.9\text{ V}}$$

---

19. ***REASONING*** We can use the equations for a series RCL circuit to solve this problem provided that we set $X_C = 0$ since there is no capacitor in the circuit. The current in the circuit can be found from Equation 23.6, $V_{\text{rms}} = I_{\text{rms}} Z$, once the impedance of the circuit has been obtained. Equation 23.8, $\tan\phi = (X_L - X_C)/R$, can then be used (with $X_C = 0\ \Omega$) to find the phase angle between the current and the voltage.

SOLUTION The inductive reactance is (Equation 23.4)

$$X_L = 2\pi f L = 2\pi(106\text{ Hz})(0.200\text{ H}) = 133\ \Omega$$

The impedance of the circuit is

$$Z = \sqrt{R^2 + (X_L - X_C)^2} = \sqrt{R^2 + X_L^2} = \sqrt{(215\ \Omega)^2 + (133\ \Omega)^2} = 253\ \Omega$$

a. The current through each circuit element is, using Equation 23.6,

$$I_{\text{rms}} = \frac{V_{\text{rms}}}{Z} = \frac{234\text{ V}}{253\ \Omega} = \boxed{0.925\text{ A}}$$

b. The phase angle between the current and the voltage is, according to Equation 23.8 (with $X_C = 0\ \Omega$),

$$\tan\phi = \frac{X_L - X_C}{R} = \frac{X_L}{R} = \frac{133\ \Omega}{215\ \Omega} = 0.619 \qquad \text{or} \qquad \phi = \tan^{-1}(0.619) = \boxed{31.8°}$$

---

21. ***REASONING*** We can use the equations for a series RCL circuit to solve this problem, provided that we set $X_L = 0$ since there is no inductance in the circuit. Thus, according to Equations 23.6 and 23.7, the current in the circuit is $I_{\text{rms}} = V_{\text{rms}} / \sqrt{R^2 + X_C^2}$. When the frequency f is very large, the capacitive reactance is zero, or $X_C = 0$, in which case the current becomes $I_{\text{rms}}(\text{large } f) = V_{\text{rms}} / R$. When the current I_{rms} in the circuit is one-half the value of $I_{\text{rms}}(\text{large } f)$ that exists when the frequency is very large, we have

$$\frac{I_{\text{rms}}}{I_{\text{rms}}(\text{large } f)} = \frac{1}{2}$$

We can use these expressions to write the ratio above in terms of the resistance and the capacitive reactance. Once the capacitive reactance is known, the frequency can be determined.

SOLUTION The ratio of the currents is

$$\frac{I_{\text{rms}}}{I_{\text{rms}}(\text{large } f)} = \frac{V_{\text{rms}} / \sqrt{R^2 + X_C^2}}{V_{\text{rms}} / R} = \frac{R}{\sqrt{R^2 + X_C^2}} = \frac{1}{2} \quad \text{or} \quad \frac{R^2}{R^2 + X_C^2} = \frac{1}{4}$$

Taking the reciprocal of this result gives

$$\frac{R^2 + X_C^2}{R^2} = 4 \quad \text{or} \quad 1 + \frac{X_C^2}{R^2} = 4$$

Therefore,

$$\frac{X_C}{R} = \sqrt{3}$$

According to Equation 23.2, $X_C = 1/(2\pi f C)$, so it follows that

$$\frac{X_C}{R} = \frac{1/(2\pi f C)}{R} = \sqrt{3}$$

Thus,

$$f = \frac{1}{2\pi RC\sqrt{3}} = \frac{1}{2\pi(85\,\Omega)(4.0 \times 10^{-6}\text{ F})\sqrt{3}} = \boxed{270\text{ Hz}}$$

---

27. ***REASONING AND SOLUTION*** The resonant frequency is given by Equation 23.10 as

$$f_0 = \frac{1}{2\pi\sqrt{LC}} = 9.3\text{ kHz}$$

If the inductance and capacitance of the circuit are each tripled, the new resonant frequency is

$$f_0' = \frac{1}{2\pi\sqrt{(3L)(3C)}} = \frac{1}{2\pi\sqrt{9LC}} = \frac{1}{3}\left(\frac{1}{2\pi\sqrt{LC}}\right) = \tfrac{1}{3}f_0 = \tfrac{1}{3}(9.3\text{ kHz}) = \boxed{3.1\text{ kHz}}$$

---

29. ***REASONING*** The current in an RCL circuit is given by Equation 23.6, $I_{\text{rms}} = V_{\text{rms}} / Z$, where the impedance Z of the circuit is given by Equation 23.7 as $Z = \sqrt{R^2 + (X_L - X_C)^2}$. The current is a maximum when the impedance is a minimum for a given generator voltage.

The minimum impedance occurs when the frequency is f_0, corresponding to the condition that $X_L = X_C$, or $2\pi f_0 L = 1/(2\pi f_0 C)$. Solving for the frequency f_0, called the resonant frequency, we find that

$$f_0 = \frac{1}{2\pi\sqrt{LC}}$$

Note that the resonant frequency depends on the inductance and the capacitance, but does not depend on the resistance.

SOLUTION

a. The frequency at which the current is a maximum is

$$f_0 = \frac{1}{2\pi\sqrt{LC}} = \frac{1}{2\pi\sqrt{(17.0\times10^{-3}\text{ H})(12.0\times10^{-6}\text{ F})}} = \boxed{352\text{ Hz}}$$

b. The maximum value of the current occurs when $f = f_0$. This occurs when $X_L = X_C$, so that $Z = R$. Therefore, according to Equation 23.6, we have

$$I_{\text{rms}} = \frac{V_{\text{rms}}}{Z} = \frac{V_{\text{rms}}}{R} = \frac{155\text{ V}}{10.0\ \Omega} = \boxed{15.5\text{ A}}$$

---

33. ***REASONING*** At the resonant frequency f_0, we have $C = 1/(4\pi^2 f_0^2 L)$. We want to determine some series combination of capacitors whose equivalent capacitance C_s' is such that $f_0' = 3f_0$. Thus,

$$C_s' = \frac{1}{4\pi^2 f_0'^2 L} = \frac{1}{4\pi^2 (3f_0)^2 L} = \frac{1}{9}\left(\frac{1}{4\pi^2 f_0^2 L}\right) = \frac{1}{9}C$$

The equivalent capacitance of a series combination of capacitors is $1/C_s' = 1/C_1 + 1/C_2 + \ldots$ If we require that all the capacitors have the same capacitance C, the equivalent capacitance is

$$\frac{1}{C_s'} = \frac{1}{C} + \frac{1}{C} + \cdots = \frac{n}{C}$$

where n is the total number of identical capacitors. Using the result above, we find that

$$\frac{1}{C_s'} = \frac{1}{\frac{1}{9}C} = \frac{n}{C} \qquad \text{or} \qquad n = 9$$

Therefore, the number of *additional* capacitors that must be inserted in series in the circuit so that the frequency triples is $n' = n - 1 = \boxed{8}$.

---

37. ***REASONING*** The voltage across the capacitor reaches its maximum instantaneous value when the generator voltage reaches its maximum instantaneous value. The maximum value of the capacitor voltage first occurs one-fourth of the way, or one-quarter of a period, through a complete cycle (see the voltage curve in Figure 23.4).

SOLUTION The period of the generator is $T = 1/f = 1/(5.00\text{ Hz}) = 0.200\text{ s}$. Therefore, the least amount of time that passes before the instantaneous voltage across the capacitor reaches its maximum value is $\frac{1}{4}T = \frac{1}{4}(0.200\text{ s}) = \boxed{5.00 \times 10^{-2}\text{ s}}$.

---

41. ***REASONING*** Since the capacitor and the inductor are connected in parallel, the voltage across each of these elements is the same or $V_L = V_C$. Using Equations 23.3 and 23.1, respectively, this becomes $I_{rms}X_L = I_{rms}X_C$. Since the currents in the inductor and capacitor are equal, this relation simplifies to $X_L = X_C$. Therefore, we can find the value of the inductance by equating the expressions (Equations 23.4 and 23.2) for the inductive reactance and the capacitive reactance, and solving for *L*.

SOLUTION Since $X_L = X_C$, we have

$$2\pi f L = \frac{1}{2\pi f C}$$

Therefore, the value of the inductance is

$$L = \frac{1}{4\pi^2 f^2 C} = \frac{1}{4\pi^2 (60.0\text{ Hz})^2 (40.0 \times 10^{-6}\text{ F})} = 0.176\text{ H} = \boxed{176\text{ mH}}$$

---

43. ***REASONING*** Since we know the values of the resonant frequency of the circuit, the capacitance, and the generator voltage, we can find the value of the inductance from Equation 23.10, the expression for the resonant frequency. The resistance can be found from energy considerations at resonance; the power factor is given by $\cos\phi$, where the phase angle ϕ is given by Equation 23.8, $\tan\phi = (X_L - X_C)/R$.

SOLUTION

a. Solving Equation 23.10 for the inductance *L*, we find that

$$L = \frac{1}{4\pi^2 f_0^2 C} = \frac{1}{4\pi^2 (1.30 \times 10^3\text{ Hz})^2 (5.10 \times 10^{-6}\text{ F})} = \boxed{2.94 \times 10^{-3}\text{ H}}$$

b. At resonance, $f = f_0$, and the current is a maximum. This occurs when $X_L = X_C$, so that $Z = R$. Thus, the average power $\overline{P}$ provided by the generator is $\overline{P} = V_{rms}^2 / R$, and solving for R we find

$$R = \frac{V_{rms}^2}{\overline{P}} = \frac{(11.0\text{ V})^2}{25.0\text{ W}} = \boxed{4.84\ \Omega}$$

c. When the generator frequency is 2.31 kHz, the individual reactances are

$$X_C = \frac{1}{2\pi f C} = \frac{1}{2\pi(2.31\times10^3\text{ Hz})(5.10\times10^{-6}\text{ F})} = 13.5\ \Omega$$

$$X_L = 2\pi f L = 2\pi(2.31\times10^3\text{ Hz})(2.94\times10^{-3}\text{ H}) = 42.7\ \Omega$$

The phase angle ϕ is, from Equation 23.8,

$$\phi = \tan^{-1}\left(\frac{X_L - X_C}{R}\right) = \tan^{-1}\left(\frac{42.7\ \Omega - 13.5\ \Omega}{4.84\ \Omega}\right) = 80.6°$$

The power factor is then given by

$$\cos\phi = \cos 80.6° = \boxed{0.163}$$

CHAPTER 24 | *ELECTROMAGNETIC WAVES*

CONCEPTUAL QUESTIONS

---

3. ***REASONING AND SOLUTION*** A transmitting antenna is located at the origin of an x, y, z axis system, and broadcasts an electromagnetic wave whose electric field oscillates along the y axis. The wave travels along the $+x$ axis. Three possible loops can be used with an LC-tuned circuit to detect the wave: One loop lies in the xy plane, another in the xz plane, and the third in the yz plane.

   The loop that will detect the electromagnetic wave must be oriented so that the normal to the loop is parallel to the magnetic field. Then, as the wave passes by the loop, the changing magnetic field penetrates the loop and results in an induced emf and current, as predicted by Faraday's law. Since the electromagnetic wave travels along the $+x$ direction, and the electric field oscillations of the electromagnetic wave are along the y axis, the magnetic field oscillations will be along the z axis. For optimum reception, therefore, the loop should lie in the xy plane so that the normal to the loop is in the z direction.

---

6. ***REASONING AND SOLUTION*** The same Doppler effect arises for electromagnetic waves when either the source or the observer of the waves moves; it is only the relative motion of the source and the observer with respect to one another that is important (see Section 24.5). Therefore, when an astronomer measures the Doppler change in frequency for the light reaching earth from a distant star, the astronomer *cannot* tell whether the star is moving away from the earth or whether the earth is moving away from the star.

---

11. ***REASONING AND SOLUTION*** Light from the sun is unpolarized; however, when the sunlight is reflected from horizontal surfaces such as the surface of a swimming pool, lake, or ocean, the reflected light is partially polarized in the horizontal direction. Polaroid sunglasses are constructed with lenses made of Polaroid (a polarizing material) with the transmission axis oriented vertically. Thus, the horizontally polarized light that is reflected from horizontal surfaces is blocked from the eyes.

    Suppose you are sitting on the beach near a lake on a sunny day, wearing Polaroid sunglasses. When you are sitting upright, the horizontally polarized light that is reflected from the lake is blocked from your eyes, as discussed above. When you lay down on your side, facing the lake, the transmission axis of the Polaroid sunglasses is now oriented in a nearly horizontal direction. Most of the horizontally polarized light that is reflected from the lake is transmitted through the sunglasses and reaches your eyes. Therefore, the sunglasses don't work as well as they did when you were sitting in an upright position.

CHAPTER 24 | *ELECTROMAGNETIC WAVES*

PROBLEMS

---

3. ***REASONING AND SOLUTION*** This is a standard exercise in units conversion. We first determine the number of meters in one light-year. The distance that light travels in one year is

$$d = ct = (3.00\times10^8 \text{ m/s})(1.00 \text{ year})\left(\frac{365.25 \text{ days}}{1 \text{ year}}\right)\left(\frac{24 \text{ hours}}{1.0 \text{ day}}\right)\left(\frac{3600 \text{ s}}{1 \text{ hour}}\right) = 9.47\times10^{15} \text{ m}$$

Thus, 1 light year $= 9.47\times10^{15}$ m. Then, the distance to Alpha Centauri is

$$(4.3 \text{ light-years})\left(\frac{9.47\times10^{15} \text{ m}}{1 \text{ light-year}}\right) = \boxed{4.1\times10^{16} \text{ m}}$$

---

5. ***REASONING*** The equation that represents the wave mathematically is $y = A\sin(2\pi ft - 2\pi x/\lambda)$. In this expression the amplitude is $A = 156$ N/C. The wavelength λ can be calculated using Equation 16.1, and we obtain

$$\lambda = \frac{c}{f} = \frac{3.00\times10^8 \text{ m/s}}{1.50\times10^8 \text{ Hz}} = 2.00 \text{ m}$$

SOLUTION

a. For $t = 0$ s, the wave expression becomes

$$y = A\sin(2\pi ft - 2\pi x/\lambda) = 156\sin\left[2\pi f(0) - \frac{2\pi x}{2.00}\right] = -156\sin\left(\frac{2\pi x}{2.00}\right) = -156\sin(\pi x)$$

In this result, the units are suppressed for convenience. The following table gives the values of the electric field obtained using this version of the wave expression with the given values of the position x. The term πx is in radians when x is in meters, and conversion from radians to degrees is accomplished using the fact that 2π rad = 360°.

x	$y = -156 \sin(\pi x)$
0 m	$-156 \sin(0) = -156 \sin(0°) = 0$
0.50 m	$-156 \sin(0.50\,\pi) = -156 \sin(90°) = -156$
1.00 m	$-156 \sin(1.00\,\pi) = -156 \sin(180°) = 0$
1.50 m	$-156 \sin(1.50\,\pi) = -156 \sin(270°) = +156$
2.00 m	$-156 \sin(2.00\,\pi) = -156 \sin(360°) = 0$

These values for the electric field are plotted in the graph shown at the right.

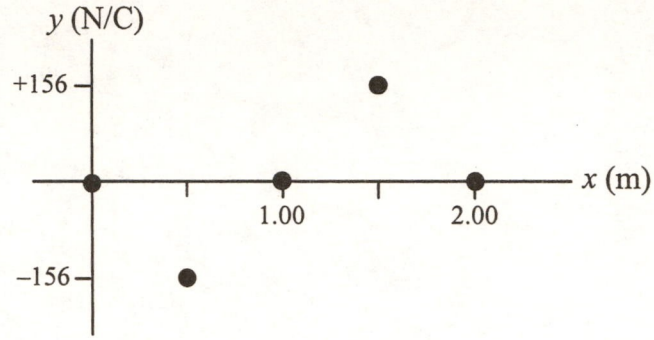

b. For $t = T/4$, we use the fact that $f = 1/T$, and the wave expression becomes

$$y = A\sin(2\pi f t - 2\pi x/\lambda) = 156\sin\left[2\pi\left(\frac{1}{T}\right)\left(\frac{T}{4}\right) - \frac{2\pi x}{2.00}\right] = 156\sin\left(\frac{\pi}{2} - \pi x\right) = 156\cos(\pi x)$$

In this result, the units are suppressed for convenience. The following table gives the values of the electric field obtained using this version of the wave expression with the given values of the position x. The term πx is in radians when x is in meters, and conversion from radians to degrees is accomplished using the fact that 2π rad = 360°.

x	$y = 156 \cos(\pi x)$
0 m	$156 \cos(0) = 156 \cos(0°) = +156$
0.50 m	$156 \cos(0.50\,\pi) = 156 \cos(90°) = 0$
1.00 m	$156 \cos(1.00\,\pi) = 156 \cos(180°) = -156$
1.50 m	$156 \cos(1.50\,\pi) = 156 \cos(270°) = 0$
2.00 m	$156 \cos(2.00\,\pi) = 156 \cos(360°) = +156$

These values for the electric field are plotted in the graph shown at the right.

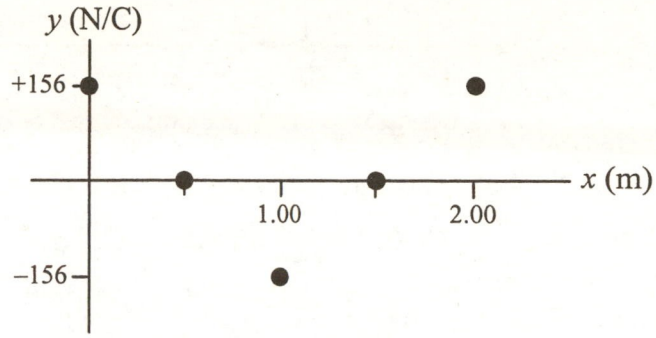

---

9. ***REASONING AND SOLUTION*** According to Equation 16.1, the wavelength is $\lambda = c/f$. Since the rods can be adjusted so that each one has a length of $\lambda/4$, the length of each rod is

$$L = \frac{\lambda}{4} = \frac{c/f}{4} = \frac{\left(3.00\times10^{8}\text{ m/s}\right)/\left(60\times10^{6}\text{ Hz}\right)}{4} = \boxed{1.25\text{ m}}$$

---

11. ***REASONING AND SOLUTION*** The number of wavelengths that can fit across the width W of your thumb is W/λ. From Equation 16.1, we know that $\lambda = c/f$, so

$$\text{No. of wavelengths} = \frac{W}{\lambda} = \frac{Wf}{c} = \frac{(2.0\times10^{-2}\text{ m})(5.5\times10^{14}\text{ Hz})}{3.0\times10^{8}\text{ m/s}} = \boxed{3.7\times10^{4}}$$

---

15. ***REASONING*** We proceed by first finding the time t for sound waves to travel between the astronauts. Since this is the same time it takes for the electromagnetic waves to travel to earth, the distance between earth and the spaceship is $d_{\text{earth-ship}} = ct$.

SOLUTION The time it takes for sound waves to travel at 343 m/s through the air between the astronauts is

$$t = \frac{d_{\text{astronaut}}}{v_{\text{sound}}} = \frac{1.5\text{ m}}{343\text{ m/s}} = 4.4\times10^{-3}\text{ s}$$

Therefore, the distance between the earth and the spaceship is

$$d_{\text{earth-ship}} = ct = (3.0\times10^{8}\text{ m/s})(4.4\times10^{-3}\text{ s}) = \boxed{1.3\times10^{6}\text{ m}}$$

---

19. ***REASONING AND SOLUTION*** If d is the distance between the mirrors, the time required for the echo to be heard is $t = 2d/v_{\text{sound}}$. In this time, the light will travel a total distance D where $D = ct = c(2d/v_{\text{sound}})$. Therefore, the number of times that the flash travels the round-trip distance between the mirrors in the time t is

$$N = \frac{D}{2d} = \frac{2cd/v_{\text{sound}}}{2d} = \frac{c}{v_{\text{sound}}} = \frac{3.00 \times 10^8 \text{ m/s}}{343 \text{ m/s}} = \boxed{8.75 \times 10^5}$$

---

23. ***REASONING*** The rms value E_{rms} of the electric field is related to the average energy density $\bar{u}$ of the microwave radiation according to Equation 24.2b: $\bar{u} = \varepsilon_0 E_{\text{rms}}^2$.

SOLUTION Solving for E_{rms} gives

$$E_{\text{rms}} = \sqrt{\frac{\bar{u}}{\varepsilon_0}} = \sqrt{\frac{4 \times 10^{-14} \text{ J/m}^3}{8.85 \times 10^{-12} \text{ C}^2/(\text{N} \cdot \text{m}^2)}} = \boxed{0.07 \text{ N/C}}$$

---

25. ***REASONING AND SOLUTION*** The energy is equal to the power P multiplied by the time t. The power, on the other hand, is equal to product of the intensity S of the wave and the area A through which the wave passes.

$$\text{Energy} = P\,t = (SA)t = (1390 \text{ W/m}^2)(25 \text{ m} \times 45 \text{ m})(3600 \text{ s}) = \boxed{5.6 \times 10^9 \text{ J}}$$

---

33. ***REASONING*** The Doppler effect for electromagnetic radiation is given by Equation 24.6, $f_o = f_s\left(1 \pm v_{\text{rel}}/c\right)$, where f_o is the observed frequency, f_s is the frequency emitted by the source, and v_{rel} is the speed of the source relative to the observer. As discussed in the text, the plus sign applies when the source and the observer are moving toward one another, while the minus sign applies when they are moving apart.

SOLUTION

a. At location A, the galaxy is moving away from the earth with a relative speed of

$$v_{\text{rel}} = (1.6 \times 10^6 \text{ m/s}) - (0.4 \times 10^6 \text{ m/s}) = 1.2 \times 10^6 \text{ m/s}$$

Therefore, the minus sign in Equation 24.6 applies and the observed frequency for the light from region A is

$$f_o = f_s\left(1 - \frac{v_{\text{rel}}}{c}\right) = (6.200 \times 10^{14} \text{ Hz})\left(1 - \frac{1.2 \times 10^6 \text{ m/s}}{3.0 \times 10^8 \text{ m/s}}\right) = \boxed{6.175 \times 10^{14} \text{ Hz}}$$

b. Similarly, at location B, the galaxy is moving away from the earth with a relative speed of

$$v_{\text{rel}} = (1.6 \times 10^6 \text{ m/s}) + (0.4 \times 10^6 \text{ m/s}) = 2.0 \times 10^6 \text{ m/s}$$

The observed frequency for the light from region B is

$$f_o = f_s\left(1 - \frac{v_{rel}}{c}\right) = (6.200 \times 10^{14}\ \text{Hz})\left(1 - \frac{2.0 \times 10^6\ \text{m/s}}{3.0 \times 10^8\ \text{m/s}}\right) = \boxed{6.159 \times 10^{14}\ \text{Hz}}$$

---

35. ***REASONING AND SOLUTION*** The average intensity of light leaving the polarizing material is given by Malus' Law (Equation 24.7). Therefore, using Malus' law, we obtain the following results:

a. $$\frac{\bar{S}}{\bar{S}_0} = \cos^2 25° = \boxed{0.82}$$

b. $$\frac{\bar{S}}{\bar{S}_0} = \cos^2 65° = \boxed{0.18}$$

---

39. ***REASONING*** The average intensity of light leaving each analyzer is given by Malus' Law (Equation 24.7). Thus, intensity of the light transmitted through the first analyzer is

$$\bar{S}_1 = \bar{S}_0 \cos^2 27°$$

Similarly, the intensity of the light transmitted through the second analyzer is

$$\bar{S}_2 = \bar{S}_1 \cos^2 27° = \bar{S}_0 \cos^4 27°$$

And the intensity of the light transmitted through the third analyzer is

$$\bar{S}_3 = \bar{S}_2 \cos^2 27° = \bar{S}_0 \cos^6 27°$$

If we generalize for the Nth analyzer, we deduce that

$$\bar{S}_N = \bar{S}_{N-1} \cos^2 27° = \bar{S}_0 \cos^{2N} 27°$$

Since we want the light reaching the photocell to have an intensity that is reduced by a least a factor of one hundred relative to the first analyzer, we want $\bar{S}_N / \bar{S}_0 = 0.010$. Therefore, we need to find N such that $\cos^{2N} 27° = 0.010$. This expression can be solved for N.

SOLUTION Taking the common logarithm of both sides of the last expression gives

$$2N \log(\cos 27°) = \log 0.010 \quad \text{or} \quad N = \frac{\log 0.010}{2 \log(\cos 27°)} = \boxed{20}$$

---

43. ***REASONING*** The electromagnetic wave will be picked up by the radio when the resonant frequency f_0 of the circuit in Figure 24.5 is equal to the frequency of the broadcast wave, or f_0 = 1400 kHz. This frequency, in turn, is related to the capacitance C and inductance L of the circuit through Equation 23.10, $f_0 = 1/\left(2\pi\sqrt{LC}\right)$. Since C is known, we can use this relation to find the inductance.

SOLUTION Solving the relation $f_0 = 1/\left(2\pi\sqrt{LC}\right)$ for the inductance L, we find that

$$L = \frac{1}{4\pi^2 f_0^2 C} = \frac{1}{4\pi^2\left(1400\times10^3\ \text{Hz}\right)^2\left(8.4\times10^{-11}\ \text{F}\right)} = \boxed{1.5\times10^{-4}\ \text{H}}$$

---

45. ***REASONING AND SOLUTION***

a. According to Equation 24.5b, the average intensity is $\overline{S} = c\varepsilon_0 E_{\text{rms}}^2$. In addition, the average intensity is the average power $\overline{P}$ divided by the area A. Therefore,

$$E_{\text{rms}} = \sqrt{\frac{\overline{S}}{c\varepsilon_0}} = \sqrt{\frac{\overline{P}}{c\varepsilon_0 A}} = \sqrt{\frac{1.20\times10^4\ \text{W}}{(3.00\times10^8\ \text{m/s})[8.85\times10^{-12}\text{C}^2/(\text{N}\cdot\text{m}^2)](135\ \text{m}^2)}} = \boxed{183\ \text{N/C}}$$

b. Then, from Equation 24.3 ($E_{\text{rms}} = cB_{\text{rms}}$), we have

$$B_{\text{rms}} = \frac{E_{\text{rms}}}{c} = \frac{183\ \text{N/C}}{3.00\times10^8\ \text{m/s}} = \boxed{6.10\times10^{-7}\ \text{T}}$$

---

47. ***REASONING AND SOLUTION*** The average intensity of light leaving each polarizer is given by Malus' Law (Equation 24.7): $\overline{S} = \overline{S}_0\cos^2\theta$. Solving for the angle θ and noting that $\overline{S} = 0.100\,\overline{S}_0$ (since 90.0% of the intensity is absorbed), we have

$$\theta = \cos^{-1}\sqrt{\frac{0.100\,\overline{S}_0}{\overline{S}_0}} = \boxed{71.6°}$$

---

51. ***REASONING AND SOLUTION*** The sun radiates sunlight (electromagnetic waves) uniformly in all directions, so the intensity at a distance r from the sun is given by Equation 16.9 as $S = P/(4\pi r^2)$, where P is the power radiated by the sun. The power that strikes an area $A_\perp$ oriented perpendicular to the direction in which the sunlight is radiated is

$P' = SA_\perp$, according to Equation 16.8. The 0.75-m^2 patch of flat land on the equator at point Q is not perpendicular to the direction of the sunlight, however.

The figure at the right shows that

$$A_\perp = (0.75\ \text{m}^2)\ \cos\ 27°$$

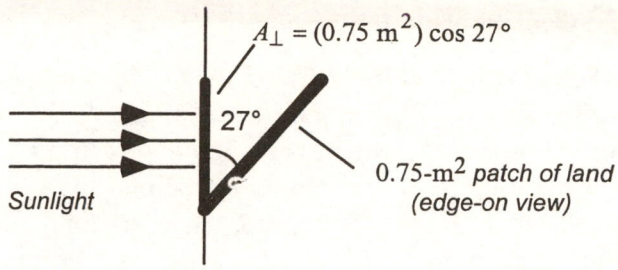

Therefore, the power striking the patch of land is

$$P' = SA_\perp = \left(\frac{P}{4\pi r^2}\right)(0.75\ \text{m}^2)\ \cos\ 27°$$

$$= \left[\frac{3.9\times10^{26}\ \text{W}}{4\pi(1.5\times10^{11}\ \text{m})^2}\right](0.75\ \text{m}^2)\ \cos\ 27° = \boxed{920\ \text{W}}$$

CHAPTER 25 | *THE REFLECTION OF LIGHT: MIRRORS*

CONCEPTUAL QUESTIONS

---

1. ***REASONING AND SOLUTION*** A sign painted on a store window is reversed when viewed from inside the store. A person inside the store views the sign in a plane mirror. As discussed in the text, the image of an object formed in a plane mirror is upright, has the same size as the object, is located as far behind the mirror as the object is in front of it, and is reversed left to right. Therefore, since the image is reversed left to right, the image of the sign painted on the store window, when viewed in a plane mirror in the store, will appear as it does when viewed from outside the store.

---

8. ***REASONING AND SOLUTION*** If you stand between two parallel plane mirrors, you see an infinite number of images of yourself. This occurs because an image in one mirror is reflected in the other mirror to produce another image, which is then re-reflected, and so-forth. The multiple images are equally spaced.

   The image produced by a convex mirror is reduced in size relative to the object. Like a plane mirror, the image is virtual and lies behind the mirror; however, the virtual image in a convex mirror is closer to the mirror than it would be if the mirror were planar. If you stand facing a convex mirror, with a plane mirror behind you, you will see an infinite number of images of yourself, as you do in the case of two parallel plane mirrors. The reason is that the first image in the convex mirror is reflected in the plane mirror, to produce an identical image behind it. The image behind the plane mirror is re-reflected in the convex mirror to produce another smaller image closer to the focal point of the convex mirror. This second image in the convex mirror is smaller than the first one, because the image behind the plane mirror is farther from the convex mirror than you are. Further reflections and re-reflections occur, leading to a series of images in the convex mirror that decrease in size as they occur closer and closer to the focal point. The images "pile up" so to speak, becoming closer and closer together as they approach the focal point. The size of the image becomes zero at the focal point. Thus, the series of images appears to vanish at the focal point of the convex mirror.

---

13. ***REASONING AND SOLUTION*** Suppose you stand in front of a spherical mirror (concave or convex). As shown in Figures 25.19 and 25.20, a concave mirror can form a real, inverted image (Figures 25.19*a* and 25.19*b*) or a virtual, upright image (Figure 25.20). As shown in Figure 25.22, a convex mirror can only form a virtual, upright image.

    a. Therefore, it is not possible for your image to be real and upright.

    b. Similarly, it is not possible for your image to be virtual and inverted.

CHAPTER 25 THE REFLECTION OF LIGHT: MIRRORS

PROBLEMS

1. ***REASONING AND SOLUTION*** The drawing at the right shows a ray diagram in which the reflected rays have been projected behind the mirror. We can see by inspection of this drawing that, after the rays reflect from the plane mirror, the angle α between them is still $\boxed{10°}$.

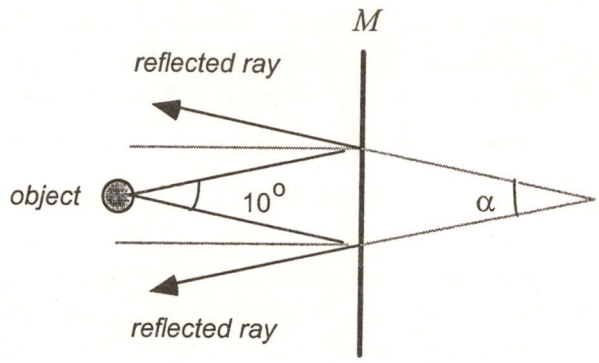

5. ***REASONING*** The geometry is shown below. According to the law of reflection, the incident ray, the reflected ray, and the normal to the surface all lie in the same plane, and the angle of reflection θ_r equals the angle of incidence θ_i. We can use the law of reflection and the properties of triangles to determine the angle θ at which the ray leaves M_2.

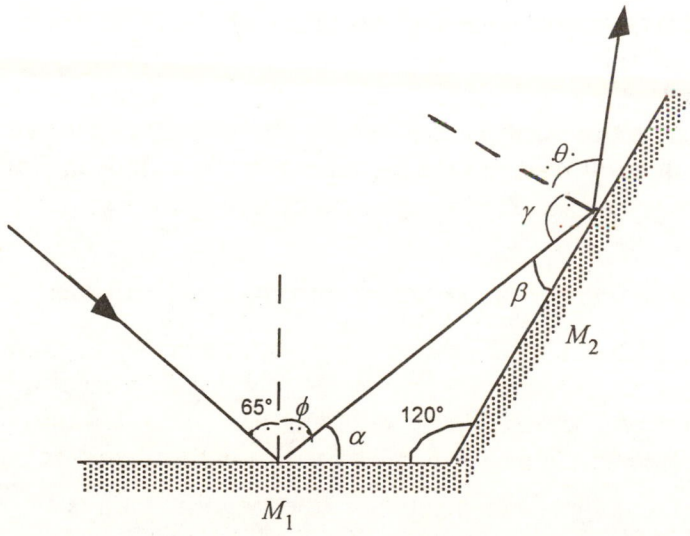

SOLUTION From the law of reflection, we know that $\phi = 65°$. We see from the figure that $\phi + \alpha = 90°$, or $\alpha = 90° - \phi = 90° - 65° = 25°$. From the figure and the fact that the sum of the interior angles in any triangle is 180°, we have $\alpha + \beta + 120° = 180°$. Solving for β, we find that $\beta = 180° - (120° + 25°) = 35°$. Therefore, since $\beta + \gamma = 90°$, we find that the angle γ is given by $\gamma = 90° - \beta = 90° - 35° = 55°$. Since γ is the angle of incidence of the ray on mirror M_2, we know from the law of reflection that $\boxed{\theta = 55°}$.

---

13. ***REASONING AND SOLUTION*** The ray diagram is shown below. (Note: $f = -50.0$ cm and $d_o = 25.0$ cm)

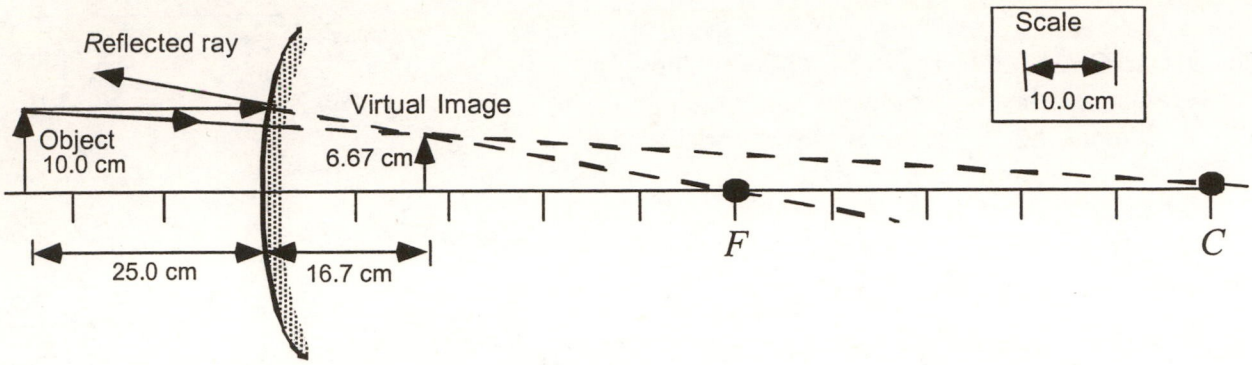

a. The ray diagram indicates that $\boxed{\text{the image is 16.7 cm behind the mirror}}$.

b. The ray diagram indicates that $\boxed{\text{the image height is 6.67 cm}}$.

---

17. ***REASONING*** We have seen that a convex mirror always forms a *virtual image* as shown in Figure 25.22*a* of the text, where the image is *upright* and *smaller* than the object. These characteristics should bear out in the results of our calculations.

SOLUTION The radius of curvature of the convex mirror is 68 cm. Therefore, the focal length is, from Equation 25.2, $f = -(1/2)R = -34$ cm. Since the image is virtual, we know that $d_i = -22$ cm.

a. With $d_i = -22$ cm and $f = -34$ cm, the mirror equation gives

$$\frac{1}{d_o} = \frac{1}{f} - \frac{1}{d_i} = \frac{1}{-34\text{ cm}} - \frac{1}{-22\text{ cm}} \qquad \text{or} \qquad \boxed{d_o = +62\text{ cm}}$$

b. According to the magnification equation, the magnification is

$$m = -\frac{d_i}{d_o} = -\frac{-22\text{ cm}}{62\text{ cm}} = \boxed{+0.35}$$

c. Since the magnification m is positive, the image is $\boxed{\text{upright}}$.

d. Since the magnification m is less than one, the image is $\boxed{\text{smaller}}$ than the object.

---

21. ***REASONING*** This problem can be solved by using the mirror equation, Equation 25.3, and the magnification equation, Equation 25.4.

SOLUTION

a. Using the mirror equation with $d_i = d_o$ and $f = R/2$, we have

$$\frac{1}{d_o} = \frac{1}{f} - \frac{1}{d_i} = \frac{1}{R/2} - \frac{1}{d_o} \qquad \text{or} \qquad \frac{2}{d_o} = \frac{2}{R}$$

Therefore, we find that $\boxed{d_o = R}$.

b. According to the magnification equation, the magnification is

$$m = -\frac{d_i}{d_o} = -\frac{d_o}{d_o} = \boxed{-1}$$

c. Since the magnification m is negative, the image is $\boxed{\text{inverted}}$.

---

23. ***REASONING*** When paraxial light rays that are parallel to the principal axis strike a convex mirror, the reflected rays diverge after being reflected, and appear to originate from the focal point F behind the mirror (see Figure 25.16). We can treat the sun as being infinitely far from the mirror, so it is reasonable to treat the incident rays as paraxial rays that are parallel to the principal axis.

SOLUTION

a. Since the sun is infinitely far from the mirror and its image is a virtual image that lies *behind* the mirror, we can conclude that the mirror is a $\boxed{\text{convex mirror}}$.

b. With $d_i = -12.0$ cm and $d_o = \infty$, the mirror equation (Equation 25.3) gives

$$\frac{1}{f} = \frac{1}{d_o} + \frac{1}{d_i} = \frac{1}{\infty} + \frac{1}{d_i} = \frac{1}{d_i}$$

Therefore, the focal length f lies 12.0 cm behind the mirror (this is consistent with the reasoning above that states that, after being reflected, the rays appear to originate from the focal point behind the mirror). In other words, $f = -12.0$ cm. Then, according to Equation 25.2, $f = -\frac{1}{2}R$, and the radius of curvature is

$$R = -2f = -2(-12.0\text{ cm}) = \boxed{24.0\text{ cm}}$$

---

27. ***REASONING AND SOLUTION*** The magnification equation, Equation 25.4, indicates that $d_i = -md_o$. Substituting this result into the mirror equation, we find that

$$\frac{1}{d_o} + \frac{1}{d_i} = \frac{1}{d_o} + \frac{1}{-md_o} = \frac{1}{f}$$

Solving for the object distance d_o, we find

$$d_o = \frac{(m-1)f}{m}$$

After the object is moved to its new position, its object distance becomes d_o' and its magnification becomes $m' = 2m$. Therefore, the amount by which the object is moved is

$$d_o - d_o' = \frac{(m-1)f}{m} - \frac{(m'-1)f}{m'} = \frac{(m-1)f}{m} - \frac{(2m-1)f}{2m}$$

$$= -\frac{f}{2m} = -\frac{-24.0\text{ cm}}{2(0.150)} = \boxed{+80.0\text{ cm}}$$

The positive answer means that the initial object distance is larger than the final object distance, so that $\boxed{\text{the object is moved toward the mirror}}$.

---

31. ***REASONING*** This problem can be solved using the mirror equation, Equation 25.3.

SOLUTION Using the mirror equation with $d_i = +26\text{ cm}$ and $f = 12\text{ cm}$, we find

$$\frac{1}{d_o} = \frac{1}{f} - \frac{1}{d_i} = \frac{1}{12\text{ cm}} - \frac{1}{26\text{ cm}} \qquad \text{or} \qquad \boxed{d_o = +22\text{ cm}}$$

---

35. ***REASONING*** This problem can be solved using the mirror equation, Equation 25.3.

SOLUTION Using the mirror equation with $d_i = +97\text{ cm}$ and $f = 42\text{ cm}$, we find

$$\frac{1}{d_o} = \frac{1}{f} - \frac{1}{d_i} = \frac{1}{42\text{ cm}} - \frac{1}{97\text{ cm}} \qquad \text{or} \qquad \boxed{d_o = +74\text{ cm}}$$

---

39. ***REASONING*** We need to know the focal length of the mirror and can obtain it from the mirror equation, Equation 25.3, as applied to the first object:

$$\frac{1}{d_{o1}}+\frac{1}{d_{i1}}=\frac{1}{14.0\text{ cm}}+\frac{1}{-7.00\text{ cm}}=\frac{1}{f} \qquad \text{or} \qquad f=-14.0\text{ cm}$$

According to the magnification equation, Equation 25.4, the image height h_i is related to the object height h_o as follows: $h_i = mh_o = (-d_i / d_o)h_o$.

SOLUTION Applying this result to each object, we find that $h_{i2} = h_{i1}$, or

$$\left(\frac{-d_{i2}}{d_{o2}}\right)h_{o2}=\left(\frac{-d_{i1}}{d_{o1}}\right)h_{o1}$$

Therefore,

$$d_{i2}=d_{o2}\left(\frac{d_{i1}}{d_{o1}}\right)\left(\frac{h_{o1}}{h_{o2}}\right)$$

Using the fact that $h_{o2} = 2h_{o1}$, we have

$$d_{i2}=d_{o2}\left(\frac{d_{i1}}{d_{o1}}\right)\left(\frac{h_{o1}}{h_{o2}}\right)=d_{o2}\left(\frac{-7.00\text{ cm}}{14.0\text{ cm}}\right)\left(\frac{h_{o1}}{2h_{o1}}\right)=-0.250\,d_{o2}$$

Using this result in the mirror equation, as applied to the second object, we find that

$$\frac{1}{d_{o2}}+\frac{1}{d_{i2}}=\frac{1}{f}$$

or

$$\frac{1}{d_{o2}}+\frac{1}{-0.250\,d_{o2}}=\frac{1}{-14.0\text{ cm}}$$

Therefore,

$$\boxed{d_{o2}=+42.0\text{ cm}}$$

41. ***REASONING*** The time of travel is proportional to the total distance for each path. Therefore, using the distances identified in the drawing, we have

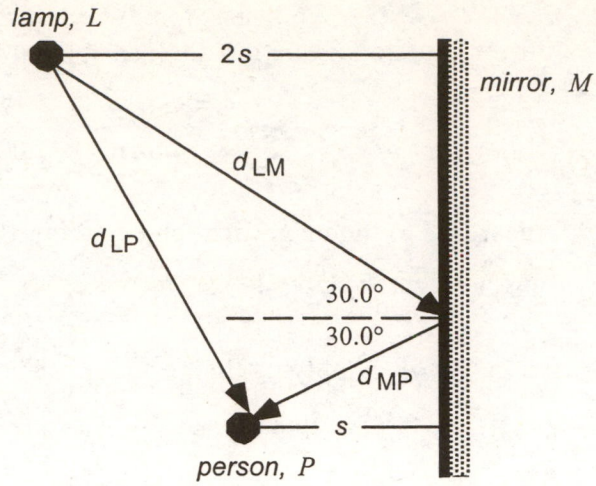

$$\frac{t_{\text{reflected}}}{t_{\text{direct}}} = \frac{d_{LM} + d_{MP}}{d_{LP}} \qquad (1)$$

We know that the angle of incidence is equal to the angle of reflection. We also know that the lamp L is a distance $2s$ from the mirror, while the person P is a distance s from the mirror.

Therefore, it follows that if $d_{MP} = d$, then $d_{LM} = 2d$. We may use the law of cosines (see Appendix E) to express the distance d_{LP} as

$$d_{LP} = \sqrt{(2d)^2 + d^2 - 2(2d)(d)\cos 2(30.0°)}$$

SOLUTION Substituting the expressions for d_{MP}, d_{LM}, and d_{LP} into Equation (1), we find that the ratio of the travel times is

$$\frac{t_{\text{reflected}}}{t_{\text{direct}}} = \frac{2d + d}{\sqrt{(2d)^2 + d^2 - 2(2d)(d)\cos 2(30.0°)}} = \frac{3}{\sqrt{5 - 4\cos 60.0°}} = \boxed{1.73}$$

CHAPTER 26 | *THE REFRACTION OF LIGHT: LENSES AND OPTICAL INSTRUMENTS*

CONCEPTUAL QUESTIONS

---

4. ***REASONING AND SOLUTION*** Two identical containers, one filled with water ($n = 1.33$) and the other filled with ethyl alcohol ($n = 1.36$) are viewed from directly above. According to Equation 26.3, when viewed from directly above in a medium of refractive index n_2, the apparent depth d' in a medium of refractive index n_1 is related to the actual depth d by the relation $d' = d(n_2 / n_1)$. Assuming that the observer is in air, $n_2 = 1.00$. Since n_1 refers to the refractive index of the liquid in the containers, we see that the apparent depth in each liquid is inversely proportional to the refractive index of the liquid. The index of refraction of water is smaller than that of ethyl alcohol; therefore, the container filled with water appears to have the greater depth of fluid.

---

14. ***REASONING AND SOLUTION*** A person is floating on an air mattress in the middle of a swimming pool. His friend is sitting on the side of the pool. The person on the air mattress claims that there is a light shining up from the bottom of the pool directly beneath him. His friend insists, however, that she cannot see any light from where she sits on the side.

Rays from a light source on the bottom of the pool will radiate outward from the source in all directions. However, only rays for which the angle of incidence is less than the critical angle will emerge from the water. Rays with an angle of incidence equal to, or greater than, the critical angle will undergo total internal reflection back into the water, as shown in the following figure.

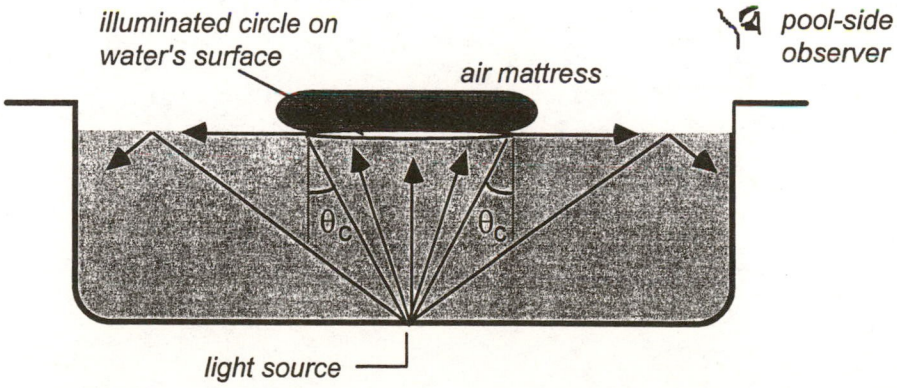

Because of the geometry, the rays that leave the water lie within a cone whose apex lies at the light source. Thus, rays of light that leave the water emerge from within an illuminated circle just above the source. If the mattress is just over the source, it could cover the area through which the light would emerge. A person sitting on the side of the pool would not see any light emerging. Therefore, the statements made by both individuals are correct.

---

21. ***REASONING AND SOLUTION*** The expert claims that the height of the window can be calculated from only two pieces of information : (1) the measured height on the film, and (2) the focal length of the camera. The expert is not correct. According to the thin-lens equation (Equation 26.6), $1/d_o + 1/d_i = 1/f$, where d_o is the object distance, d_i is the image distance, and f is the focal length of the lens. The magnification equation (Equation 26.7), relates the image and object heights to the image and object distances: $h_o/h_i = -d_o/d_i$. These two equations contain five unknowns. To determine any one of the unknowns, three of the other unknowns must be known. In this case, all that we know is the height of the image, h_i, and the focal length of the camera, f. Therefore, we do not have enough information given to determine the distance from the ground to the window (the height of the object in this case), h_o. We still need to know either the distance from the photographer to the house (the object distance, d_o), or the distance from the center of the lens to the film (the image distance, d_i). We can conclude, therefore, that the expert is incorrect.

CHAPTER 26 THE REFRACTION OF LIGHT: LENSES AND OPTICAL INSTRUMENTS

PROBLEMS

1. ***REASONING AND SOLUTION*** The speed of light in benzene v is related to the speed of light in vacuum c by the index of refraction n. The index of refraction is defined by Equation 26.1 ($n = c/v$). According to Table 26.1, the index of refraction of benzene is 1.501. Therefore, solving for v, we have

$$v = \frac{c}{n} = \frac{3.00 \times 10^8 \text{ m/s}}{1.501} = \boxed{2.00 \times 10^8 \text{ m/s}}$$

3. ***REASONING*** Since the light will travel in glass at a constant speed v, the time it takes to pass perpendicularly through the glass is given by $t = d/v$, where d is the thickness of the glass. The speed v is related to the vacuum value c by Equation 26.1: $n = c/v$.

SOLUTION Substituting for v from Equation 26.1 and substituting values, we obtain

$$t = \frac{d}{v} = \frac{nd}{c} = \frac{(1.5)(4.0 \times 10^{-3} \text{ m})}{3.00 \times 10^8 \text{ m/s}} = \boxed{2.0 \times 10^{-11} \text{ s}}$$

9. ***REASONING AND SOLUTION***

a. We know from the law of reflection (Section 25.2), that the angle of reflection is equal to the angle of incidence, so the reflected ray is reflected at $\boxed{43°}$.

b. Snell's law of refraction (Equation 26.2: $n_1 \sin\theta_1 = n_2 \sin\theta_2$ can be used to find the angle of refraction. Table 26.1 indicates that the index of refraction of water is 1.333. Solving for θ_2 and substituting values, we find that

$$\sin\theta_2 = \frac{n_1 \sin\theta_1}{n_2} = \frac{(1.000)(\sin 43°)}{1.333} = 0.51 \qquad \text{or} \qquad \theta_2 = \sin^{-1} 0.51 = \boxed{31°}$$

13. ***REASONING*** We begin by using Snell's law (Equation 26.2: $n_1 \sin\theta_1 = n_2 \sin\theta_2$) to find the index of refraction of the material. Then we will use Equation 26.1, the definition of the index of refraction ($n = c/v$) to find the speed of light in the material.

SOLUTION From Snell's law, the index of refraction of the material is

$$n_2 = \frac{n_1 \sin\theta_1}{\sin\theta_2} = \frac{(1.000)\sin 63.0°}{\sin 47.0°} = 1.22$$

Then, from Equation 26.1, we find that the speed of light v in the material is

$$v = \frac{c}{n_2} = \frac{3.00 \times 10^8 \text{ m/s}}{1.22} = \boxed{2.46 \times 10^8 \text{ m/s}}$$

---

19. The drawing at the right shows the geometry of the situation using the same notation as that in Figure 26.6. In addition to the text's notation, let t represent the thickness of the pane, let L represent the length of the ray in the pane, let x (shown twice in the figure) equal the displacement of the ray, and let the difference in angles $\theta_1 - \theta_2$ be given by ϕ.

We wish to find the amount x by which the emergent ray is displaced relative to the incident ray. This can be done by applying Snell's law at each interface, and then making use of the geometric and trigonometric relations in the drawing.

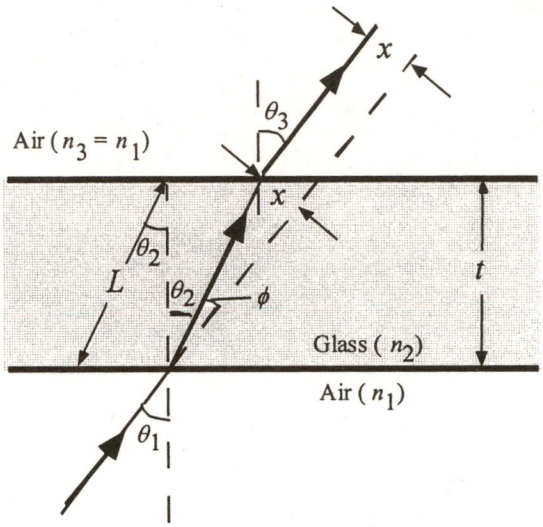

SOLUTION If we apply Snell's law (see Equation 26.2) to the bottom interface we obtain $n_1 \sin\theta_1 = n_2 \sin\theta_2$. Similarly, if we apply Snell's law at the top interface where the ray emerges, we have $n_2 \sin\theta_2 = n_3 \sin\theta_3 = n_1 \sin\theta_3$. Comparing this with Snell's law at the bottom face, we see that $n_1 \sin\theta_1 = n_1 \sin\theta_3$, from which we can conclude that $\theta_3 = \theta_1$. Therefore, the emerging ray is parallel to the incident ray.

From the geometry of the ray and thickness of the pane, we see that $L\cos\theta_2 = t$, from which it follows that $L = t/\cos\theta_2$. Furthermore, we see that $x = L\sin\phi = L\sin(\theta_1 - \theta_2)$. Substituting for L, we find

$$x = L\sin(\theta_1 - \theta_2) = \frac{t\sin(\theta_1 - \theta_2)}{\cos\theta_2}$$

Before we can use this expression to determine a numerical value for x, we must find the value of θ_2. Solving the expression for Snell's law at the bottom interface for θ_2, we have

$$\sin\theta_2 = \frac{n_1 \sin\theta_1}{n_2} = \frac{(1.000)\,(\sin 30.0°)}{1.52} = 0.329 \qquad \text{or} \qquad \theta_2 = \sin^{-1} 0.329 = 19.2°$$

Therefore, the amount by which the emergent ray is displaced relative to the incident ray is

$$x = \frac{t\,\sin(\theta_1 - \theta_2)}{\cos\theta_2} = \frac{(6.00\text{ mm})\sin(30.0° - 19.2°)}{\cos 19.2°} = \boxed{1.19\text{ mm}}$$

---

21. ***REASONING*** The drawing at the right shows the situation. As discussed in the text, when the observer is directly above, the apparent depth d' of the object is related to the actual depth by Equation 26.3:

$$d' = d\left(\frac{n_2}{n_1}\right)$$

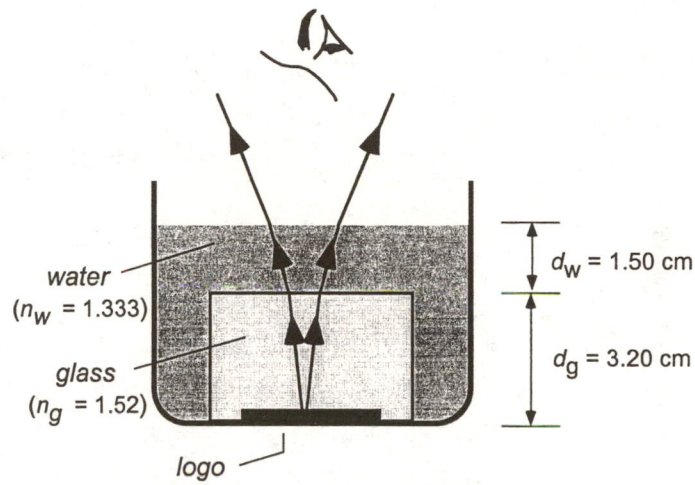

In this case, we must apply Equation 26.3 twice; once for the rays in the glass, and once again for the rays in the water.

SOLUTION We refer to the drawing for our notation and begin at the logo. To an observer *in the water* directly above the logo, the apparent depth of the logo is $d'_g = d_g\left(n_w / n_g\right)$. When viewed directly from above in air, the logo's apparent depth is, $d'_w = \left(d_w + d'_g\right)\left(n_{air} / n_w\right)$, where we have used the fact that when viewed from air, the logo's actual depth appears to be $d_w + d'_g$. Substituting the expression for d'_g into the expression for d'_w, we obtain

$$d'_w = (d_w + d'_g)\left(\frac{n_{air}}{n_w}\right) = d_w\left(\frac{n_{air}}{n_w}\right) + d_g\left(\frac{n_w}{n_g}\right)\left(\frac{n_{air}}{n_w}\right) = d_w\left(\frac{n_{air}}{n_w}\right) + d_g\left(\frac{n_{air}}{n_g}\right)$$

$$= (1.50\text{ cm})\left(\frac{1.000}{1.333}\right) + (3.20\text{ cm})\left(\frac{1.000}{1.52}\right) = \boxed{3.23\text{ cm}}$$

---

23. ***REASONING AND SOLUTION*** According to Equation 26.4, the critical angle is related to the refractive indices n_1 and n_2 by $\sin\theta_c = n_2 / n_1$, where $n_1 > n_2$. Solving for n_1, we find

$$n_1 = \frac{n_2}{\sin\theta_c} = \frac{1.000}{\sin 40.5°} = \boxed{1.54}$$

---

31. ***REASONING*** Since the light reflected from the coffee table is completely polarized parallel to the surface of the glass, the angle of incidence must be the Brewster angle ($\theta_B = 56.7°$) for the air-glass interface. We can use Brewster's law (Equation 26.5: $\tan\theta_B = n_2 / n_1$) to find the index of refraction n_2 of the glass.

SOLUTION Solving Brewster's law for n_2, we find that the refractive index of the glass is

$$n_2 = n_1 \tan\theta_B = (1.000)(\tan 56.7°) = \boxed{1.52}$$

---

35. ***REASONING*** Brewster's law (Equation 26.5: $\tan\theta_B = n_2 / n_1$) relates the angle of incidence θ_B at which the reflected ray is completely polarized parallel to the surface to the indices of refraction n_1 and n_2 of the two media forming the interface. We can use Brewster's law for light incident from above to find the ratio of the refractive indices n_2/n_1. This ratio can then be used to find the Brewster angle for light incident from below on the same interface.

SOLUTION The index of refraction for the medium in which the incident ray occurs is designated by n_1. For the light striking from above $n_2 / n_1 = \tan\theta_B = \tan 65.0° = 2.14$. The same equation can be used when the light strikes from below if the indices of refraction are interchanged

$$\theta_B = \tan^{-1}\left(\frac{n_1}{n_2}\right) = \tan^{-1}\left(\frac{1}{n_2/n_1}\right) = \tan^{-1}\left(\frac{1}{2.14}\right) = \boxed{25.0°}$$

---

41. ***REASONING*** Because the refractive index of the glass depends on the wavelength (i.e., the color) of the light, the rays corresponding to different colors are bent by different amounts in the glass. We can use Snell's law (Equation 26.2: $n_1 \sin\theta_1 = n_2 \sin\theta_2$) to find the angle of refraction for the violet ray and the red ray. The angle between these rays can be found by the subtraction of the two angles of refraction.

SOLUTION In Table 26.2 the index of refraction for violet light in crown glass is 1.538, while that for red light is 1.520. According to Snell's law, then, the sine of the angle of refraction for the violet ray in the glass is $\sin\theta_2 = (1.000/1.538)\ \sin 45.00° = 0.4598$, so that

$$\theta_2 = \sin^{-1}(0.4598) = 27.37°$$

Similarly, for the red ray, $\sin\theta_2 = (1.000/1.520)\ \sin 45.00° = 0.4652$, from which it follows that

$$\theta_2 = \sin^{-1}(0.4652) = 27.72°$$

Therefore, the angle between the violet ray and the red ray in the glass is

$$27.72° - 27.37° = \boxed{0.35°}$$

---

43. ***REASONING*** We can use Snell's law (Equation 26.2: $n_1 \sin\theta_1 = n_2 \sin\theta_2$) at each face of the prism. At the first interface where the ray enters the prism, $n_1 = 1.000$ for air and $n_2 = n_g$ for glass. Thus, Snell's law gives

$$(1)\sin 60.0° = n_g \sin\theta_2 \qquad \text{or} \qquad \sin\theta_2 = \frac{\sin 60.0°}{n_g} \tag{1}$$

We will represent the angles of incidence and refraction at the second interface as θ_1' and θ_2', respectively. Since the triangle is an equilateral triangle, the angle of incidence at the second interface, where the ray emerges back into air, is $\theta_1' = 60.0° - \theta_2$. Therefore, at the second interface, where $n_1 = n_g$ and $n_2 = 1.000$, Snell's law becomes

$$n_g \sin\ (60.0° - \theta_2) = (1)\sin\theta_2' \tag{2}$$

We can now use Equations (1) and (2) to determine the angles of refraction θ_2' at which the red and violet rays emerge into the air from the prism.

SOLUTION

Red Ray The index of refraction of flint glass at the wavelength of red light is $n_g = 1.662$. Therefore, using Equation (1), we can find the angle of refraction for the red ray as it enters the prism:

$$\sin\theta_2 = \frac{\sin 60.0°}{1.662} = 0.521 \qquad \text{or} \qquad \theta_2 = \sin^{-1} 0.521 = 31.4°$$

Substituting this value for θ_2 into Equation (2), we can find the angle of refraction at which the red ray emerges from the prism:

$$\sin\theta_2' = 1.662\sin(60.0° - 31.4°) = 0.796 \qquad \text{or} \qquad \theta_2' = \sin^{-1} 0.796 = \boxed{52.7°}$$

Violet Ray For violet light, the index of refraction for glass is $n_g = 1.698$. Again using Equation (1), we find

$$\sin\theta_2 = \frac{\sin 60.0°}{1.698} = 0.510 \qquad \text{or} \qquad \theta_2 = \sin^{-1} 0.510 = 30.7°$$

Using Equation (2), we find

$$\sin\theta_2' = 1.698\sin(60.0° - 30.7°) = 0.831 \qquad \text{or} \qquad \theta_2' = \sin^{-1} 0.831 = \boxed{56.2°}$$

---

45. ***REASONING AND SOLUTION*** Equation 26.6 gives the thin-lens equation which relates the object and image distances d_o and d_i, respectively, to the focal length f of the lens: $(1/d_o) + (1/d_i) = (1/f)$.

The optical arrangement is similar to that in Figure 26.26. The problem statement gives values for the focal length ($f = 50.0$ mm) and the maximum lens-to-film distance ($d_i = 275$ mm). Therefore, the maximum distance that the object can be located in front of the lens is

$$\frac{1}{d_o} = \frac{1}{f} - \frac{1}{d_i} = \frac{1}{50.0\text{ mm}} - \frac{1}{275\text{ mm}} \qquad \text{or} \qquad \boxed{d_o = 61.1\text{ mm}}$$

---

53. ***REASONING*** The optical arrangement is similar to that in Figure 26.26. We begin with the thin-lens equation, [Equation 26.6: $(1/d_o) + (1/d_i) = (1/f)$]. Since the distance between the moon and the camera is so large, the object distance d_o is essentially infinite, and $1/d_o = 1/\infty = 0$. Therefore the thin-lens equation becomes $1/d_i = 1/f$ or $d_i = f$. The diameter of the moon's imagine on the slide film is equal to the image height h_i, as given by the magnification equation (Equation 26.7: $h_i / h_o = -d_i / d_o$).

When the slide is projected onto a screen, the situation is similar to that in Figure 26.27. In this case, the thin-lens and magnification equations can be used in their usual forms.

SOLUTION

a. Solving the magnification equation for h_i gives

$$h_i = -h_o \frac{d_i}{d_o} = (-3.48 \times 10^6 \text{ m})\left(\frac{50.0 \times 10^{-3} \text{ m}}{3.85 \times 10^8 \text{ m}}\right) = -4.52 \times 10^{-4} \text{ m}$$

The diameter of the moon's image on the slide film is, therefore, $\boxed{4.52 \times 10^{-4} \text{ m}}$.

b. From the magnification equation, $h_i = -h_o\left(d_i / d_o\right)$. We need to find the ratio d_i / d_o. Beginning with the thin-lens equation, we have

$$\frac{1}{d_o} + \frac{1}{d_i} = \frac{1}{f} \quad \text{or} \quad \frac{1}{d_o} = \frac{1}{f} - \frac{1}{d_i} \quad \text{which leads to} \quad \frac{d_i}{d_o} = \frac{d_i}{f} - \frac{d_i}{d_i} = \frac{d_i}{f} - 1$$

Therefore,

$$h_i = -h_o\left(\frac{d_i}{f} - 1\right) = -\left(4.52 \times 10^{-4} \text{ m}\right)\left(\frac{15.0 \text{ m}}{110.0 \times 10^{-3} \text{ m}} - 1\right) = -6.12 \times 10^{-2} \text{ m}$$

The diameter of the image on the screen is $\boxed{6.12 \times 10^{-2} \text{ m}}$.

---

55. ***REASONING*** The magnification equation (Equation 26.7) relates the object and image distances d_o and d_i, respectively, to the relative size of the of the image and object: $m = -(d_i / d_o)$. We consider two cases: in case 1, the object is placed 18 cm in front of a diverging lens. The magnification for this case is given by m_1. In case 2, the object is moved so that the magnification m_2 is reduced by a factor of 2 compared to that in case 1. In other words, we have $m_2 = \frac{1}{2} m_1$. Using Equation 26.7, we can write this as

$$-\frac{d_{i2}}{d_{o2}} = -\frac{1}{2}\left(\frac{d_{i1}}{d_{o1}}\right) \tag{1}$$

This expression can be solved for d_{o2}. First, however, we must find a numerical value for d_{i1}, and we must eliminate the variable d_{i2}.

SOLUTION

The image distance for case 1 can be found from the thin-lens equation [Equation 26.6: $(1/d_o) + (1/d_i) = (1/f)$]. The problem statement gives the focal length as $f = -12$ cm. Since the object is 18 cm in front of the diverging lens, $d_{o1} = 18$ cm. Solving for d_{i1}, we find

$$\frac{1}{d_{i1}} = \frac{1}{f} - \frac{1}{d_{o1}} = \frac{1}{-12 \text{ cm}} - \frac{1}{18 \text{ cm}} \quad \text{or} \quad d_{i1} = -7.2 \text{ cm}$$

where the minus sign indicates that the image is virtual. Solving Equation (1) for d_{o2}, we have

$$d_{o2} = 2d_{i2}\left(\frac{d_{o1}}{d_{i1}}\right) \tag{2}$$

To eliminate d_{i2} from this result, we note that the thin-lens equation applied to case 2 gives

$$\frac{1}{d_{i2}} = \frac{1}{f} - \frac{1}{d_{o2}} = \frac{d_{o2} - f}{f d_{o2}} \qquad \text{or} \qquad d_{i2} = \frac{f d_{o2}}{d_{o2} - f}$$

Substituting this expression for d_{i2} into Equation (2), we have

$$d_{o2} = \left(\frac{2 f d_{o2}}{d_{o2} - f}\right)\left(\frac{d_{o1}}{d_{i1}}\right) \qquad \text{or} \qquad d_{o2} - f = 2f\left(\frac{d_{o1}}{d_{i1}}\right)$$

Solving for d_{o2}, we find

$$d_{o2} = f\left[2\left(\frac{d_{o1}}{d_{i1}}\right) + 1\right] = (-12\text{cm})\left[2\left(\frac{18\text{ cm}}{-7.2\text{ cm}}\right) + 1\right] = \boxed{48\text{ cm}}$$

---

63. ***REASONING*** The problem can be solved using the thin-lens equation [Equation 26.6: $(1/d_o) + (1/d_i) = (1/f)$] twice in succession. We begin by using the thin lens-equation to find the location of the image produced by the converging lens; this image becomes the object for the diverging lens.

SOLUTION

a. The image distance for the converging lens is determined as follows:

$$\frac{1}{d_{i1}} = \frac{1}{f} - \frac{1}{d_{o1}} = \frac{1}{12.0\text{ cm}} - \frac{1}{36.0\text{ cm}} \qquad \text{or} \qquad d_{i1} = 18.0\text{ cm}$$

This image acts as the object for the diverging lens. Therefore,

$$\frac{1}{d_{i2}} = \frac{1}{f} - \frac{1}{d_{o2}} = \frac{1}{-6.00\text{ cm}} - \frac{1}{(30.0\text{ cm} - 18.0\text{ cm})} \qquad \text{or} \qquad d_{i2} = -4.00\text{ cm}$$

Thus, the final image is located $\boxed{\text{4.00 cm to the left of the diverging lens}}$.

b. The magnification equation (Equation 26.7: $h_i / h_o = -d_i / d_o$) gives

$$\underbrace{m_c = -\frac{d_{i1}}{d_{o1}} = -\frac{18.0\text{ cm}}{36.0\text{ cm}} = -0.500}_{\text{Converging lens}} \qquad \underbrace{m_d = -\frac{d_{i2}}{d_{o2}} = -\frac{-4.00\text{ cm}}{12.0\text{ cm}} = 0.333}_{\text{Diverging lens}}$$

Therefore, the overall magnification is given by the product $m_c m_d = \boxed{-0.167}$.

c. Since the final image distance is negative, we can conclude that the image is $\boxed{\text{virtual}}$.

d. Since the overall magnification of the image is negative, the image is $\boxed{\text{inverted}}$.

e. The magnitude of the overall magnification is less than one; therefore, the final image is $\boxed{\text{smaller}}$.

---

65. ***REASONING*** We begin by using the thin-lens equation [Equation 26.6: $(1/d_o)+(1/d_i)=(1/f)$] to locate the image produced by the lens. This image is then treated as the object for the mirror.

SOLUTION

a. The image distance from the diverging lens can be determined as follows:

$$\frac{1}{d_i} = \frac{1}{f} - \frac{1}{d_o} = \frac{1}{-8.00\text{ cm}} - \frac{1}{20.0\text{ cm}} \qquad \text{or} \qquad d_i = -5.71\text{ cm}$$

The image produced by the lens is 5.71 cm to the left of the lens. The distance between this image and the concave mirror is 5.71 cm + 30.0 cm = 35.7 cm. The mirror equation [Equation 25.3: $(1/d_o)+(1/d_i)=(1/f)$] gives the image distance from the mirror:

$$\frac{1}{d_i} = \frac{1}{f} - \frac{1}{d_o} = \frac{1}{12.0\text{ cm}} - \frac{1}{35.7\text{ cm}} \qquad \text{or} \qquad \boxed{d_i = 18.1\text{ cm}}$$

b. The image is $\boxed{\text{real}}$, because d_i is a positive number, indicating that the final image lies to the left of the concave mirror.

c. The image is $\boxed{\text{inverted}}$, because a diverging lens always produces an upright image, and the concave mirror produces an inverted image when the object distance is greater than the focal length of the mirror.

---

69. ***REASONING*** The far point is 5.0 m from the right eye, and 6.5 m from the left eye. For an object infinitely far away ($d_o = \infty$), the image distances for the corrective lenses are then –5.0 m for the right eye and –6.5 m for the left eye, the negative sign indicating that the images are virtual images formed to the left of the lenses. The thin-lens equation [Equation 26.6: $(1/d_o)+(1/d_i)=(1/f)$] can be used to find the focal length. Then, Equation 26.8 can be used to find the refractive power for the lens for each eye.

SOLUTION Since the object distance d_o is essentially infinite, $1/d_o = 1/\infty = 0$, and the thin-lens equation becomes $1/d_i = 1/f$, or $d_i = f$. Therefore, for the right eye, $f = -5.0\text{ m}$, and the refractive power is (see Equation 26.8)

[Right eye]

$$\begin{matrix}\text{Refractive power}\\ \text{(in diopters)}\end{matrix} = \frac{1}{f} = \frac{1}{(-5.0\text{ m})} = \boxed{-0.20\text{ diopters}}$$

Similarly, for the left eye, $f = -6.5\text{ m}$, and the refractive power is

[Left eye]

$$\begin{matrix}\text{Refractive power}\\ \text{(in diopters)}\end{matrix} = \frac{1}{f} = \frac{1}{(-6.5\text{ m})} = \boxed{-0.15\text{ diopters}}$$

---

71. ***REASONING AND SOLUTION*** An optometrist prescribes contact lenses that have a focal length of 55.0 cm.

a. The focal length is positive (+55.0 cm); therefore, we can conclude that the lenses are $\boxed{\text{converging}}$.

b. As discussed in the text (see Section 26.10), farsightedness is corrected by converging lenses. Therefore, the person who wears these lens is $\boxed{\text{farsighted}}$.

c. If the lenses are designed so that objects no closer than 35.0 cm can be seen clearly, we have $d_o = 35.0\text{ cm}$. The thin-lens equation (Equation 26.6) gives the image distance:

$$\frac{1}{d_i} = \frac{1}{f} - \frac{1}{d_o} = \frac{1}{55.0\text{ cm}} - \frac{1}{35.0\text{ cm}} \qquad \text{or} \qquad d_i = -96.3\text{ cm}$$

Thus, the near point is located $\boxed{96.3\text{ cm}}$ from the eyes.

---

77. ***REASONING*** The angular size of a distant object in radians is approximately equal to the diameter of the object divided by the distance from the eye. We will use this definition to calculate the angular size of the quarter, and then, calculate the angular size of the sun; we can then form the ratio $\theta_{\text{quarter}}/\theta_{\text{sun}}$.

SOLUTION The angular sizes are

$$\theta_{\text{quarter}} \approx \frac{2.4\text{ cm}}{70.0\text{ cm}} = 0.034\text{ rad} \qquad \text{and} \qquad \theta_{\text{sun}} \approx \frac{1.39\times10^{9}\text{ m}}{1.50\times10^{11}\text{ m}} = 0.0093\text{ rad}$$

Therefore, the ratio of the angular sizes is

$$\frac{\theta_{\text{quarter}}}{\theta_{\text{sun}}} = \frac{0.034\text{ rad}}{0.0093\text{ rad}} = \boxed{3.7}$$

---

81. ***REASONING*** The angular magnification of a magnifying glass is given by Equation 26.10: $M \approx \left[(1/f)-(1/d_{\text{i}})\right]N$, where N is the distance from the eye to the near-point. For maximum magnification, the closest to the eye that the image can be is at the near point, with $d_{\text{i}} = -N$ (where the minus sign indicates that the image lies to the left of the lens and is virtual). In this case, Equation 26.10 becomes $M_{\text{max}} \approx N/f + 1$. At minimum magnification, the image is as far from the eye as it can be ($d_{\text{i}} = -\infty$); this occurs when the object is placed at the focal point of the lens. In this case, Equation 26.10 simplifies to $M_{\text{min}} \approx N/f$.

Since the woman observes that for clear vision, the maximum angular magnification is 1.25 times larger than the minimum angular magnification, we have $M_{\text{max}} = 1.25M_{\text{min}}$. This equation can be written in terms of N and f using the above expressions, and then solved for f.

SOLUTION We have

$$\underbrace{\frac{N}{f}+1}_{M_{\text{max}}} = 1.25\underbrace{\frac{N}{f}}_{M_{\text{min}}}$$

Solving for f, we find that

$$f = 0.25N = (0.25)(25\text{ cm}) = \boxed{6.3\text{ cm}}$$

---

83. ***REASONING*** The angular magnification of a compound microscope is given by Equation 26.11:

$$M \approx -\frac{(L-f_{\text{e}})N}{f_{\text{o}}f_{\text{e}}}$$

where f_{o} is the focal length of the objective, f_{e} is the focal length of the eyepiece, and L is the separation between the two lenses. This expression can be solved for f_{o}, the focal length of the objective.

SOLUTION Solving for f_o, we find that the focal length of the objective is

$$f_o = -\frac{(L - f_e)N}{f_e M} = -\frac{(16.0\text{ cm} - 1.4\text{ cm})(25\text{ cm})}{(1.4\text{ cm})(-320)} = \boxed{0.81\text{ cm}}$$

---

91. ***REASONING AND SOLUTION*** The angular magnification of an astronomical telescope, is given by Equation 26.12 as $M \approx -f_o / f_e$. Solving for the focal length of the eyepiece, we find

$$f_e \approx -\frac{f_o}{M} = -\frac{48.0\text{ cm}}{(-184)} = \boxed{0.261\text{ cm}}$$

---

93. ***REASONING AND SOLUTION***

a. The lens with the largest focal length should be used for the objective of the telescope. Since the refractive power is the reciprocal of the focal length (in meters), the lens with the smallest refractive power is chosen as the objective, namely, the $\boxed{\text{1.3 - diopter lens}}$.

b. According to Equation 26.8, the refractive power is related to the focal length f by Refractive power (in diopters) $=1/[f\text{(in meters)}]$. Since we know the refractive powers of the two lenses, we can solve Equation 26.8 for the focal lengths of the objective and the eyepiece. We find that $f_o = 1/(1.3\text{ diopters}) = 0.77\text{ m}$. Similarly, for the eyepiece, $f_e = 1/(11\text{ diopters}) = 0.091\text{ m}$. Therefore, the distance between the lenses should be

$$L \approx f_o + f_e = 0.77\text{ m} + 0.091\text{ m} = \boxed{0.86\text{ m}}$$

c. The angular magnification of the telescope is given by Equation 26.12 as

$$M \approx -\frac{f_o}{f_e} = -\frac{0.77\text{ m}}{0.091\text{ m}} = \boxed{-8.5}$$

---

97. ***REASONING*** The ray diagram is constructed by drawing the paths of two rays from a point on the object. For convenience, we choose the top of the object. The ray that is parallel to the principal axis will be refracted by the lens and pass through the focal point on the right side. The ray that passes through the center of the lens passes through undeflected. The image is formed at the intersection of these two rays on the right side of the lens.

SOLUTION The following ray diagram (to scale) shows that $\boxed{d_i = 18\text{ cm}}$ and reveals a real, inverted, and enlarged image.

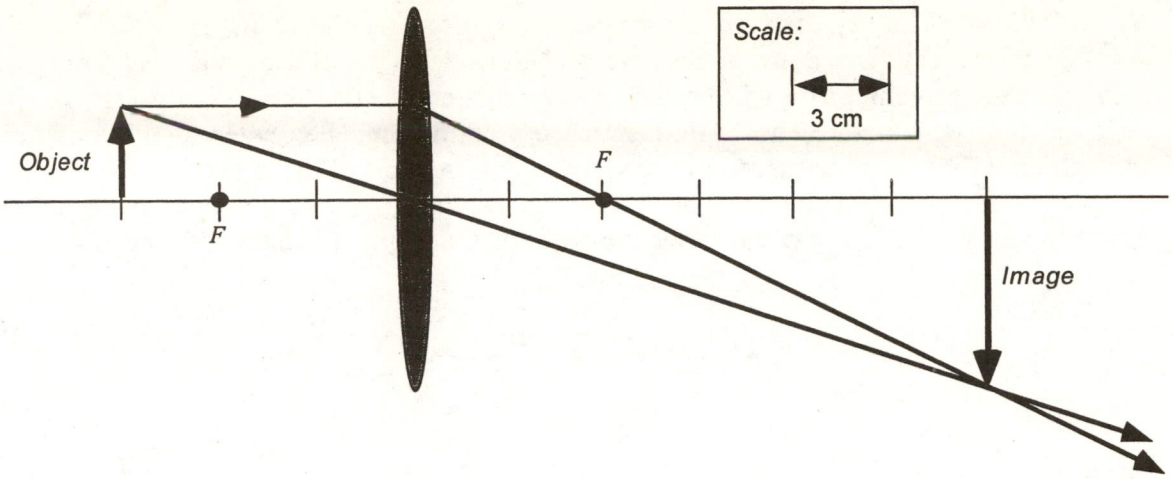

99. ***REASONING AND SOLUTION***
a. Using Equation 26.4 and the refractive index for crown glass given in Table 26.1, we find that the critical angle for a crown glass-air interface is

$$\theta_c = \sin^{-1}\left(\frac{1.00}{1.523}\right) = 41.0°$$

The light will be totally reflected at point A since the incident angle of 60.0° is greater than θ_c. The incident angle at point B, however, is 30.0° and smaller than θ_c. Thus, the light will exit first at $\boxed{\text{point B}}$.

b. The critical angle for a crown glass-water interface is

$$\theta_c = \sin^{-1}\left(\frac{1.333}{1.523}\right) = 61.1°$$

The incident angle at point A is less than this, so the light will first exit at $\boxed{\text{point A}}$.

101. ***REASONING AND SOLUTION*** Since the far point is 220 cm, we know that the image distance is $d_i = -220$ cm, when the object is infinitely far from the lens ($d_o = \infty$). Thus, the thin-lens equation (Equation 26.6) becomes

$$\frac{1}{d_o} + \frac{1}{d_i} = \frac{1}{\infty} + \frac{1}{(-220\text{ cm})} = \frac{1}{(-220\text{ cm})} = \frac{1}{f} \qquad \text{or} \qquad \boxed{f = -220\text{cm}}$$

103. ***REASONING*** The ray diagram is constructed by drawing the paths of two rays from a point on the object. For convenience, we will choose the top of the object. The ray that is parallel to the principal axis will be refracted by the lens so that it passes through the focal point on the right of the lens. The ray that passes through the center of the lens passes through undeflected. The image is formed at the intersection of these two rays. In this case, the rays do not intersect on the right of the lens. However, if they are extended backwards they intersect on the left of the lens, locating a virtual, upright, and enlarged image.

SOLUTION

a. The ray-diagram, drawn to scale, is shown below.

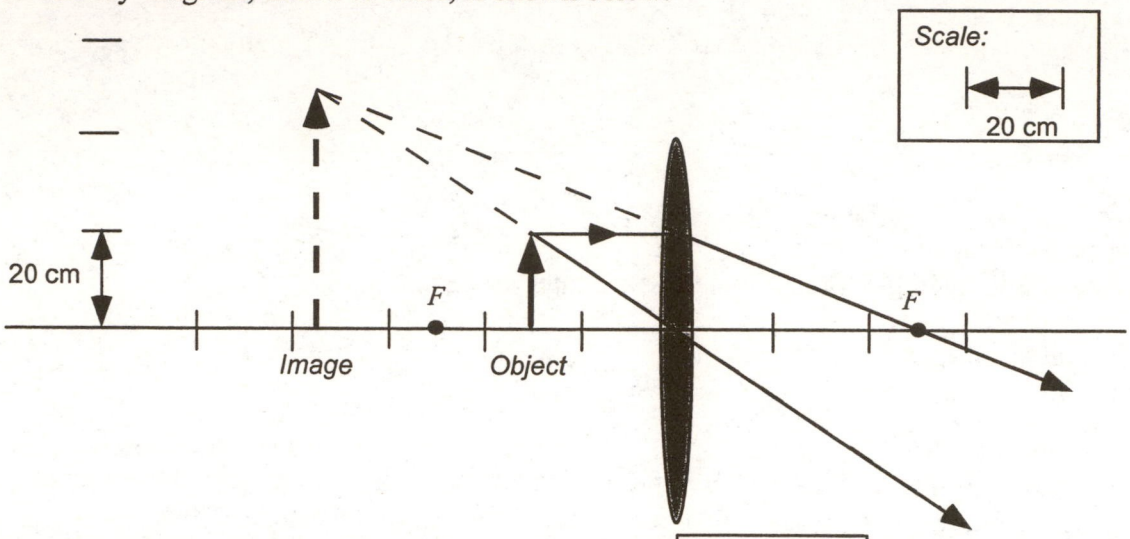

From the diagram, we see that the image distance is $\boxed{d_i = -75 \text{ cm}}$ and the magnification is $\boxed{+2.5}$. The negative image distance indicates that the image is virtual. The positive magnification indicates that the image is larger than the object.

b. From the thin-lens equation [Equation 26.6: $(1/d_o) + (1/d_i) = (1/f)$], we obtain

$$\frac{1}{d_i} = \frac{1}{f} - \frac{1}{d_o} = \frac{1}{50.0 \text{ m}} - \frac{1}{30.0 \text{ cm}} \quad \text{or} \quad \boxed{d_i = -75.0 \text{ cm}}$$

The magnification equation (Equation 26.7) gives the magnification to be

$$m = -\frac{d_i}{d_o} = -\frac{-75.0 \text{ cm}}{30.0 \text{ cm}} = \boxed{+2.50}$$

---

105. ***REASONING*** We will use the geometry of the situation to determine the angle of incidence. Once the angle of incidence is known, we can use Snell's law to find the index of refraction of the unknown liquid. The speed of light v in the liquid can then be determined.

SOLUTION From the drawing in the text, we see that the angle of incidence at the liquid-air interface is

$$\theta_1 = \tan^{-1}\left(\frac{5.00\text{ cm}}{6.00\text{ cm}}\right) = 39.8°$$

The drawing also shows that the angle of refraction is 90.0°. Thus, according to Snell's law (Equation 26.2: $n_1 \sin\theta_1 = n_2 \sin\theta_2$), the index of refraction of the unknown liquid is

$$n_1 = \frac{n_2 \sin\theta_2}{\sin\theta_1} = \frac{(1.000)\,(\sin 90.0°)}{\sin 39.8°} = 1.56$$

From Equation 26.1 ($n = c/v$), we find that the speed of light in the unknown liquid is

$$v = \frac{c}{n_1} = \frac{3.00\times10^8\text{ m/s}}{1.56} = \boxed{1.92\times10^8\text{ m/s}}$$

---

111. ***REASONING AND SOLUTION***

a. A real image must be projected on the drum; therefore, the lens in the copier must be a $\boxed{\text{converging lens}}$.

b. If the document and its copy have the same size, but are inverted with respect to one another, the magnification equation (Equation 26.7) indicates that $m = -d_i/d_o = -1$. Therefore, $d_i/d_o = 1$ or $d_i = d_o$. Then, the thin-lens equation (Equation 26.6) gives

$$\frac{1}{d_i} + \frac{1}{d_o} = \frac{1}{f} = \frac{2}{d_o} \qquad \text{or} \qquad d_o = d_i = 2f$$

Therefore the document is located at a distance $\boxed{2f}$ from the lens.

c. Furthermore, the image is located at a distance of $\boxed{2f}$ from the lens.

---

113. ***REASONING*** A contact lens is placed directly on the eye. Therefore, the object distance, which is the distance from the book to the lens, is 25.0 cm. The near point can be determined from the thin-lens equation [Equation 26.6: $(1/d_o) + (1/d_i) = (1/f)$].

SOLUTION

a. Using the thin-lens equation, we have

$$\frac{1}{d_i} = \frac{1}{f} - \frac{1}{d_o} = \frac{1}{65.0\text{ cm}} - \frac{1}{25.0\text{ cm}} \qquad \text{or} \qquad d_i = -40.6\text{ cm}$$

In other words, at age 40, the man's near point is 40.6 cm. Similarly, when the man is 45, we have

$$\frac{1}{d_i} = \frac{1}{f} - \frac{1}{d_o} = \frac{1}{65.0\text{ cm}} - \frac{1}{29.0\text{ cm}} \qquad \text{or} \qquad d_i = -52.4\text{ cm}$$

and his near point is 52.4 cm. Thus, the man's near point has changed by $52.4\text{ cm} - 40.6\text{ cm} = \boxed{11.8\text{ cm}}$.

b. With $d_o = 25.0$ cm and $d_i = -52.4$ cm, the focal length of the lens is found as follows:

$$\frac{1}{f} = \frac{1}{d_o} + \frac{1}{d_i} = \frac{1}{25.0\text{ cm}} + \frac{1}{(-52.4\text{ cm})} \qquad \text{or} \qquad \boxed{f = 47.8\text{ cm}}$$

---

115. ***REASONING*** The angular magnification of a refracting telescope is 32 800 times larger when you look through the correct end of the telescope than when you look through the wrong end. We wish to find the angular magnification, $M = -f_o / f_e$ (see Equation 26.12) of the telescope. Thus, we proceed by finding the ratio of the focal lengths of the objective and the eyepiece and using Equation 26.12 to find M.

SOLUTION When you look through the correct end of the telescope, the angular magnification of the telescope is $M_c = -f_o / f_e$. If you look through the wrong end, the roles of the objective and eyepiece lenses are interchanged, so that the angular magnification would be $M_w = -f_e / f_o$. Therefore,

$$\frac{M_c}{M_w} = \frac{-f_o / f_e}{-f_e / f_o} = \left(\frac{f_o}{f_e}\right)^2 = 32\,800 \qquad \text{or} \qquad \frac{f_o}{f_e} = \pm\sqrt{32\,800} = \pm 181$$

The angular magnification of the telescope is negative, so we choose the positive root and obtain $M = -f_o / f_e = -(+181) = \boxed{-181}$.

CHAPTER 27 | *INTERFERENCE AND THE WAVE NATURE OF LIGHT*

CONCEPTUAL QUESTIONS

---

3. ***REASONING AND SOLUTION***

a. The light waves coming from *both* slits in a Young's double slit experiment have their phases shifted by an amount equivalent to half a wavelength. Since the light from both slits is changed by the same amount, the relative phase difference between the light from the two slits is zero when the light leaves the slits. When the light reaches the screen, the relative phase difference between light waves from the two slits will be the same as if the phase of the waves had not been shifted at the slits. Therefore, the pattern will be exactly the same as that described in the text.

b. Light coming from *only one* of the slits in a Young's double slit experiment has its phase shifted by an amount equivalent to half of a wavelength. Now, there is a relative phase difference of one half of a wavelength between the light leaving the slits. When the light reaches the screen, there will be a relative phase difference due to the fact that light from each slit, in general, traveled along different paths. In addition, there will be the initial phase difference that is equivalent to half of a wavelength. Therefore, the pattern will be similar to that described in the text; however, the points of constructive interference will be points of destructive interference, and the points of destructive interference will be points of constructive interference. In other words, the positions of the light and dark fringes are interchanged.

---

10. ***REASONING AND SOLUTION*** A thin film of material is floating on water (n = 1.33). When the material has a refractive index of n = 1.20, the film looks bright in reflected light as its thickness approaches zero. But when the material has a refractive index of n = 1.45, the film looks black in reflected light as its thickness approaches zero.

When the material has a refractive index of n = 1.20, both the light reflected from the top of the film and the light reflected from the bottom occur under conditions where light is traveling through a material with a smaller refractive index toward a material with a larger refractive index. Therefore, both the light reflected from the top and the bottom undergoes a phase change upon reflection. Both phase changes are equivalent to half of a wavelength. Therefore, the reflections introduce no net phase difference between the light reflected from the top and bottom. As the thickness of the film approaches zero, the path difference between the light reflected from the top and the bottom of the film approaches zero. Since the path difference is close to zero, and there is no relative phase difference due to reflection, the light reflected from the top will be in phase with the light reflected from the bottom of the film; constructive interference will occur, and the film looks bright in reflected light.

When the material has a refractive index of n = 1.45, only the light reflected from the top of the film is traveling through a material with lower refractive index toward a material with a larger refractive index. Therefore, only the light reflected from the top undergoes a phase change upon reflection. This phase change is equivalent to half of a wavelength. Therefore, there is a net phase difference equivalent to half of a wavelength between the light reflected from the top and bottom of the film. As the thickness of the film approaches zero, the path difference between the light reflected from the top and the light reflected from the bottom of the film approaches zero. Since the path difference is close to zero, and there is a relative phase difference due to reflection that is equal to half of a wavelength, the light reflected from the top will be exactly out of phase with the light reflected from the bottom of the film, and destructive interference will occur. Therefore, the film looks black in reflected light.

---

17. ***REASONING AND SOLUTION*** Rayleigh's criterion states that two point objects are just resolved when the first dark fringe in the diffraction pattern of one falls directly on the central bright fringe in the diffraction pattern of the other. Equation 27.6 approximates the minimum angular displacement θ_{min} between the two objects so that they are just resolved by an optical instrument: $\theta_{min} \approx 1.22\lambda / D$, where λ is the wavelength of light used and D is the diameter of the aperture of the optical instrument. According to Equation 27.6, optical instruments with the highest resolution should have the largest possible diameter D and utilize light of the shortest possible wavelength.

Suppose you were designing an eye and could select the size of the pupil and the wavelength of the electromagnetic waves to which the eye is sensitive. As far as the limitation created by diffraction is concerned, the following list ranks the design choices in order of decreasing resolving power (greatest first): (a) large pupil (large D) and ultraviolet wavelengths (small λ); (c) small pupil (small D) and ultraviolet wavelengths (small λ); (b) small pupil (small D) and infrared wavelength (larger λ).

CHAPTER 27 | *INTERFERENCE AND THE WAVE NATURE OF LIGHT*

PROBLEMS

---

1. ***REASONING AND SOLUTION*** The angular position θ of the bright fringes of a double slit are given by Equation 27.1 as $\sin\theta = m\lambda / d$, with the order of the fringe specified by $m = 0, 1, 2, 3, \ldots$. Solving for λ, we have

$$\lambda = \frac{d \sin\theta}{m} = \frac{(3.8\times10^{-5}\text{ m})\sin 2.0°}{2} = 6.6\times10^{-7}\text{ m} = \boxed{660\text{ nm}}$$

---

5. ***REASONING*** The slit separation d is given by Equation 27.1 with $m = 1$; namely $d = \lambda / \sin\theta$. As shown in Example 1 in the text, the angle θ is given by $\theta = \tan^{-1}(y / L)$.

SOLUTION The angle θ is

$$\theta = \tan^{-1}\left(\frac{0.037\text{ m}}{4.5\text{ m}}\right) = 0.47°$$

Therefore, the slit separation d is

$$d = \frac{\lambda}{\sin\theta} = \frac{490\times10^{-9}\text{ m}}{\sin 0.47°} = \boxed{6.0\times10^{-5}\text{ m}}$$

---

9. ***REASONING*** The light that travels through the plastic has a different path length than the light that passes through the unobstructed slit. Since the center of the screen now appears dark, rather than bright, destructive interference, rather than constructive interference occurs there. This means that the difference between the number of wavelengths in the plastic sheet and that in a comparable thickness of air is $\frac{1}{2}$.

SOLUTION The wavelength of the light in the plastic sheet is given by Equation 27.3 as

$$\lambda_{\text{plastic}} = \frac{\lambda_{\text{vacuum}}}{n} = \frac{586\times10^{-9}\text{ m}}{1.60} = 366\times10^{-9}\text{ m}$$

The number of wavelengths contained in a plastic sheet of thickness t is

$$N_{\text{plastic}} = \frac{t}{\lambda_{\text{plastic}}} = \frac{t}{366\times10^{-9}\text{ m}}$$

The number of wavelengths contained in an equal thickness of air is

$$N_{\text{air}} = \frac{t}{\lambda_{\text{air}}} = \frac{t}{586 \times 10^{-9}\ \text{m}}$$

where we have used the fact that $\lambda_{\text{air}} \approx \lambda_{\text{vacuum}}$. Destructive interference occurs when the difference, $N_{\text{plastic}} - N_{\text{air}}$, in the number of wavelengths is $\frac{1}{2}$:

$$N_{\text{plastic}} - N_{\text{air}} = \frac{1}{2}$$

$$\frac{t}{366 \times 10^{-9}\ \text{m}} - \frac{t}{586 \times 10^{-9}\ \text{m}} = \frac{1}{2}$$

Solving this equation for t yields $t = 487 \times 10^{-9}\ \text{m} = \boxed{487\ \text{nm}}$.

---

11. ***REASONING*** When the light strikes the film from above, the wave reflected from the top surface of the film undergoes a phase shift that is equivalent to one-half of a wavelength, since the light travels from a smaller refractive index ($n_{\text{air}} = 1.00$) toward a larger refractive index ($n_{\text{film}} = 1.33$). On the other hand, there is no phase shift when the light reflects from the bottom surface of the film, since the light travels from a larger refractive index ($n_{\text{film}} = 1.33$) toward a smaller refractive index ($n_{\text{air}} = 1.00$). Thus, the net phase change due to reflection from the two surfaces is equivalent to one-half of a wavelength in the film. This half-wavelength must be combined with the extra distance $2t$ traveled by the wave reflected from the bottom surface, where t is the film thickness. Thus, the condition for constructive interference is

$$\underbrace{2t}_{\substack{\text{Extra distance}\\ \text{traveled by wave}\\ \text{in the film}}} + \underbrace{\tfrac{1}{2}\lambda_{\text{film}}}_{\substack{\text{Half-wavelength}\\ \text{net phase change}\\ \text{due to reflection}}} = \underbrace{\lambda_{\text{film}},\ 2\lambda_{\text{film}},\ 3\lambda_{\text{film}},\ \ldots}_{\substack{\text{Condition for}\\ \text{constructive interference}}}$$

We will use this relation to find the two smallest non-zero film thicknesses for which constructive interference occurs in the reflected light.

SOLUTION The smallest film thickness occurs when the condition for constructive interference is λ_{film}. Then, the relation above becomes

$$2t + \tfrac{1}{2}\lambda_{\text{film}} = \lambda_{\text{film}} \qquad \text{or} \qquad t = \tfrac{1}{4}\lambda_{\text{film}}$$

Since $\lambda_{\text{film}} = \dfrac{\lambda_{\text{vacuum}}}{n_{\text{film}}}$ (Equation 27.3), we have that

$$t = \tfrac{1}{4}\lambda_{\text{film}} = \tfrac{1}{4}\left(\frac{\lambda_{\text{vacuum}}}{n_{\text{film}}}\right) = \tfrac{1}{4}\left(\frac{691\text{ nm}}{1.33}\right) = \boxed{1.30\times10^2\text{ nm}}$$

The next smallest film thickness occurs when the condition for constructive interference is $2\lambda_{\text{film}}$. Then, we have

$$2t + \tfrac{1}{2}\lambda_{\text{film}} = 2\lambda_{\text{film}} \qquad \text{or} \qquad t = \tfrac{3}{4}\lambda_{\text{film}} = \tfrac{3}{4}\left(\frac{\lambda_{\text{vacuum}}}{n_{\text{film}}}\right) = \tfrac{3}{4}\left(\frac{691\text{ nm}}{1.33}\right) = \boxed{3.90\times10^2\text{ nm}}$$

---

15. ***REASONING*** To solve this problem, we must express the condition for constructive interference in terms of the film thickness t and the wavelength λ_{film} of the light in the soap film. We must also take into account any phase changes that occur upon reflection.

SOLUTION For the reflection at the top film surface, the light travels from air, where the refractive index is smaller ($n = 1.00$), toward the film, where the refractive index is larger ($n = 1.33$). Associated with this reflection there is a phase change that is equivalent to one-half of a wavelength. For the reflection at the bottom film surface, the light travels from the film, where the refractive index is larger ($n = 1.33$), toward air, where the refractive index is smaller ($n = 1.00$). Associated with this reflection, there is no phase change. As a result of these two reflections, there is a net phase change that is equivalent to one-half of a wavelength. To obtain the condition for constructive interference, this net phase change must be added to the phase change that arises because of the film thickness t, which is traversed twice by the light that penetrates it. For constructive interference we find that

$$2t + \tfrac{1}{2}\lambda_{\text{film}} = \lambda_{\text{film}}, 2\lambda_{\text{film}}, 3\lambda_{\text{film}}, \ldots$$

or

$$2t = \left(m + \tfrac{1}{2}\right)\lambda_{\text{film}}, \qquad \text{where } m = 0, 1, 2, \ldots$$

Equation 27.3 indicates that $\lambda_{\text{film}} = \lambda_{\text{vacuum}} / n$. Using this expression and the fact that $m = 0$ for the minimum thickness t, we find that the condition for constructive interference becomes

$$2t = \left(m + \tfrac{1}{2}\right)\lambda_{\text{film}} = \left(0 + \tfrac{1}{2}\right)\left(\frac{\lambda_{\text{vacuum}}}{n}\right)$$

or

$$t = \frac{\lambda_{\text{vacuum}}}{4n} = \frac{611\text{ nm}}{4(1.33)} = \boxed{115\text{ nm}}$$

---

19. ***REASONING*** This problem can be solved by using Equation 27.4 for the value of the angle θ when $m = 1$ (first dark fringe).

SOLUTION

a. When the slit width is $W = 1.8 \times 10^{-4}$ m and $\lambda = 675 \text{ nm} = 675 \times 10^{-9}$ m, we find, according to Equation 27.4,

$$\theta = \sin^{-1}\left(m\frac{\lambda}{W}\right) = \sin^{-1}\left[(1)\frac{675 \times 10^{-9} \text{ m}}{1.8 \times 10^{-4} \text{ m}}\right] = \boxed{0.21°}$$

b. Similarly, when the slit width is $W = 1.8 \times 10^{-6}$ m and $\lambda = 675 \times 10^{-9}$ m, we find

$$\theta = \sin^{-1}\left[(1)\frac{675 \times 10^{-9} \text{ m}}{1.8 \times 10^{-6} \text{ m}}\right] = \boxed{22°}$$

---

23. ***REASONING*** According to Equation 27.4, the angles at which the dark fringes occur are given by $\sin\theta = m\lambda / W$, where W is the slit width. In Figure 27.24, we see from the trigonometry of the situation that $\tan\theta = y / L$. Therefore, the latter expression can be used to determine the angle θ, and Equation 27.4 can be used to find the wavelength λ.

SOLUTION The angle θ is given by

$$\theta = \tan^{-1}\left(\frac{y}{L}\right) = \tan^{-1}\left(\frac{3.5 \times 10^{-3} \text{ m}}{4.0 \text{ m}}\right) = 0.050°$$

The wavelength of the light is

$$\lambda = W\sin\theta = \left(5.6 \times 10^{-4} \text{ m}\right)\sin 0.050° = 4.9 \times 10^{-7} \text{ m} = \boxed{490 \text{ nm}}$$

---

25. ***REASONING AND SOLUTION*** Using Equation 27.4 for the first-order dark fringes ($m = 1$) and referring to Figure 27.24 in the text, we see that

$$\sin\theta = (1)\frac{\lambda}{W} = \frac{y}{\sqrt{L^2 + y^2}}$$

Since the distance L between the slit and the screen equals the width $2y$ of the central bright fringe, this equation becomes

$$\frac{\lambda}{W} = \frac{y}{\sqrt{(2y)^2 + y^2}} = \frac{1}{\sqrt{5}} = \boxed{0.447}$$

31. ***REASONING AND SOLUTION*** According to Rayleigh's criterion (Equation 27.6), the minimum angular separation of the two objects is

$$\theta_{\text{min}}\,(\text{in rad}) \approx 1.22\frac{\lambda}{D} = 1.22\left(\frac{565\times10^{-9}\text{ m}}{1.02\text{ m}}\right) = 6.76\times10^{-7}\text{ rad}$$

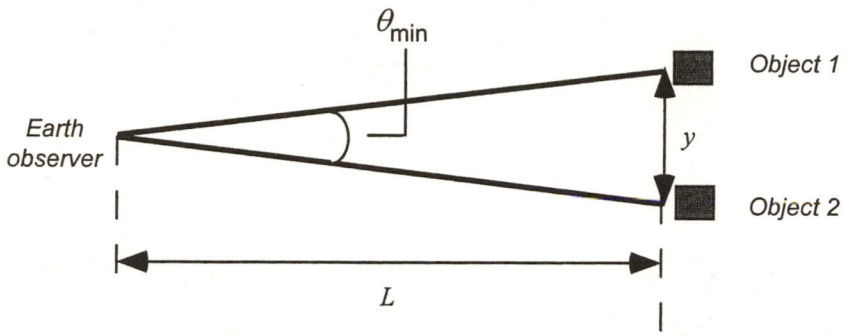

Therefore, from the figure above, the separation y of the two objects is

$$y = L\theta_{\text{min}} = (3.75\times10^{4}\text{ m})(6.76\times10^{-7}\text{ rad}) = \boxed{0.0254\text{ m}}$$

35. ***REASONING*** When light of wavelength λ passes through a circular opening of diameter D, a circular diffraction pattern results, and Equation 27.5 [$\sin\theta = 1.22(\lambda/D)$] locates the first circular dark fringe relative to the central bright spot. Therefore, Equation 27.5 can be used to find the angular separation θ of the central bright spot and the first dark fringe. The geometry of the situation can be used to find the diameter of the central bright spot on the moon.

SOLUTION From Equation 27.5,

$$\theta = \sin^{-1}\left(\frac{1.22\lambda}{D}\right) = \sin^{-1}\left[\frac{1.22(694.3\times10^{-9}\text{ m})}{0.20\text{ m}}\right] = 4.2\times10^{-6}\text{ rad}$$

From the figure below, we see that $2\theta = \dfrac{d}{R}$.

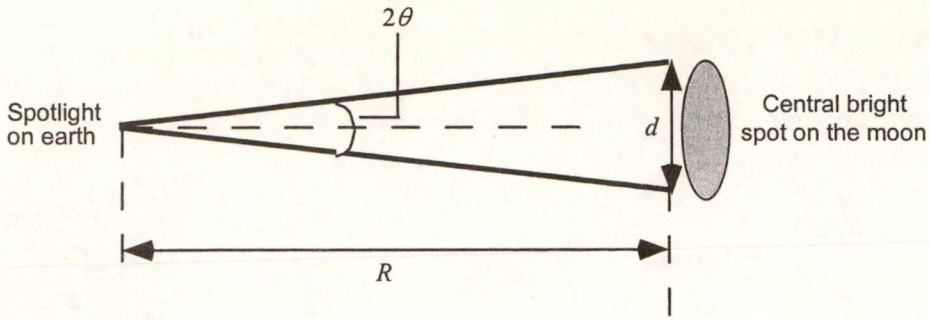

Therefore, the diameter of the spot on the moon is

$$d = 2\theta R = 2(4.2\times10^{-6}\text{ rad})(3.77\times10^{8}\text{ m}) = \boxed{3.2\times10^{3}\text{ m}}$$

---

37. ***REASONING AND SOLUTION*** The geometry of the situation is shown below.

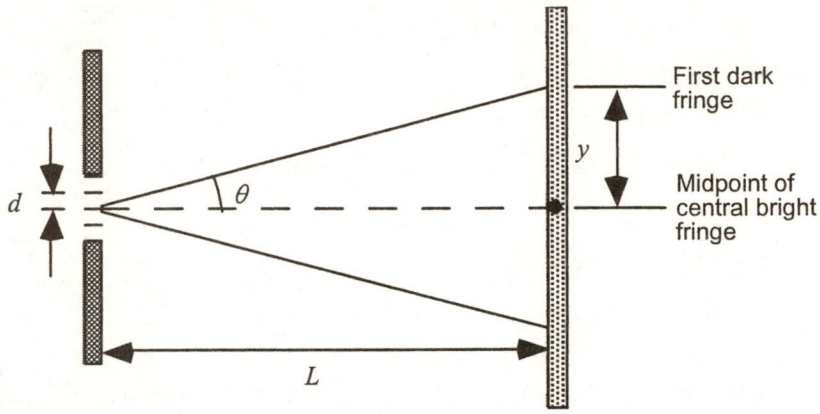

From the geometry, we have

$$\tan\theta = \frac{y}{L} = \frac{0.60\text{ mm}}{3.0\text{ mm}} = 0.20 \qquad \text{or} \qquad \theta = 11.3°$$

Then, solving Equation 27.7 with $m = 1$ for the separation d between the slits, we have

$$d = \frac{m\lambda}{\sin\theta} = \frac{(1)\left(780\times10^{-9}\text{ m}\right)}{\sin 11.3°} = \boxed{4.0\times10^{-6}\text{ m}}$$

---

41. ***REASONING AND SOLUTION*** According to Equation 27.7, the angles that correspond to the first-order ($m = 1$) maximum for the two wavelengths in question are:

a. for $\lambda = 660\text{ nm} = 660\times10^{-9}\text{ m}$,

$$\theta = \sin^{-1}\left(m\frac{\lambda}{d}\right) = \sin^{-1}\left[(1)\left(\frac{660\times10^{-9}\ \text{m}}{1.1\times10^{-6}\ \text{m}}\right)\right] = \boxed{37°}$$

b. for $\lambda = 410\ \text{nm} = 410\times10^{-9}\ \text{m}$,

$$\theta = \sin^{-1}\left(m\frac{\lambda}{d}\right) = \sin^{-1}\left[(1)\left(\frac{410\times10^{-9}\ \text{m}}{1.1\times10^{-6}\ \text{m}}\right)\right] = \boxed{22°}$$

---

43. ***REASONING*** The angle θ that locates the first-order maximum produced by a grating with 3300 lines/cm is given by Equation 27.7, $\sin\theta = m\lambda / d$, with the order of the fringes given by $m = 0, 1, 2, 3, \ldots$ Any two of the diffraction patterns will overlap when their angular positions are the same.

SOLUTION Since the grating has 3300 lines/cm, we have

$$d = \frac{1}{3300\ \text{lines/cm}} = 3.0\times10^{-4}\ \text{cm} = 3.0\times10^{-6}\ \text{m}$$

a. In first order, $m = 1$; therefore, for violet light,

$$\theta = \sin^{-1}\left(m\frac{\lambda}{d}\right) = \sin^{-1}\left[(1)\left(\frac{410\times10^{-9}\ \text{m}}{3.0\times10^{-6}\ \text{m}}\right)\right] = \boxed{7.9°}$$

Similarly for red light,

$$\theta = \sin^{-1}\left(m\frac{\lambda}{d}\right) = \sin^{-1}\left[(1)\left(\frac{660\times10^{-9}\ \text{m}}{3.0\times10^{-6}\ \text{m}}\right)\right] = \boxed{13°}$$

b. Repeating the calculation for the second order maximum ($m = 2$), we find that

$(m = 2)$	
for violet	$\theta = 16°$
for red	$\theta = 26°$

c. Repeating the calculation for the third order maximum ($m = 3$), we find that

$(m = 3)$	
for violet	$\theta = 24°$
for red	$\theta = 41°$

d. Comparisons of the values for θ calculated in parts (a), (b) and (c) show that the $\boxed{\text{second and third orders overlap}}$.

---

47. ***REASONING AND SOLUTION*** To decide whether constructive or destructive interference occurs, we need to determine the wavelength of the wave. For electromagnetic waves, Equation 16.1 can be written $f\lambda = c$, so that

$$\lambda = \frac{c}{f} = \frac{3.00 \times 10^8 \text{ m}}{536 \times 10^3 \text{ Hz}} = 5.60 \times 10^2 \text{ m}$$

Since the two wave sources are in phase, constructive interference occurs when the path difference is an integer number of wavelengths, and destructive interference occurs when the path difference is an odd number of half wavelengths. We find that the path difference is 8.12 km – 7.00 km = 1.12 km. The number of wavelengths in this path difference is $\left(1.12 \times 10^3 \text{ m}\right)/\left(5.60 \times 10^2 \text{ m}\right) = 2.00$. Therefore, $\boxed{\text{constructive interference occurs}}$.

---

51. ***REASONING*** According to Rayleigh's criterion, the two taillights must be separated by a distance s sufficient to subtend an angle $\theta_{\text{min}} \approx 1.22\lambda / D$ at the pupil of the observer's eye. Recalling that this angle must be expressed in radians, we relate θ_{min} to the distances s and L.

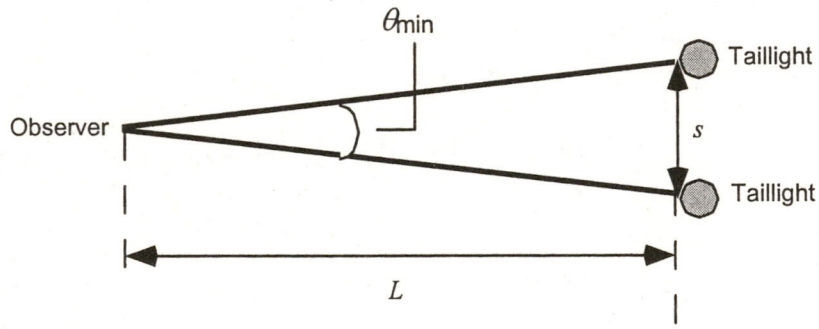

SOLUTION The wavelength λ is 660 nm. Therefore, we have from Equation 27.6

$$\theta_{\text{min}} \approx 1.22\frac{\lambda}{D} = 1.22\left(\frac{660 \times 10^{-9} \text{ m}}{7.0 \times 10^{-3} \text{ m}}\right) = 1.2 \times 10^{-4} \text{ rad}$$

According to Equation 8.1, the distance L between the observer and the taillights is

$$L = \frac{s}{\theta_{\text{min}}} = \frac{1.2 \text{ m}}{1.2 \times 10^{-4} \text{ rad}} = \boxed{1.0 \times 10^4 \text{ m}}$$

---

55. ***REASONING*** For a diffraction grating, the angular position θ of a principal maximum on the screen is given by Equation 27.7 as $\sin\theta = m\lambda / d$ with $m = 0, 1, 2, 3, \ldots$

SOLUTION When the fourth-order principal maximum of light A exactly overlaps the third-order principal maximum of light B, we have

$$\sin\theta_A = \sin\theta_B$$

$$\frac{4\lambda_A}{d} = \frac{3\lambda_B}{d} \quad \text{or} \quad \frac{\lambda_A}{\lambda_B} = \boxed{\frac{3}{4}}$$

CHAPTER 28 | *SPECIAL RELATIVITY*

CONCEPTUAL QUESTIONS

---

2. ***REASONING AND SOLUTION*** A baseball player at home plate hits a pop fly straight up (the beginning event) that is caught by the catcher at home plate (the ending event).

   a. A spectator in the stands is at rest relative to these events; therefore, this spectator would record the proper time interval between them Δt_0.

   b. A spectator sitting on the couch and watching the game on TV is at rest relative to these events; therefore, this spectator would also record the proper time interval between them Δt_0.

   c. The third baseman running in to cover the play is moving relative to the events; therefore, the third baseman *will not* record the proper time interval between them. He will record a dilated time interval Δt given by the time-dilation equation (Equation 28.1): $\Delta t = \Delta t_0 \Big/ \sqrt{1 - v^2 / c^2}$, where v is the relative speed between the observer who measures Δt_0 and the observer who measures Δt.

---

7. ***REASONING AND SOLUTION*** Light travels in water at a speed of 2.26×10^8 m/s. It is possible for a particle that has mass to travel through water at a faster speed than this. But objects that have mass *cannot reach the speed of light in a vacuum*, that is, they cannot attain a speed of $c = 3.00 \times 10^8$ m / s. It is the speed of light in a vacuum that is the ultimate speed, not the speed of light in a material like water.

---

13. ***REASONING AND SOLUTION*** The quantities listed are (a) the time interval between two events, (b) the length of an object, (c) the speed of light in a vacuum, and (d) the relative speed between the observers.

    a. The time interval between two events depends on the reference frame in which the time measurements are made. An observer in the rest frame of the two events will measure the proper time interval. In any other inertial reference frame in motion relative to the rest frame of the events, the time interval will be longer or dilated. Therefore, two observers in relative motion will not, in general, agree on the measured value of the time interval between two events.

    b. The length of an object depends on the reference frame in which the length measurement is made. An observer in the rest frame of the object will measure the proper length of the object. In any other inertial reference frame in motion relative to the rest frame of the object, the length will be shorter or contracted in the direction of motion. Therefore, two observers in relative motion will not, in general, agree on the measured value of the length of the object.

c. According to the speed of light postulate, the speed of light in a vacuum will be the same in all inertial reference frames that are in motion relative to each other. Therefore, two observers will always measure the speed of light to be the same, regardless of the relative velocity between them.

d. Since there is no preferred inertial reference frame, two observers in inertial reference frames that are in motion relative to each other will always measure the same relative speed between them, regardless of the relative velocity.

CHAPTER 28 | *SPECIAL RELATIVITY*

PROBLEMS

---

1. ***REASONING*** The expression for time dilation is, according to Equation 28.1,

$$\Delta t = \frac{\Delta t_0}{\sqrt{1 - v^2 / c^2}}$$

For a given event, it relates the proper time interval Δt_0 to the time interval Δt that would be measured by an observer moving at a speed v relative to the frame of reference in which the event takes place.

We must consider two situations; in the first situation, the Klingon spacecraft has a speed of $0.75c$ with respect to the earth. In the second situation, the craft has a speed of $0.94c$ relative to the earth. We will refer to these two situations as A and B, respectively.

Since the proper time interval always has the same value, $(\Delta t_0)_A = (\Delta t_0)_B$. We can express both sides of this expression using Equation 28.1. The result can be solved for Δt_B.

SOLUTION Use of Equation 28.1 gives

$$\Delta t_A \sqrt{1 - v_A^2 / c^2} = \Delta t_B \sqrt{1 - v_B^2 / c^2}$$

$$\Delta t_B = \Delta t_A \frac{\sqrt{1 - v_A^2 / c^2}}{\sqrt{1 - v_B^2 / c^2}} = \Delta t_A \sqrt{\frac{1 - (v_A / c)^2}{1 - (v_B / c)^2}} = (37.0 \text{ h}) \sqrt{\frac{1 - (0.75c/c)^2}{1 - (0.94c/c)^2}} = \boxed{72 \text{ h}}$$

---

3. ***REASONING*** Since the "police car" is moving relative to the earth observer, the earth observer measures a greater time interval Δt between flashes. Since both the proper time Δt_0 (as observed by the officer) and the dilated time Δt (as observed by the person on earth) are known, the speed of the "police car" relative to the observer can be determined from the time dilation relation, Equation 28.1.

SOLUTION According to Equation 28.1, the dilated time interval between flashes is $\Delta t = \Delta t_0 / \sqrt{1 - (v^2 / c^2)}$, where Δt_0 is the proper time. Solving for the speed v, we find

$$v = c\sqrt{1-\left(\frac{\Delta t_0}{\Delta t}\right)^2} = (3.0\times10^8\ \text{m/s})\sqrt{1-\left(\frac{1.5\ \text{s}}{2.5\ \text{s}}\right)^2} = \boxed{2.4\times10^8\ \text{m/s}}$$

---

9. ***REASONING AND SOLUTION*** The diameter D of the planet, as measured by a moving spacecraft, is given in terms of the proper diameter D_0 by Equation 28.2. Taking the ratio of the diameter D_A of the planet measured by spaceship A to the diameter D_B measured by spaceship B, we find

$$\frac{D_A}{D_B} = \frac{D_0\sqrt{1-\frac{v_A^2}{c^2}}}{D_0\sqrt{1-\frac{v_B^2}{c^2}}} = \frac{\sqrt{1-\frac{(0.60\,c)^2}{c^2}}}{\sqrt{1-\frac{(0.80\,c)^2}{c^2}}} = \boxed{1.3}$$

---

11. ***REASONING AND SOLUTION*** The length L_0 that the person measures for the UFO when it lands is the proper length, since the UFO is at rest with respect to the person. Therefore, from Equation 28.2 we have

$$L_0 = \frac{L}{\sqrt{1-\frac{v^2}{c^2}}} = \frac{230\ \text{m}}{\sqrt{1-\frac{(0.90\,c)^2}{c^2}}} = \boxed{530\ \text{m}}$$

---

19. ***REASONING*** The height of the woman as measured by the observer is given by Equation 28.2 as $h = h_0\sqrt{1-(v/c)^2}$, where h_0 is her proper height. In order to use this equation, we must determine the speed v of the woman relative to the observer. We are given the magnitude of her relativistic momentum, so we can determine v from p.

SOLUTION According to Equation 28.3 $p = mv/\sqrt{1-v^2/c^2}$, so $mv = p\sqrt{1-v^2/c^2}$. Squaring both sides, we have

$$m^2v^2 = p^2(1-v^2/c^2) \qquad \text{or} \qquad m^2v^2 + p^2\frac{v^2}{c^2} = p^2$$

$$v^2\left(m^2 + \frac{p^2}{c^2}\right) = p^2 \qquad \text{or} \qquad v^2 = \frac{p^2}{m^2 + \frac{p^2}{c^2}}$$

Solving for v and substituting values, we have

$$v = \frac{p}{\sqrt{m^2 + \frac{p^2}{c^2}}} = \frac{2.0\times10^{10}\ \text{kg}\cdot\text{m/s}}{\sqrt{(55\ \text{kg})^2 + \left(\frac{2.0\times10^{10}\ \text{kg}\cdot\text{m/s}}{3.00\times10^8\ \text{m/s}}\right)^2}} = 2.3\times10^8\ \text{m/s}$$

Then, the height that the observer measures for the woman is

$$h = h_0\sqrt{1-\left(\frac{v}{c}\right)^2} = (1.6\ \text{m})\sqrt{1-\left(\frac{2.3\times10^8\ \text{m/s}}{3.0\times10^8\ \text{m/s}}\right)^2} = \boxed{1.0\ \text{m}}$$

---

21. ***REASONING*** The total linear momentum of the system is conserved, since no net external force acts on the system. Therefore, the final total momentum $p_1 + p_2$ of the two fragments must equal the initial total momentum, which is zero since the particle is initially at rest. As a result, $p_1 = -p_2$, where Equation 28.3 must be used for the magnitudes of the momenta p_1 and p_2. Thus, we find

$$\frac{m_1 v_1}{\sqrt{1-(v_1^2/c^2)}} = \frac{-m_2 v_2}{\sqrt{1-(v_2^2/c^2)}}$$

SOLUTION Letting fragment 2 be the more-massive fragment, we have that $v_1 = 0.800c$, $m_1 = 1.67\times10^{-27}$ kg, and $m_2 = 5.01\times10^{-27}$ kg. Squaring both sides of the above equation, rearranging terms, substituting the known values for v_1, m_1, and m_2, we find that

$$\frac{v_2^2}{1-(v_2^2/c^2)} = \frac{m_1^2 v_1^2}{m_2^2\left[1-(v_1^2/c^2)\right]} = \frac{(1.67\times10^{-27}\ \text{kg})^2(0.800c)^2}{(5.01\times10^{-27}\ \text{kg})^2\left[1-(0.800c/c)^2\right]} = 0.1975c^2$$

Therefore,

$$v_2^2 = 0.1975c^2\left[1-(v_2^2/c^2)\right] = 0.1975c^2 - 0.1975v_2^2$$

Solving for v_2 gives

$$v_2 = \pm\sqrt{\frac{0.1975c^2}{1.1975}} = \pm0.406c$$

We reject the positive root, since then both fragments would be moving in the same direction after the break-up and the system would have a non-zero momentum. According to the principle of conservation of momentum, the total momentum after the break-up must be zero, just as it was before the break-up. The momentum of the system will be zero only if the velocity v_2 is opposite to the velocity v_1. Hence, we chose the negative root and $\boxed{v_2 = -0.406c}$.

---

23. ***REASONING AND SOLUTION***
a. In Section 28.6 it is shown that when the speed of a particle is $0.01c$ (or less), the relativistic kinetic energy becomes nearly equal to the nonrelativistic kinetic energy. Since the speed of the particle here is $0.001c$, the ratio of the relativistic kinetic energy to the nonrelativistic kinetic energy is $\boxed{1.0}$.

b. Taking the ratio of the relativistic kinetic energy, Equation 28.6, to the nonrelativistic kinetic energy, $\frac{1}{2}mv^2$, we find that

$$\frac{mc^2\left(\dfrac{1}{\sqrt{1-(v^2/c^2)}}-1\right)}{\frac{1}{2}mv^2} = 2\left(\frac{c}{v}\right)^2\left(\frac{1}{\sqrt{1-(v^2/c^2)}}-1\right)$$

$$= 2\left(\frac{c}{0.970c}\right)^2\left(\frac{1}{\sqrt{1-(0.970c)^2/c^2}}-1\right) = \boxed{6.6}$$

---

27. ***REASONING*** According to the work-energy theorem, Equation 6.3, the work that must be done on the electron to accelerate it from rest to a speed of $0.990c$ is equal to the kinetic energy of the electron when it is moving at $0.990c$.

SOLUTION Using Equation 28.6, we find that

$$\text{KE} = mc^2\left(\frac{1}{\sqrt{1-(v^2/c^2)}}-1\right)$$

$$= (9.11\times10^{-31}\text{ kg})(3.00\times10^8\text{ m/s})^2\left(\frac{1}{\sqrt{1-(0.990c)^2/c^2}}-1\right) = \boxed{5.0\times10^{-13}\text{ J}}$$

---

31. ***REASONING*** Let's define the following relative velocities, assuming that the spaceship and exploration vehicle are moving in the positive direction.

v_{ES} = velocity of **E**xploration vehicle relative to the **S**paceship.

v_{EO} = velocity of **E**xploration vehicle relative to an **O**bserver on earth = +0.70c
v_{SO} = velocity of **S**paceship relative to an **O**bserver on earth = +0.50c

The velocity v_{ES} can be determined from the velocity-addition formula, Equation 28.8:

$$v_{ES} = \frac{v_{EO} + v_{OS}}{1 + \dfrac{v_{EO}v_{OS}}{c^2}}$$

The velocity v_{OS} of the observer on earth relative to the spaceship is not given. However, we know that v_{OS} is the negative of v_{SO}, so $v_{OS} = -v_{SO} = -(+0.50c) = -0.50c$.

SOLUTION The velocity of the exploration vehicle relative to the spaceship is

$$v_{ES} = \frac{v_{EO} + v_{OS}}{1 + \dfrac{v_{EO}v_{OS}}{c^2}} = \frac{+0.70c + (-0.50c)}{1 + \dfrac{(+0.70c)(-0.50c)}{c^2}} = \boxed{+0.31c}$$

---

33. ***REASONING*** Since the crew is initially at rest relative to the escape pod, the length of 45 m is the proper length L_0 of the pod. The length of the escape pod as determined by an observer on earth can be obtained from the relation for length contraction given by Equation 28.2, $L = L_0\sqrt{1-\left(v_{PE}^2/c^2\right)}$. The quantity v_{PE} is the speed of the escape pod relative to the earth, which can be found from the velocity-addition formula, Equation 28.8. The following are the relative velocities, assuming that the direction away from the earth is the positive direction:

v_{PE} = velocity of the escape **P**od relative to **E**arth.
v_{PR} = velocity of escape **P**od relative to the **R**ocket = –0.55c. This velocity is negative because the rocket is moving away from the earth (in the positive direction), and the escape pod is moving in an opposite direction (the negative direction) relative to the rocket.
v_{RE} = velocity of **R**ocket relative to **E**arth = +0.75c

These velocities are related by the velocity-addition formula, Equation 28.8.

SOLUTION The relative velocity of the escape pod relative to the earth is

$$v_{PE} = \frac{v_{PR} + v_{RE}}{1 + \dfrac{v_{PR}v_{RE}}{c^2}} = \frac{-0.55c + 0.75c}{1 + \dfrac{(-0.55c)(+0.75c)}{c^2}} = +0.34c$$

The speed of the pod relative to the earth is $0.34c$. The length of the pod as determined by an observer on earth is

$$L = L_0\sqrt{1-\frac{v_{\text{PE}}^2}{c^2}} = (45\text{ m})\sqrt{1-\frac{(0.34c)^2}{c^2}} = \boxed{42\text{ m}}$$

---

37. ***REASONING*** All standard meter sticks at rest have a length of 1.00 m for observers who are at rest with respect to them. Thus, 1.00 m is the proper length L_0 of the meter stick. When the meter stick moves with speed v relative to an earth-observer, its length $L = 0.500$ m will be a contracted length. Since both L_0 and L are known, v can be found directly from Equation 28.2, $L = L_0\sqrt{1-\left(v^2/c^2\right)}$.

SOLUTION Solving Equation 28.2 for v, we find that

$$v = c\sqrt{1-\left(\frac{L}{L_0}\right)^2} = (3.00\times10^8\text{ m/s})\sqrt{1-\left(\frac{0.500\text{ m}}{1.00\text{ m}}\right)^2} = \boxed{2.60\times10^8\text{ m/s}}$$

---

43. ***REASONING*** Only the sides of the rectangle that lie in the direction of motion will experience length contraction. In order to make the rectangle look like a square, each side must have a length of L = 2.0 m. Thus, we move along the long side, taking the proper length to be $L_0 = 3.0$ m. We can solve for the speed using Equation 28.2. Then, with this speed, we can use the relation for length contraction to find L for the short side as we move along it.

SOLUTION From Equation 28.2, $L = L_0\sqrt{1-(v^2/c^2)}$, we find that

$$v = c\sqrt{1-\left(\frac{L}{L_0}\right)^2} = c\sqrt{1-\left(\frac{2.0\text{ m}}{3.0\text{ m}}\right)^2} = 0.75\,c$$

Moving at this speed along the short side, we take $L_0 = 2.0$ m and find L:

$$L = L_0\sqrt{1-\left(\frac{v}{c}\right)^2} = (2.0\text{ m})\sqrt{1-\left(\frac{0.75\,c}{c}\right)^2} = 1.3\text{ m}$$

The observed dimensions of the rectangle are, therefore, $\boxed{3.0\text{ m}\times 1.3\text{ m}}$, since the long side is not contracted due to motion along the short side.

CHAPTER 29 | *PARTICLES AND WAVES*

CONCEPTUAL QUESTIONS

---

6. ***REASONING AND SOLUTION*** In a photoelectric effect experiment, the intensity of the light is increased while the frequency is kept constant. The frequency is greater than the minimum frequency f_0, so that photoelectrons are emitted from the negatively charged metal plate (see Figure 29.4).

Light intensity is the energy per second per unit area that crosses a transverse surface. According to Einstein's photon hypothesis (see Equation 29.2), a beam of light is composed of individual photons, each having an energy $E = hf$, where h is Planck's constant and f is the frequency of the light. From this viewpoint, the intensity of the light is equal to the energy of each photon multiplied by the number of photons per second per unit area crossing a transverse surface. An increase in the intensity of the light beam, therefore, corresponds to an increase in the number of photons per second that strikes the negatively charged metal plate. This will result in the emission of an increased number of photoelectrons per second at the negative plate.

a. As the number of photoelectrons emitted per second from the negative plate increases, the number of photoelectrons collected per second at the positive plate increases. Since the current in the phototube is proportional to the number of photoelectrons per second collected at the positive plate, increasing the light intensity will cause the current in the phototube to increase.

b. As discussed above, an increase in intensity will cause the number of photoelectrons emitted per second from the metal surface to increase.

c. According to Equation 29.3, the maximum kinetic energy that a photoelectron could have is given by $\text{KE}_{\text{max}} = hf - W_0$, where W_0 is the work function of the metal. Since the frequency is kept constant, and W_0 is a property of the metal that is independent of the intensity, an increase in intensity will have no effect on KE_{max}. The maximum kinetic energy that an electron could have remains the same.

d. The kinetic energy of an object of mass m moving non-relativistically with speed v can be expressed in terms of its momentum p as follows:

$$\text{KE} = \tfrac{1}{2}mv^2 = \frac{(mv)^2}{2m} = \frac{p^2}{2m}$$

Therefore, treating the electron classically, the maximum momentum that a photoelectron could have is $p_{\text{max}} = \sqrt{2m\,(\text{KE}_{\text{max}})}$. Since the maximum kinetic energy that an electron could have remains the same as the intensity is increased at constant frequency, the maximum momentum that an electron could have remains the same as well.

e. According to Equation 29.8, the minimum de Broglie wavelength of an object of maximum momentum p_{max} is given by

$$\lambda_{min} = \frac{h}{p_{max}}$$

Since the maximum momentum that a photoelectron could have remains the same as the intensity is increased at constant frequency, we can conclude that the minimum de Broglie wavelength that an electron could have also remains the same.

---

10. ***REASONING AND SOLUTION*** When bright light is incident on the radiometer, photons strike both the black and the shiny surfaces. The photons are absorbed by the black surfaces and are reflected by the shiny surfaces. Linear momentum is transferred to the panels because of the photon collisions. Since the orientation of the light is arbitrary, photons will strike the panels at arbitrary angles of incidence. Only the component of the photon's momentum that is perpendicular to the face of the panel contributes to its motion. As discussed in Conceptual Example 4, the momentum transfer to the panels is a maximum for the shiny surfaces and is twice as large in magnitude than it is for the black surfaces. Following the reasoning of Conceptual Example 4, we can use the impulse-momentum theorem and Newton's third law to deduce that the force on the shiny side is greater than the force on the black side. Therefore, in bright light, the arrangement would spin in the direction from the shiny side toward the black side. The observed spinning is in the opposite direction. Thus, photon collisions with the panels *cannot* be the cause of the spinning.

---

13. ***REASONING AND SOLUTION*** Suppose that in Figure 29.1, the electrons are replaced with protons that have the same speed. The speed of the protons is the same as that of the electrons; however, the mass of a proton is greater than the mass of an electron. Therefore, the magnitude of the linear momentum of a proton is greater than the linear momentum of an electron traveling at the same speed. According to Equation 29.8, $\lambda = h / p$, the de Broglie wavelength of the "proton beam" will be smaller than that of the "electron beam." Equation 27.1 gives the condition that must be satisfied by the angle θ that locates the interference maxima in a Young's double-slit experiment: $\sin\theta = m\lambda / d$, where m takes on integer values. According to Equation 27.1, the sine of the angle θ is directly proportional to the wavelength of the incident beam. Since the proton beam has a smaller wavelength than the electron beam, the values of $\sin\theta$ that correspond to maxima will be smaller. Consequently, the difference in the corresponding values of θ will be smaller, and the angular separation between the bright fringes will *decrease* when the electrons are replaced by protons.

CHAPTER 29 | *PARTICLES AND WAVES*

PROBLEMS

---

1. ***REASONING*** According to Equation 29.3, the work function W_0 is related to the photon energy hf and the maximum kinetic energy KE_{max} by $W_0 = hf - \text{KE}_{\text{max}}$. This expression can be used to find the work function of the metal.

SOLUTION KE_{max} is 6.1 eV. The photon energy (in eV) is, according to Equation 29.2,

$$hf = \left(6.63 \times 10^{-34}\ \text{J}\cdot\text{s}\right)\left(3.00 \times 10^{15}\ \text{Hz}\right)\left(\frac{1\ \text{eV}}{1.60 \times 10^{-19}\ \text{J}}\right) = 12.4\ \text{eV}$$

The work function is, therefore,

$$W_0 = hf - \text{KE}_{\text{max}} = 12.4\ \text{eV} - 6.1\ \text{ev} = \boxed{6.3\ \text{eV}}$$

---

5. ***REASONING*** The energy of the photon is related to its frequency by Equation 29.2, $E = hf$. Equation 16.1, $v = f\lambda$, relates the frequency and the wavelength for any wave.

SOLUTION Combining Equations 29.2 and 16.1, and noting that the speed of a photon is c, the speed of light in a vacuum, we have

$$\lambda = \frac{c}{f} = \frac{c}{(E/h)} = \frac{hc}{E} = \frac{(6.63 \times 10^{-34}\ \text{J}\cdot\text{s})(3.0 \times 10^{8}\ \text{m/s})}{6.4 \times 10^{-19}\ \text{J}} = 3.1 \times 10^{-7}\ \text{m} = \boxed{310\ \text{nm}}$$

---

9. ***REASONING AND SOLUTION*** The number of photons per second, N, entering the owl's eye is $N = SA/E$, where S is the intensity of the beam, A is the area of the owl's pupil, and E is the energy of a single photon. Assuming that the owl's pupil is circular, $A = \pi r^2 = \pi\left(\frac{1}{2}d\right)^2$, where d is the diameter of the owl's pupil. Combining Equations 29.2 and 16.1, we have $E = hf = hc/\lambda$. Therefore,

$$N = \frac{SA\lambda}{hc} = \frac{(5.0 \times 10^{-13}\ \text{W/m}^2)\pi\left[\frac{1}{2}\left(8.5 \times 10^{-3}\ \text{m}\right)\right]^2 (510 \times 10^{-9}\ \text{m})}{(6.63 \times 10^{-34}\ \text{J}\cdot\text{s})(3.0 \times 10^{8}\ \text{m/s})} = \boxed{73\ \text{photons/s}}$$

---

11. ***REASONING AND SOLUTION***

a. According to Equation 24.5b, the electric field can be found from $E = \sqrt{S/(\varepsilon_0 c)}$. The intensity S of the beam is

$$S = \frac{\text{Energy per unit time}}{A} = \frac{Nhf}{A} = \frac{Nh}{A}\left(\frac{c}{\lambda}\right)$$

$$= \frac{\left(1.30\times10^{18}\text{ photons/s}\right)\left(6.63\times10^{-34}\text{ J}\cdot\text{s}\right)}{\pi\left(1.00\times10^{-3}\text{ m}\right)^2}\left(\frac{3.00\times10^{8}\text{ m/s}}{514.5\times10^{-9}\text{ m}}\right)$$

$$= 1.60\times10^{5}\text{ W/m}^2$$

where N is the number of photons per second emitted. Then,

$$E = \sqrt{S/\left(\varepsilon_0 c\right)} = \boxed{7760\text{ N/C}}$$

b. According to Equation 24.3, the average magnetic field is

$$B = E/c = \boxed{2.59\times10^{-5}\text{ T}}$$

---

15. ***REASONING*** The angle θ through which the X-rays are scattered is related to the difference between the wavelength λ' of the scattered X-rays and the wavelength λ of the incident X-rays by Equation 29.7 as

$$\lambda' - \lambda = \frac{h}{mc}(1-\cos\theta)$$

where h is Planck's constant, m is the mass of the electron, and c is the speed of light in a vacuum. We can use this relation directly to find the angle, since all the other variables are known.

SOLUTION Solving Equation 29.7 for the angle θ, we obtain

$$\cos\theta = 1 - \frac{mc}{h}(\lambda' - \lambda)$$

$$= 1 - \frac{\left(9.11\times10^{-31}\text{ kg}\right)\left(3.00\times10^{8}\text{ m/s}\right)}{6.63\times10^{-34}\text{ J}\cdot\text{s}}\left(0.2703\times10^{-9}\text{ m} - 0.2685\times10^{-9}\text{ m}\right) = 0.26$$

$$\theta = \cos^{-1}(0.26) = \boxed{75°}$$

---

19. ***REASONING*** The change in wavelength that occurs during Compton scattering is given by Equation 29.7:

$$\lambda' - \lambda = \frac{h}{mc}(1 - \cos\theta) \qquad \text{or} \qquad (\lambda' - \lambda)_{\max} = \frac{h}{mc}(1 - \cos 180°) = \frac{2h}{mc}$$

$(\lambda' - \lambda)_{\max}$ is the maximum change in the wavelength, and to calculate it we need a value for the mass m of a nitrogen molecule. This value can be obtained from the mass per mole M of nitrogen (N_2) and Avogadro's number N_A, according to $m = M / N_A$ (see Section 14.1).

SOLUTION Using a value of $M = 0.0280$ kg/mol, we obtain the following result for the maximum change in the wavelength:

$$(\lambda' - \lambda)_{\max} = \frac{2h}{mc} = \frac{2h}{\left(\frac{M}{N_A}\right)c} = \frac{2\left(6.63 \times 10^{-34}\ \text{J}\cdot\text{s}\right)}{\left(\frac{0.0280\ \text{kg/mol}}{6.02 \times 10^{23}\ \text{mol}^{-1}}\right)\left(3.00 \times 10^{8}\ \text{m/s}\right)}$$

$$= \boxed{9.50 \times 10^{-17}\ \text{m}}$$

---

23. ***REASONING*** In order for the person to diffract to the same extent as the sound wave, the de Broglie wavelength of the person must be equal to the wavelength of the sound wave.

SOLUTION a. Since the wavelengths are equal, we have that

$$\lambda_{\text{sound}} = \lambda_{\text{person}}$$

$$\lambda_{\text{sound}} = \frac{h}{m_{\text{person}} v_{\text{person}}}$$

Solving for v_{person}, and using the relation $\lambda_{\text{sound}} = v_{\text{sound}} / f_{\text{sound}}$ (Equation 16.1), we have

$$v_{\text{person}} = \frac{h}{m_{\text{person}}(v_{\text{sound}} / f_{\text{sound}})} = \frac{h f_{\text{sound}}}{m_{\text{person}} v_{\text{sound}}}$$

$$= \frac{(6.63 \times 10^{-34}\ \text{J}\cdot\text{s})(128\ \text{Hz})}{(55.0\ \text{kg})(343\ \text{m/s})} = \boxed{4.50 \times 10^{-36}\ \text{m/s}}$$

b. At the speed calculated in part (a), the time required for the person to move a distance of one meter is

$$t = \frac{x}{v} = \frac{1.0\text{ m}}{4.50\times10^{-36}\text{ m/s}}\underbrace{\left(\frac{1.0\text{ h}}{3600\text{ s}}\right)\left(\frac{1\text{ day}}{24.0\text{ h}}\right)\left(\frac{1\text{ year}}{365.25\text{ days}}\right)}_{\text{Factors to convert seconds to years}} = \boxed{7.05\times10^{27}\text{ years}}$$

---

27. ***REASONING AND SOLUTION*** The de Broglie wavelength λ of the woman is given by Equation 29.8 as $\lambda = h/p$, where p is the magnitude of her momentum. The magnitude of the momentum is $p = mv$, where m is the woman's mass and v is her speed. According to Equation 3.6b of the equations of kinematics, the speed v is given by $v = \sqrt{2a_y y}$, since the woman jumps from rest. In this expression, $a_y = -9.80\text{ m/s}^2$ and $y = -9.5\text{ m}$. With these considerations we find that

$$\lambda = \frac{h}{mv} = \frac{h}{m\sqrt{2a_y y}} = \frac{6.63\times10^{-34}\text{ J}\cdot\text{s}}{(41\text{ kg})\sqrt{2(-9.80\text{ m/s}^2)(-9.5\text{ m})}} = \boxed{1.2\times10^{-36}\text{ m}}$$

---

33. ***REASONING AND SOLUTION*** According to the uncertainty principle, the minimum uncertainty in the momentum can be determined from $\Delta p_y \Delta y = h/(4\pi)$. Since $p_y = mv_y$, it follows that $\Delta p_y = m\Delta v_y$. Thus, the minimum uncertainty in the velocity of the oxygen molecule is given by

$$\Delta v_y = \frac{h}{4\pi m\Delta y} = \frac{6.63\times10^{-34}\text{ J}\cdot\text{s}}{4\pi\left(5.3\times10^{-26}\text{ kg}\right)\left(0.25\times10^{-3}\text{ m}\right)} = \boxed{4.0\times10^{-6}\text{ m/s}}$$

---

35. ***REASONING AND SOLUTION*** If we assume that the number of particles that strikes the screen outside the bright fringe is negligible, the particles that pass through the slit will hit the screen somewhere in the central bright fringe. The angular width of the central bright fringe is equal to 2θ, where θ is shown in Figure 29.15. From this figure we see that

$$\theta = \tan^{-1}\left(\frac{\Delta p_y}{p_x}\right)$$

According to Equation 29.10, the minimum uncertainty in Δp_y occurs when

$$(\Delta p_y)(\Delta y) = \frac{h}{4\pi}$$

Since the electron can pass anywhere through the width W of the slit, the uncertainty in the y position of the electron is $\Delta y = W$. Thus, the minimum uncertainty can be written as

$$(\Delta p_y)W = \frac{h}{4\pi} \qquad \text{or} \qquad \Delta p_y = \frac{h}{4\pi W}$$

Using the fact that $p_x = h/\lambda$, we find

$$\theta = \tan^{-1}\left(\frac{h}{p_x 4\pi W}\right) = \tan^{-1}\left[\frac{h}{(h/\lambda)4\pi W}\right] = \tan^{-1}\left(\frac{\lambda}{4\pi W}\right)$$

$$= \tan^{-1}\left[\frac{633\times10^{-9}\text{ m}}{4\pi(0.200\times10^{-3}\text{ m})}\right] = 1.44\times10^{-2}\text{ degrees}$$

Therefore, the minimum range of angles over which the particles spread is $\boxed{-0.0144° \le \theta \le +0.0144°}$.

---

39. ***REASONING AND SOLUTION*** The de Broglie wavelength λ is given by Equation 29.8 as $\lambda = h/p$, where p is the magnitude of the momentum of the particle. The magnitude of the momentum is $p = mv$, where m is the mass and v is the speed of the particle. Using this expression in Equation 29.8, we find that $\lambda = h/(mv)$, or

$$v = \frac{h}{m\lambda} = \frac{6.63\times10^{-34}\text{ J}\cdot\text{s}}{(1.67\times10^{-27}\text{ kg})(1.30\times10^{-14}\text{ m})} = 3.05\times10^{7}\text{ m/s}$$

The kinetic energy of the proton is

$$\text{KE} = \tfrac{1}{2}mv^2 = \tfrac{1}{2}(1.67\times10^{-27}\text{ kg})(3.05\times10^{7}\text{ m/s})^2 = \boxed{7.77\times10^{-13}\text{ J}}$$

---

43. ***REASONING*** The de Broglie wavelength λ of the electron is related to the magnitude p of its momentum by $\lambda = h/p$ (Equation 29.8), where h is Planck's constant. If the speed of the electron is much less than the speed of light, the magnitude of the electron's momentum is given by $p = mv$ (Equation 7.2). Thus, the de Broglie wavelength can be written as $\lambda = h/(mv)$.

When the electron is at rest, it has electric potential energy, but no kinetic energy. The electric potential energy EPE is given by EPE = eV (Equation 19.3), where e is the magnitude of the charge on the electron and V is the potential difference. When the electron reaches its maximum speed, it has no potential energy, but its kinetic energy is $\frac{1}{2}mv^2$. The conservation of energy states that the final total energy of the electron equals the initial total energy:

$$\underbrace{\tfrac{1}{2}mv^2}_{\text{Final total energy}} = \underbrace{eV}_{\text{Initial total energy}}$$

Solving this equation for the final speed gives $v = \sqrt{2eV/m}$. Substituting this expression for v into $\lambda = h/(mv)$ gives $\lambda = h/\sqrt{2meV}$.

SOLUTION After accelerating through the potential difference, the electron has a de Broglie wavelength of

$$\lambda = \frac{h}{\sqrt{2meV}} = \frac{6.63\times10^{-34}\ \text{J}\cdot\text{s}}{\sqrt{2(9.11\times10^{-31}\ \text{kg})(1.60\times10^{-19}\ \text{C})(418\ \text{V})}} = \boxed{6.01\times10^{-11}\ \text{m}}$$

CHAPTER 30 | *THE NATURE OF THE ATOM*

CONCEPTUAL QUESTIONS

---

3. ***REASONING AND SOLUTION*** According to Equation 30.13, the energy E_n of the nth atomic state in a Bohr atom is given by $E_n = -(13.6\text{ eV})Z^2/n^2$, where n = 1, 2, 3, . . ., and Z is equal to the number of protons in the nucleus. The energy that must be absorbed by the electron to cause an upward transition from the initial state n_i to the final state n_f is $\Delta E_{fi} = E_f - E_i$. Using Equation 30.13, we find that the required energy is $\Delta E_{fi} = -(13.6\text{ eV})Z^2\left(1/n_f^2 - 1/n_i^2\right)$, with n_i, n_f = 1, 2, 3, . . . and $n_f > n_i$. The energy required to ionize the atom from when the outermost electron is in the state n_i can be found by letting the value of n_f approach infinity. The ionization energy is then

$$\Delta E_\infty = -(13.6\text{ eV})Z^2\left(1/\infty - 1/n_i^2\right) = -(13.6\text{ eV})Z^2\left(0 - 1/n_i^2\right) = (13.6\text{ eV})Z^2/n_i^2$$

From this expression, we see that the ionization energy is inversely proportional to the square of the principal quantum number n_i of the initial state of the electron; thus, less energy is required to remove the outermost electron in an atom when the atom is in an excited state. Therefore, when the atom is in an excited state, it is more easily ionized than when it is in the ground state.

---

8. ***REASONING AND SOLUTION*** An electronic configuration for manganese (Z = 25) is written as $1s^2\,2s^2\,2p^6\,3s^2\,3p^6\,3d^4\,4s^2\,4p^1$. According to Figure 30.17, the ground state electronic configuration for manganese is written as $1s^2\,2s^2\,2p^6\,3s^2\,3p^6\,3d^5\,4s^2$. Clearly, the electronic configuration given in the question corresponds to *an excited state* in which the last electron in the 3d subshell has been excited to the 4p subshell.

---

10. ***REASONING AND SOLUTION*** Bremsstrahlung X-rays are produced when the electrons decelerate upon hitting the target. In the production of X-rays, it is possible to create Bremsstrahlung X-rays without producing the characteristic X-rays. The wavelengths of the characteristic lines depend on the target material; the cutoff wavelength, however, is independent of the target material and depends only on the energy of the impinging electrons. In order to produce Bremsstrahlung X-rays without producing the characteristic X-rays, the potential difference V applied to the tube must be chosen so that the cutoff wavelength is greater than the wavelengths of the characteristic X-rays for the target material. Therefore, using Equation 30.17, the value of V must be chosen so that $V = (hc)/(e\lambda_0)$, where $\lambda_0 > \lambda_{K_\alpha}$.

CHAPTER 30 | *THE NATURE OF THE ATOM*

PROBLEMS

---

1. ***REASONING*** Assuming that the hydrogen atom is a sphere of radius r_{atom}, its volume V_{atom} is given by $\frac{4}{3}\pi r_{\text{atom}}^3$. Similarly, if the radius of the nucleus is r_{nucleus}, the volume V_{nucleus} is given by $\frac{4}{3}\pi r_{\text{nucleus}}^3$.

SOLUTION

a. According to the given data, the nuclear dimensions are much smaller than the orbital radius of the electron; therefore, we can treat the nucleus as a point about which the electron orbits. The electron is normally at a distance of about 5.3×10^{-11} m from the nucleus, so we can treat the atom as a sphere of radius $r_{\text{atom}} = 5.3 \times 10^{-11}$ m. The volume of the atom is

$$V_{\text{atom}} = \frac{4}{3}\pi r_{\text{atom}}^3 = \frac{4}{3}\pi\left(5.3\times10^{-11}\text{ m}\right)^3 = \boxed{6.2\times10^{-31}\text{ m}^3}$$

b. Similarly, since the nucleus has a radius of approximately $r_{\text{nucleus}} = 1\times10^{-15}$ m, its volume is

$$V_{\text{nucleus}} = \frac{4}{3}\pi r_{\text{nucleus}}^3 = \frac{4}{3}\pi\left(1\times10^{-15}\text{ m}\right)^3 = \boxed{4\times10^{-45}\text{ m}^3}$$

c. The percentage of the atomic volume occupied by the nucleus is

$$\frac{V_{\text{nucleus}}}{V_{\text{atom}}}\times100\% = \frac{\frac{4}{3}\pi r_{\text{nucleus}}^3}{\frac{4}{3}\pi r_{\text{atom}}^3}\times100\% = \left(\frac{1\times10^{-15}\text{ m}}{5.3\times10^{-11}\text{ m}}\right)^3\times100\% = \boxed{7\times10^{-13}\ \%}$$

---

5. ***REASONING*** The copper nucleus has a total charge $Ze = 29\times1.6\times10^{-19}$ C, and radius $r_{\text{nucleus}} = 4.8\times10^{-15}$ m. The required work can be found from Equation 19.1: $W_{\text{AB}} = \text{EPE}_{\text{A}} - \text{EPE}_{\text{B}}$. According to Equation 19.3, EPE $= Vq$, where $q = e$ for a proton. Therefore, $W_{\text{AB}} = V_{\text{A}}e - V_{\text{B}}e$. The proton is initially at infinity, so $V_{\text{A}} = 0$ V and $V_{\text{B}} = kZe / r_{\text{nucleus}}$, according to Equation 19.6. Thus, $W_{\text{AB}} = -kZe^2 / r_{\text{nucleus}}$.

SOLUTION The work done by the electric force is, therefore,

$$W_{AB} = \frac{-kZe^2}{r_{\text{nucleus}}} = \frac{-(8.99\times10^9\ \text{N}\cdot\text{m}^2/\text{C}^2)(29)(1.6\times10^{-19}\ \text{C})^2}{4.8\times10^{-15}\ \text{m}} = -1.39\times10^{-12}\ \text{J}$$

$$W_{AB} = \left(-1.39\times10^{-12}\ \text{J}\right)\underbrace{\left(\frac{1.0\ \text{eV}}{1.6\times10^{-19}\ \text{J}}\right)}_{\substack{\text{Converts from joules}\\ \text{to electron volts}}} = \boxed{-8.7\times10^6\ \text{eV}}$$

---

7. ***REASONING AND SOLUTION*** The first excited state occurs for $n = 2$ and has an energy given by Equation 30.13:

$$E_2 = -(13.6\ \text{eV})\frac{Z^2}{n^2} = -(13.6\ \text{eV})\frac{1^2}{2^2} = -3.40\ \text{eV}$$

After absorbing 2.86 eV of energy, the electron has an energy

$$E_n = -3.40\ \text{eV} + 2.86\ \text{eV} = -0.54\ \text{eV}$$

Again using Equation 30.13, we find $E_n = -0.54$ eV $= -(13.6\ \text{eV})(1^2/n^2)$. Solving for n, we obtain

$$n = \sqrt{\frac{13.6\ \text{eV}}{0.54\ \text{eV}}} = \boxed{5}$$

---

11. ***REASONING*** According to Equation 30.14, the wavelength λ emitted by the hydrogen atom when it makes a transition from the level with n_i to the level with n_f is given by

$$\frac{1}{\lambda} = \frac{2\pi^2 mk^2e^4}{h^3c}(Z^2)\left(\frac{1}{n_f^2} - \frac{1}{n_i^2}\right) \quad \text{with} \quad n_i, n_f = 1, 2, 3, \ldots \quad \text{and} \quad n_i > n_f$$

where $2\pi^2 mk^2e^4/(h^3c) = 1.097\times10^7\ \text{m}^{-1}$ and $Z = 1$ for hydrogen. Once the wavelength for the particular transition in question is determined, Equation 29.2 ($E = hf = hc/\lambda$) can be used to find the energy of the emitted photon.

SOLUTION In the Paschen series, $n_f = 3$. Using the above expression with $Z = 1$, $n_i = 7$ and $n_f = 3$, we find that

$$\frac{1}{\lambda} = \left(1.097\times10^7\ \text{m}^{-1}\right)\left(1^2\right)\left(\frac{1}{3^2} - \frac{1}{7^2}\right) \qquad \text{or} \qquad \lambda = 1.005\times10^{-6}\ \text{m}$$

The photon energy is

$$E = \frac{hc}{\lambda} = \frac{\left(6.63\times10^{-34}\ \text{J}\cdot\text{s}\right)\left(3.00\times10^{8}\ \text{m/s}\right)}{1.005\times10^{-6}\ \text{m}} = \boxed{1.98\times10^{-19}\ \text{J}}$$

---

15. ***REASONING*** A wavelength of 410.2 nm is emitted by the hydrogen atoms in a high-voltage discharge tube. This transition lies in the visible region (380–750 nm) of the hydrogen spectrum. Thus, we can conclude that the transition is in the Balmer series and, therefore, that $n_f = 2$. The value of n_i can be found using Equation 30.14, according to which the Balmer series transitions are given by

$$\frac{1}{\lambda} = R\left(1^2\right)\left(\frac{1}{2^2} - \frac{1}{n_i^2}\right) \qquad n = 3, 4, 5, \ldots$$

This expression may be solved for n_i for the energy transition that produces the given wavelength.

SOLUTION Solving for n_i, we find that

$$n_i = \frac{1}{\sqrt{\dfrac{1}{2^2} - \dfrac{1}{R\lambda}}} = \frac{1}{\sqrt{\dfrac{1}{2^2} - \dfrac{1}{(1.097\times10^{7}\ \text{m}^{-1})(410.2\times10^{-9}\ \text{m})}}} = 6$$

Therefore, the initial and final states are identified by $\boxed{n_i = 6 \text{ and } n_f = 2}$.

---

17. ***REASONING AND SOLUTION*** For the Paschen series, $n_f = 3$. The range of wavelengths occurs for values of $n_i = 4$ to $n_i = \infty$. Using Equation 30.14, we find that the shortest wavelength occurs for $n_i = \infty$ and is given by

$$\frac{1}{\lambda} = \left(1.097\times10^{7}\ \text{m}^{-1}\right)(1)^2\left(\frac{1}{n_f^2} - \frac{1}{n_i^2}\right) = \left(1.097\times10^{7}\ \text{m}^{-1}\right)\left(\frac{1}{3^2}\right) \quad \text{or} \quad \underbrace{\lambda = 8.204\times10^{-7}\ \text{m}}_{\text{Shortest wavelength in Paschen series}}$$

The longest wavelength in the Paschen series occurs for $n_i = 4$ and is given by

$$\frac{1}{\lambda} = \left(1.097 \times 10^7 \text{ m}^{-1}\right)\left(\frac{1}{3^2} - \frac{1}{4^2}\right) \quad \text{or} \quad \underbrace{\lambda = 1.875 \times 10^{-6} \text{ m}}_{\text{Longest wavelength in Paschen series}}$$

For the Brackett series, $n_f = 4$. The range of wavelengths occurs for values of $n_i = 5$ to $n_i = \infty$. Using Equation 30.14, we find that the shortest wavelength occurs for $n_i = \infty$ and is given by

$$\frac{1}{\lambda} = \left(1.097 \times 10^7 \text{ m}^{-1}\right)(1)^2\left(\frac{1}{n_f^2} - \frac{1}{n_i^2}\right) = \left(1.097 \times 10^7 \text{ m}^{-1}\right)\left(\frac{1}{4^2}\right) \quad \text{or} \quad \underbrace{\lambda = 1.459 \times 10^{-6} \text{ m}}_{\text{Shortest wavelength in Brackett series}}$$

The longest wavelength in the Brackett series occurs for $n_i = 5$ and is given by

$$\frac{1}{\lambda} = \left(1.097 \times 10^7 \text{ m}^{-1}\right)\left(\frac{1}{4^2} - \frac{1}{5^2}\right) \quad \text{or} \quad \underbrace{\lambda = 4.051 \times 10^{-6} \text{ m}}_{\text{Longest wavelength in Brackett series}}$$

Since the longest wavelength in the Paschen series falls within the Brackett series, the wavelengths of the two series overlap.

---

23. ***REASONING*** The values that ℓ can have depend on the value of n, and only the following integers are allowed: $\ell = 0, 1, 2, \ldots (n-1)$. The values that m_ℓ can have depend on the value of ℓ, with only the following positive and negative integers being permitted: $m_\ell = -\ell, \ldots -2, -1, 0, +1, +2, \ldots +\ell$.

SOLUTION Thus, when $n = 6$, the possible values of ℓ are 0, 1, 2, 3, 4, 5. Now when $m_\ell = 2$, the possible values of ℓ are 2, 3, 4, 5, . . . These two series of integers overlap for the integers 2, 3, 4, and 5. Therefore, the possible values for the orbital quantum number ℓ that this electron could have are $\boxed{\ell = 2, 3, 4, 5}$.

---

27. ***REASONING*** Let θ denote the angle between the angular momentum L and its z-component L_z. We can see from the figure at the right that $L_z = L\cos\theta$. Using Equation 30.16 for L_z and Equation 30.15 for L, we have

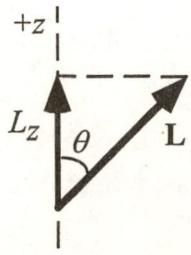

$$\cos\theta = \frac{L_z}{L} = \frac{m_\ell h/(2\pi)}{\sqrt{\ell(\ell+1)}\left[h/(2\pi)\right]} = \frac{m_\ell}{\sqrt{\ell(\ell+1)}}$$

The smallest value for θ corresponds to the largest value of cos θ. For a given value of ℓ, the largest value for cos θ corresponds to the largest value for m_ℓ. But the largest possible value for m_ℓ is $m_\ell = \ell$. Therefore, we find that

$$\cos\theta = \frac{\ell}{\sqrt{\ell(\ell+1)}} = \sqrt{\frac{\ell}{\ell+1}}$$

SOLUTION The smallest value for θ corresponds to the largest value for ℓ. When the electron is in the $n = 5$ state, the largest allowed value of ℓ is $\ell = 4$; therefore, we see that

$$\cos\theta = \sqrt{\frac{\ell}{\ell+1}} = \sqrt{\frac{4}{4+1}} \qquad \text{or} \qquad \theta = \cos^{-1}\left(\sqrt{4/5}\right) = \boxed{26.6°}$$

---

33. ***REASONING*** This problem is similar to Example 11 in the text. We use Equation 30.14 with the initial value of n being $n_i = 2$, and the final value being $n_f = 1$. As in Example 11, we use a value of Z that is one less than the atomic number of the atom in question (in this case, a value of $Z = 41$ rather than 42); this accounts approximately for the shielding effect of the single K-shell electron in canceling out the attraction of one nuclear proton.

SOLUTION Using Equation 30.14, we obtain

$$\frac{1}{\lambda} = (1.097\times10^7\ \text{m}^{-1})(41)^2\left(\frac{1}{1^2}-\frac{1}{2^2}\right) \qquad \text{or} \qquad \boxed{\lambda = 7.230\times10^{-11}\ \text{m}}$$

---

37. ***REASONING*** In the spectrum of X-rays produced by the tube, the cutoff wavelength λ_0 and the voltage V of the tube are related according to Equation 30.17, $V = hc/(e\lambda_0)$. Since the voltage is increased from zero until the K_α X-ray just appears in the spectrum, it follows that $\lambda_0 = \lambda_\alpha$ and $V = hc/(e\lambda_\alpha)$. Using Equation 30.14 for $1/\lambda_\alpha$, we find that

$$V = \frac{hc}{e\lambda_\alpha} = \frac{hcR(Z-1)^2}{e}\left(\frac{1}{1^2}-\frac{1}{2^2}\right)$$

In this expression we have replaced Z with $Z-1$, in order to account for shielding, as explained in Example 11 in the text.

SOLUTION The desired voltage is, then,

$$V = \frac{\left(6.63\times10^{-34}\ \text{J}\cdot\text{s}\right)\left(3.00\times10^{8}\ \text{m/s}\right)\left(1.097\times10^{7}\ \text{m}^{-1}\right)(47-1)^2}{\left(1.60\times10^{-19}\ \text{C}\right)}\left(\frac{1}{1^2}-\frac{1}{2^2}\right) = \boxed{21\,600\ \text{V}}$$

---

39. ***REASONING*** The number of photons emitted by the laser will be equal to the total energy carried in the beam divided by the energy per photon.

SOLUTION The total energy carried in the beam is, from the definition of power,

$$E_{\text{total}} = Pt = (1.5\ \text{W})(0.050\ \text{s}) = 0.075\ \text{J}$$

The energy of a single photon is given by Equations 29.2 and 16.1 as

$$E_{\text{photon}} = hf = \frac{hc}{\lambda} = \frac{\left(6.63\times10^{-34}\ \text{J}\cdot\text{s}\right)\left(3.00\times10^{8}\ \text{m/s}\right)}{514\times10^{-9}\ \text{m}} = 3.87\times10^{-19}\ \text{J}$$

where we have used the fact that $514\ \text{nm} = 514\times10^{-9}\ \text{m}$. Therefore, the number of photons emitted by the laser is

$$\frac{E_{\text{total}}}{E_{\text{photon}}} = \frac{0.075\ \text{J}}{3.87\times10^{-19}\ \text{J/photon}} = \boxed{1.9\times10^{17}\ \text{photons}}$$

---

45. ***REASONING AND SOLUTION***

a. The longest wavelength in the Pfund series occurs for the transition $n = 6$ to $n = 5$, so that

$$\frac{1}{\lambda} = R\left(\frac{1}{5^2}-\frac{1}{n^2}\right) = \left(1.097\times10^{7}\ \text{m}^{-1}\right)\left(\frac{1}{5^2}-\frac{1}{6^2}\right) \quad \text{or} \quad \boxed{\lambda = 7458\ \text{nm}}$$

b. The shortest wavelength occurs when $1/n^2 = 0$, so that

$$\frac{1}{\lambda} = R\left(\frac{1}{5^2}-\frac{1}{n^2}\right) = \left(1.097\times10^{7}\ \text{m}^{-1}\right)\left(\frac{1}{5^2}\right) \quad \text{or} \quad \boxed{\lambda = 2279\ \text{nm}}$$

c. The lines in the Pfund series occur in the $\boxed{\text{infrared region}}$.

---

47. ***REASONING*** In the theory of quantum mechanics, there is a selection rule that restricts the initial and final values of the orbital quantum number ℓ. The selection rule states that when

an electron makes a transition between energy levels, the value of ℓ may not remain the same or increase or decrease by more than one. In other words, the rule requires that $\Delta\ell = \pm 1$.

SOLUTION

a. For the transition $2s \rightarrow 1s$, the electron makes a transition from the 2s state ($n = 2$, $\ell = 0$) to the 1s state ($n = 1$, $\ell = 0$). Since the value of ℓ is the same in both states, $\Delta\ell = 0$, and we can conclude that this energy level transition is $\boxed{\text{not allowed}}$.

b. For the transition $2p \rightarrow 1s$, the electron makes a transition from the 2p state ($n = 2$, $\ell = 1$) to the 1s state ($n = 1$, $\ell = 0$). The value of ℓ changes so that $\Delta\ell = 0 - 1 = -1$, and we can conclude that this energy level transition is $\boxed{\text{allowed}}$.

c. For the transition $4p \rightarrow 2p$, the electron makes a transition from the 4p state ($n = 4$, $\ell = 1$) to the 2p state ($n = 2$, $\ell = 1$). Since the value of ℓ is the same in both states, $\Delta\ell = 0$, and we can conclude that this energy level transition is $\boxed{\text{not allowed}}$.

d. For the transition $4s \rightarrow 2p$, the electron makes a transition from the 4s state ($n = 4$, $\ell = 0$) to the 2p state ($n = 2$, $\ell = 1$). The value of ℓ changes so that $\Delta\ell = 1 - 0 = +1$, and we can conclude that this energy level transition is $\boxed{\text{allowed}}$.

e. For the transition $3d \rightarrow 3s$, the electron makes a transition from the 3d state ($n = 3$, $\ell = 2$) to the 3s state ($n = 3$, $\ell = 0$). The value of ℓ changes so that $\Delta\ell = 0 - 2 = -2$, and we can conclude that this energy level transition is $\boxed{\text{not allowed}}$.

---

49. ***REASONING*** According to the Bohr model, the energy E_n (in eV) of the electron in an orbit is given by Equation 30.13: $E_n = -13.6\left(Z^2/n^2\right)$. In order to find the principal quantum number of the state in which the electron in a doubly ionized lithium atom Li^{2+} has the same total energy as a ground state electron in a hydrogen atom, we equate the right hand sides of Equation 30.13 for the hydrogen atom and the lithium ion. This gives

$$-(13.6)\left(\frac{Z^2}{n^2}\right)_{\text{H}} = -(13.6)\left(\frac{Z^2}{n^2}\right)_{\text{Li}} \qquad \text{or} \qquad n_{\text{Li}}^2 = \left(\frac{n^2}{Z^2}\right)_{\text{H}} Z_{\text{Li}}^2$$

This expression can be evaluated to find the desired principal quantum number.

SOLUTION For hydrogen, $Z = 1$, and $n = 1$ for the ground state. For lithium Li^{2+}, $Z = 3$. Therefore,

$$n_{Li}^2 = \left(\frac{n^2}{Z^2}\right)_H Z_{Li}^2 = \left(\frac{1^2}{1^2}\right)(3^2) \quad \text{or} \quad \boxed{n_{Li} = 3}$$

---

53. ***REASONING*** Singly ionized helium, He^+, is a hydrogen-like species with $Z = 2$. The wavelengths of the series of lines produced when the electron makes a transition from higher energy levels into the $n_f = 4$ level are given by Equation 30.14 with $Z = 2$ and $n_f = 4$:

$$\frac{1}{\lambda} = \left(1.097 \times 10^7 \text{ m}^{-1}\right)(2^2)\left(\frac{1}{4^2} - \frac{1}{n_i^2}\right)$$

SOLUTION Solving this expression for n_i gives

$$n_i = \left[\frac{1}{4^2} - \frac{1}{4\lambda(1.097 \times 10^7 \text{ m}^{-1})}\right]^{-1/2}$$

Evaluating this expression at the limits of the range for λ, we find that $n_i = 19.88$ for $\lambda = 380$ nm, and $n_i = 5.58$ for $\lambda = 750$ nm. Therefore, the values of n_i for energy levels from which the electron makes the transitions that yield wavelengths in the range between 380 nm and 750 nm are $\boxed{6 \le n_i \le 19}$.

CHAPTER 31 | *NUCLEAR PHYSICS AND RADIOACTIVITY*

CONCEPTUAL QUESTIONS

---

7. ***REASONING AND SOLUTION*** In order for a nuclear decay to occur, it must bring the parent nucleus toward a more stable state by allowing the release of energy. The total mass of the decay products must be less than the mass of the parent nucleus. The difference in mass between the products and the parent nucleus represents an equivalent amount of energy that is given by Equation 28.5: $E_0 = mc^2$. This is the energy that is released during the decay.

As discussed in Example 4, uranium ${}^{238}_{92}\text{U}$ decays into thorium ${}^{234}_{90}\text{Th}$ by means of α decay. The ${}^{238}_{92}\text{U}$ nucleus never decays by just emitting a single proton, instead of an alpha particle. The hypothetical decay scheme for the unobserved decay, is given below along with the pertinent atomic masses:

$$\underbrace{{}^{238}_{92}\text{U}}_{\substack{\text{Uranium}\\ 238.050\,78\text{ u}}} \rightarrow \underbrace{{}^{237}_{91}\text{Pa}}_{\substack{\text{Protactinium}\\ 237.051\,14\text{ u}}} + \underbrace{{}^{1}_{1}\text{H}}_{\substack{\text{Proton}\\ 1.007\,83\text{ u}}}$$

The total mass of the decay products is (237.051 14 u) + (1.007 83 u) = 238.058 97 u. This is greater, not smaller, than the mass of the parent nucleus ${}^{238}_{92}\text{U}$; therefore, the daughter nucleus is less stable than the parent nucleus. Consequently, the hypothetical decay never occurs.

---

13. ***REASONING AND SOLUTION*** In radiocarbon dating, the age of the sample is determined from the ratio of the present activity to the original activity. Suppose that there were a greater number of ${}^{14}_{6}\text{C}$ atoms in a plant living 5000 yr ago than is currently believed. According to Equation 31.4: $\Delta N / \Delta t = -\lambda N$, and the original activity would be greater than is currently believed. A longer time would be required for the activity of a larger number of ${}^{14}_{6}\text{C}$ nuclei to decrease to its present value. Therefore, the age of the seeds obtained from radiocarbon dating would be too small.

---

16. ***REASONING AND SOLUTION*** According to Equation 21.2, the radius r of curvature for a particle with charge q and mass m, that moves perpendicular to a magnetic field of magnitude B, is given by

$$r = \frac{mv}{|q|B}$$

Thus, the ratio of the radii of curvature for the motion of the α particle and the β^- particle is

$$\frac{r_\beta}{r_\alpha} = \frac{m_\beta v / |q_\beta| B}{m_\alpha v / |q_\alpha| B} = \frac{m_\beta / |q_\beta|}{m_\alpha / |q_\alpha|} = \frac{m_\beta}{m_\alpha} \frac{|q_\alpha|}{|q_\beta|}$$

The mass of the α particle is 1.0026 u, while the mass of a β^- particle (an electron) is

$$(9.109\ 389\ 731 \times 10^{-31}\ \text{kg}) \left(\frac{1.0073\ \text{u}}{1.6726 \times 10^{-27}\ \text{kg}} \right) = 5.486 \times 10^{-4}\ \text{u}$$

Since the magnitudes of the charge on the α and β^- particles are $2e$ and e, respectively, the ratio of the radii of curvature is

$$\frac{r_\beta}{r_\alpha} = \left(\frac{5.486 \times 10^{-4}\ \text{u}}{4.0026\ \text{u}} \right) \left(\frac{2e}{e} \right) = 2.7 \times 10^{-4}$$

Therefore, the radius of curvature of the path of the β^- particle is smaller than the radius of curvature of the path of the α particle. Therefore, the trajectory of the β^- particle has a greater curvature than that of the α particle.

CHAPTER 31 | *NUCLEAR PHYSICS AND RADIOACTIVITY*

PROBLEMS

---

1. ***REASONING AND SOLUTION*** Solving Equation 31.2 for A gives

$$A = r^3/(1.2 \times 10^{-15}\ \text{m})^3$$

If r is doubled, then A will increase by a factor of $2^3 = \boxed{8}$

---

5. ***REASONING AND SOLUTION*** The surface area of a sphere is $\text{Area} = 4\pi r^2$. But according to Equation 31.2, the radius of a nucleus in meters is $r = (1.2 \times 10^{-15}\ \text{m})A^{1/3}$, where A is the nucleon number. With this expression for r, the surface area becomes $\text{Area} = 4\pi(1.2 \times 10^{-15}\ \text{m})^2 A^{2/3}$. The ratio of the largest to the smallest surface area is, then,

$$\frac{\text{Largest area}}{\text{Smallest area}} = \frac{4\pi(1.2 \times 10^{-15}\ \text{m})^2 A_{\text{largest}}^{2/3}}{4\pi(1.2 \times 10^{-15}\ \text{m})^2 A_{\text{smallest}}^{2/3}} = \frac{209^{2/3}}{1^{2/3}} = \boxed{35.2}$$

---

9. ***REASONING*** According to Equation 31.2, the radius of a nucleus in meters is $r = (1.2 \times 10^{-15}\ \text{m})A^{1/3}$, where A is the nucleon number. If we treat the neutron star as a uniform sphere, its density (Equation 11.1) can be written as

$$\rho = \frac{M}{V} = \frac{M}{\frac{4}{3}\pi r^3}$$

Solving for the radius r, we obtain,

$$r = \sqrt[3]{\frac{M}{\frac{4}{3}\pi\rho}}$$

This expression can be used to find the radius of a neutron star of mass M and density ρ.

SOLUTION As discussed in Conceptual Example 1, nuclear densities have the same approximate value in all atoms. If we consider a uniform spherical nucleus, then the density of nuclear matter is approximately given by

$$\rho = \frac{M}{V} \approx \frac{A \times (\text{mass of a nucleon})}{\frac{4}{3}\pi r^3} = \frac{A \times (\text{mass of a nucleon})}{\frac{4}{3}\pi\left[(1.2\times10^{-15}\text{ m})A^{1/3}\right]^3}$$

$$= \frac{1.67\times10^{-27}\text{ kg}}{\frac{4}{3}\pi(1.2\times10^{-15}\text{ m})^3} = 2.3\times10^{17}\text{ kg/m}^3$$

The mass of the sun is 1.99×10^{30} kg (see inside of the front cover of the text). Substituting values into the expression for r determined above, we find

$$r = \sqrt[3]{\frac{(0.40)(1.99\times10^{30}\text{ kg})}{\frac{4}{3}\pi(2.3\times10^{17}\text{ kg/m}^3)}} = \boxed{9.4\times10^3\text{ m}}$$

---

11. ***REASONING AND SOLUTION*** The earth revolves around the sun, and the two represent a bound system that has a binding energy of 2.6×10^{33} J. Therefore, to completely separate the earth and the sun so that they are infinitely far apart requires an amount of energy equal to the binding energy, namely $\Delta E_0 = 2.6\times10^{33}$ J.

Any change in the energy of a system causes a change in the mass of the system according to $\Delta E_0 = (\Delta m)c^2$ (see Section 28.6). Therefore, the change in mass of the earth-sun system when its energy changes by an amount $\Delta E_0 = 2.6\times10^{33}$ J is given by

$$\Delta m = \frac{\Delta E_0}{c^2} = \frac{2.6\times10^{33}\text{ J}}{(3.0\times10^8\text{ m/s})^2} = 2.9\times10^{16}\text{ kg}$$

Therefore, the difference between the mass of the separated system and that of the bound system is $\boxed{2.9\times10^{16}\text{ kg}}$.

---

15. ***REASONING*** Since we know the difference in binding energies for the two isotopes, we can determine the corresponding mass defect. Also knowing that the isotope with the larger binding energy contains one more neutron than the other isotope gives us enough information to calculate the atomic mass difference between the two isotopes.

SOLUTION The mass defect corresponding to a binding energy difference of 5.03 MeV is

$$(5.03\text{ MeV})\left(\frac{1\text{ u}}{931.5\text{ MeV}}\right) = 0.005\,40\text{ u}$$

Since the isotope with the larger binding energy has one more neutron ($m = 1.008\,665$ u) than the other isotope, the difference in atomic mass between the two isotopes is

$$1.008\ 665\ \text{u} - 0.005\ 40\ \text{u} = \boxed{1.003\ 27\ \text{u}}$$

---

17. ***REASONING AND SOLUTION*** The general form for β^- decay is

$$\underbrace{{}^{A}_{Z}\text{P}}_{\substack{\text{Parent}\\ \text{nucleus}}} \rightarrow \underbrace{{}^{A}_{Z+1}\text{D}}_{\substack{\text{Daughter}\\ \text{nucleus}}} + \underbrace{{}^{0}_{-1}\text{e}}_{\substack{\beta^- \text{ particle}\\ \text{(electron)}}}$$

Therefore, the β^- decay process for ${}^{35}_{16}\text{S}$ is $\boxed{{}^{35}_{16}\text{S} \rightarrow {}^{35}_{17}\text{Cl} + {}^{0}_{-1}\text{e}}$.

---

19. ***REASONING AND SOLUTION*** The general form for β^+ decay is

$$\underbrace{{}^{A}_{Z}\text{P}}_{\substack{\text{Parent}\\ \text{nucleus}}} \rightarrow \underbrace{{}^{A}_{Z-1}\text{D}}_{\substack{\text{Daughter}\\ \text{nucleus}}} + \underbrace{{}^{0}_{+1}\text{e}}_{\substack{\beta^+ \text{ particle}\\ \text{(positron)}}}$$

a. Therefore, the β^+ decay process for ${}^{18}_{9}\text{F}$ is $\boxed{{}^{18}_{9}\text{F} \rightarrow {}^{18}_{8}\text{O} + {}^{0}_{+1}\text{e}}$.

b. Similarly, the β^+ decay process for ${}^{15}_{8}\text{O}$ is $\boxed{{}^{15}_{8}\text{O} \rightarrow {}^{15}_{7}\text{N} + {}^{0}_{+1}\text{e}}$.

---

27. ***REASONING*** Energy is released during the β decay. To find the energy released, we determine how much the mass has decreased because of the decay and then calculate the equivalent energy. The reaction and masses are shown below:

$$\underbrace{{}^{22}_{11}\text{Na}}_{21.994\ 434\ \text{u}} \rightarrow \underbrace{{}^{22}_{10}\text{Ne}}_{21.991\ 383\ \text{u}} + \underbrace{{}^{0}_{+1}\text{e}}_{5.485\ 799 \times 10^{-4}\ \text{u}}$$

SOLUTION The decrease in mass is

$$21.994\ 434\ \text{u} - \left(21.991\ 383\ \text{u} + 5.485\ 799 \times 10^{-4}\ \text{u} + 5.485\ 799 \times 10^{-4}\ \text{u}\right) = 0.001\ 954\ \text{u}$$

where the extra electron mass takes into account the fact that the atomic mass for sodium includes the mass of 11 electrons, whereas the atomic mass for neon includes the mass of only 10 electrons.

Since 1 u is equivalent to 931.5 MeV, the released energy is

$$(0.001\,954\text{ u})\left(\frac{931.5\text{ MeV}}{1\text{ u}}\right) = \boxed{1.82\text{ MeV}}$$

---

31. ***REASONING AND SOLUTION*** The number of radioactive nuclei that remains in a sample after a time t is given by Equation 31.5, $N = N_0 e^{-\lambda t}$, where λ is the decay constant. From Equation 31.6, we know that the decay constant is related to the half-life by $T_{1/2} = 0.693/\lambda$; therefore, $\lambda = 0.693/T_{1/2}$ and we can write

$$\frac{N}{N_0} = e^{-(0.693/T_{1/2})t} \qquad \text{or} \qquad \frac{t}{T_{1/2}} = -\frac{1}{0.693}\ln\left(\frac{N}{N_0}\right)$$

When the number of radioactive nuclei decreases to one-millionth of the initial number, $N/N_0 = 1.00\times10^{-6}$; therefore, the number of half-lives is

$$\frac{t}{T_{1/2}} = -\frac{1}{0.693}\ln\left(1.00\times10^{-6}\right) = \boxed{19.9}$$

---

35. ***REASONING AND SOLUTION*** According to Equation 31.5, $N = N_0 e^{-\lambda t}$, the decay constant is

$$\lambda = -\frac{1}{t}\ln\left(\frac{N}{N_0}\right) = -\frac{1}{20\text{ days}}\ln\left(\frac{8.14\times10^{14}}{4.60\times10^{15}}\right) = 0.0866\text{ days}^{-1}$$

The half-life is, from Equation 31.6,

$$T_{1/2} = \frac{0.693}{\lambda} = \frac{0.693}{0.0866\text{ days}^{-1}} = \boxed{8.00\text{ days}}$$

---

39. ***REASONING*** According to Equation 31.5, the number of nuclei remaining after a time t is $N = N_0 e^{-\lambda t}$. Using this expression, we find the ratio N_A/N_B as follows:

$$\frac{N_A}{N_B} = \frac{N_{0A}e^{-\lambda_A t}}{N_{0B}e^{-\lambda_B t}} = e^{-(\lambda_A - \lambda_B)t}$$

where we have used the fact that initially the numbers of the two types of nuclei are equal $(N_{0A} = N_{0B})$. Taking the natural logarithm of both sides of the equation above shows that

$$\ln\left(N_A / N_B\right) = -\left(\lambda_A - \lambda_B\right)t \qquad \text{or} \qquad \lambda_A - \lambda_B = \frac{-\ln\left(N_A / N_B\right)}{t}$$

SOLUTION Since $N_A / N_B = 3.00$ when $t = 3.00$ days, it follows that

$$\lambda_A - \lambda_B = \frac{-\ln(3.00)}{3.00 \text{ days}} = -0.366 \text{ days}^{-1}$$

But we need to find the half-life of species B, so we use Equation 31.6, which indicates that $\lambda = 0.693 / T_{1/2}$. With this expression for λ, the result for $\lambda_A - \lambda_B$ becomes

$$0.693\left(\frac{1}{T_{1/2}^A} - \frac{1}{T_{1/2}^B}\right) = -0.366 \text{ days}^{-1}$$

Since $T_{1/2}^B = 1.50$ days, the result above can be solved to show that $\boxed{T_{1/2}^A = 7.23 \text{ days}}$.

---

41. ***REASONING AND SOLUTION*** The answer can be obtained directly from Equation 31.5, combined with Equation 31.6:

$$\frac{N}{N_0} = e^{-\lambda t} = e^{-(0.693)t/T_{1/2}} = e^{-(0.693)(41\,000 \text{ yr})/(5730 \text{ yr})} = 0.0070$$

The percent of atoms remaining is $\boxed{0.70\ \%}$.

---

45. ***REASONING*** According to Equation 31.5, $N = N_0 e^{-\lambda t}$. If we multiply both sides by the decay constant λ, we have

$$\lambda N = \lambda N_0 e^{-\lambda t} \qquad \text{or} \qquad A = A_0 e^{-\lambda t}$$

where A_0 is the initial activity and A is the activity after a time t. The decay constant λ is related to the half-life through Equation 31.6: $\lambda = 0.693 / T_{1/2}$. We can find the age of the fossils by solving for the time t. The maximum error can be found by evaluating the limits of the accuracy as given in the problem statement.

SOLUTION The age of the fossils is

$$t = -\frac{T_{1/2}}{0.693}\ln\left(\frac{A}{A_0}\right) = -\frac{5730\text{ yr}}{0.693}\ln\left(\frac{0.100\text{ Bq}}{0.23\text{ Bq}}\right) = \boxed{6900\text{ yr}}$$

The maximum error can be found as follows:

When there is an error of +10.0 %, $A = 0.100\text{ Bq} + 0.0100\text{ Bq} = 0.110\text{ Bq}$, and we have

$$t = -\frac{5730\text{ yr}}{0.693}\ln\left(\frac{0.110\text{ Bq}}{0.23\text{ Bq}}\right) = 6100\text{ yr}$$

Similarly, when there is an error of –10.0 %, $A = 0.100\text{ Bq} - 0.0100\text{ Bq} = 0.090\text{ Bq}$, and we have

$$t = -\frac{5730\text{ yr}}{0.693}\ln\left(\frac{0.090\text{ Bq}}{0.23\text{ Bq}}\right) = 7800\text{ yr}$$

The maximum error in the age of the fossils is $7800\text{ yr} - 6900\text{ yr} = \boxed{900\text{ yr}}$.

---

49. ***REASONING*** We can find the decay constant from Equation 31.5, $N = N_0e^{-\lambda t}$. If we multiply both sides by the decay constant λ, we have

$$\lambda N = \lambda N_0 e^{-\lambda t} \qquad \text{or} \qquad A = A_0 e^{-\lambda t}$$

where A_0 is the initial activity and A is the activity after a time t. Once the decay constant is known, we can use the same expression to determine the activity after a total of six days.

SOLUTION Solving the expression above for the decay constant λ, we have

$$\lambda = -\frac{1}{t}\ln\left(\frac{A}{A_0}\right) = -\frac{1}{2\text{ days}}\ln\left(\frac{285\text{ disintegrations/min}}{398\text{ disintegrations/min}}\right) = 0.167\text{ days}^{-1}$$

Then the activity four days after the second day is

$$A = (285\text{ disintegrations/min})e^{-(0.167\text{ days}^{-1})(4.00\text{ days})} = \boxed{146\text{ disintegrations/min}}$$

---

53. ***REASONING AND SOLUTION*** As shown in Figure 31.18, if the first dynode produces 3 electrons, the second produces 9 electrons (3^2), the third produces 27 electrons (3^3), so the Nth produces 3^N electrons. The number of electrons that leaves the 14th dynode and strikes the 15th dynode is

$$3^{14} = \boxed{4\ 782\ 969 \text{ electrons}}$$

CHAPTER 32 | *IONIZING RADIATION, NUCLEAR ENERGY, AND ELEMENTARY PARTICLES*

CONCEPTUAL QUESTIONS

---

4. ***REASONING AND SOLUTION*** In any nuclear reaction, the following quantities are conserved: (1) mass / energy; (2) electric charge, (3) linear momentum, (4) angular momentum, and (5) nucleon number.

a. ${}^{60}_{28}\text{Ni}\ (\alpha, p)\ {}^{62}_{29}\text{Cu}$ corresponds to the reaction

$$ {}^{4}_{2}\text{He} + {}^{60}_{28}\text{Ni} \rightarrow {}^{62}_{29}\text{Cu} + {}^{1}_{1}\text{H} $$

The nucleon number for the reactants is 4 + 60 = 64, while the nucleon number of the products is 62 + 1 = 63. Therefore, this reaction does *not* occur, because it violates the conservation of nucleon number.

b. ${}^{27}_{13}\text{Al}\ (n, n)\ {}^{28}_{13}\text{Al}$ corresponds to the reaction

$$ {}^{1}_{0}\text{n} + {}^{27}_{13}\text{Al} \rightarrow {}^{28}_{13}\text{Al} + {}^{1}_{0}\text{n} $$

The nucleon number for the reactants is 1 + 27 = 28, while the nucleon number of the products is 28 + 1 = 29. Therefore, this reaction does *not* occur, because it violates the conservation of nucleon number.

c. ${}^{39}_{19}\text{K}\ (p, \alpha)\ {}^{36}_{17}\text{Cl}$ corresponds to the reaction

$$ {}^{1}_{1}\text{H} + {}^{39}_{19}\text{K} \rightarrow {}^{36}_{17}\text{Cl} + {}^{4}_{2}\text{He} $$

The nucleon number for the reactants is 1 + 39 = 40, while the nucleon number of the products is 36 + 4 = 40. Nucleon number is conserved. The proton number on the left is $1 + 19 = 20$, while the proton number on the right is $17 + 2 = 19$. Since each proton carries a unit of charge $+e$, the total electric charge on the left is $+20\ e$, while the total electric charge on the right is $+19\ e$. Therefore, this reaction does *not* occur because it violates the conservation of electric charge.

---

9. ***REASONING AND SOLUTION*** Figure 32.9 shows that the binding energy of a nucleus with $A = 60$ is about 9.0 MeV per nucleon. If two such nuclei fuse, the daughter nucleus would have $A = 120$. According to Figure 32.9, the binding energy of the daughter nucleus *decreases* to about 8.3 MeV per nucleon. Consequently, the difference between the binding energy per nucleon of the initial nuclei and the binding energy per nucleon of the fused nucleus is about $8.3\text{ MeV} - 9.0\text{ MeV} = -0.7\text{ MeV}$ of energy per nucleon, where the minus sign indicates that the energy is absorbed. Therefore, the fusion of two nuclei, each with $A = 60$, would *not* release energy.

CHAPTER 32 *IONIZING RADIATION, NUCLEAR ENERGY, AND ELEMENTARY PARTICLES*

PROBLEMS

---

1. ***REASONING AND SOLUTION*** The energy absorbed is given by the absorbed dose in grays times the mass of the person in kilograms, according to Equation 32.2. To find the absorbed dose from the biologically equivalent dose (BED), we turn to Equation 32.4:

$$\text{Absorbed dose in rad} = \frac{\text{BED}}{\text{RBE}} = \frac{45 \times 10^{-3}\ \text{rem}}{12}$$

Using the fact that 0.01 Gy = 1 rad, we find that

$$\text{Absorbed dose in grays} = \left(\frac{45 \times 10^{-3}}{12}\ \text{rad}\right)\left(\frac{0.01\ \text{Gy}}{1\ \text{rad}}\right)$$

$$\text{Energy} = (\text{Absorbed dose in grays})(\text{Mass}) = \left(\frac{45 \times 10^{-5}}{12}\ \text{Gy}\right)(75\ \text{kg}) = \boxed{2.8 \times 10^{-3}\ \text{J}}$$

---

5. ***REASONING AND SOLUTION*** According to Equation 32.2, the absorbed dose (AD) is equal to the energy absorbed by the tumor divided by its mass:

$$\text{AD} = \frac{\text{Energy absorbed}}{\text{Mass}} = \frac{(25\ \text{s})\left(1.6 \times 10^{10}\ \text{s}^{-1}\right)\left(4.0 \times 10^{6}\ \text{eV}\right)\left(\frac{1.60 \times 10^{-19}\ \text{J}}{1\ \text{eV}}\right)}{0.015\ \text{kg}}$$

$$= 1.7 \times 10^{1}\ \text{Gy} = 1.7 \times 10^{3}\ \text{rd}$$

The biologically equivalent dose (BED) is equal to the product of the absorbed dose (AD) and the RBE (see Equation 32.4):

$$\text{BED} = \text{AD} \times \text{RBE} = (1.7 \times 10^{3}\ \text{rd})(14) = \boxed{2.4 \times 10^{4}\ \text{rem}} \qquad (32.4)$$

---

9. ***REASONING*** The number of nuclei in the beam is equal to the energy absorbed by the tumor divided by the energy per nucleus (130 MeV). According to Equation 32.2, the energy (in joules) absorbed by the tumor is equal to the absorbed dose (expressed in grays)

times the mass of the tumor. The absorbed dose (expressed in rads) is equal to the biologically equivalent dose divided by the RBE of the radiation (see Equation 32.4). We can use these concepts to determine the number of nuclei in the beam.

SOLUTION The number N of nuclei in the beam is equal to the energy E absorbed by the tumor divided by the energy per nucleus. Since the energy absorbed is equal to the absorbed dose (in Gy) times the mass m (see Equation 32.2) we have

$$N = \frac{E}{\text{Energy per nucleus}} = \frac{[\text{Absorbed dose (in Gy)}]m}{\text{Energy per nucleus}}$$

We can express the absorbed dose in terms of rad units, rather than Gy units, by noting that 1 rad = 0.01 Gy. Therefore,

$$\text{Absorbed dose (in Gy)} = \text{Absorbed dose (in rad)}\left(\frac{0.01\text{ Gy}}{1\text{ rad}}\right)$$

The number of nuclei can now be written as

$$N = \frac{E}{\text{Energy per nucleus}} = \frac{[\text{Absorbed dose (in rad)}]\left(\frac{0.01\text{ Gy}}{1\text{ rad}}\right)m}{\text{Energy per nucleus}}$$

We know from Equation 32.4 that the Absorbed dose (in rad) is equal to the Biologically equivalent dose divided by the RBE, so that

$$N = \frac{E}{\text{Energy per nucleus}} = \frac{\left[\frac{\text{Biologically equivalent dose}}{\text{RBE}}\right]\left(\frac{0.01\text{ Gy}}{1\text{ rad}}\right)m}{\text{Energy per nucleus}}$$

$$= \frac{\left[\frac{180\text{ rem}}{16}\right]\left(\frac{0.01\text{ Gy}}{1\text{ rad}}\right)(0.17\text{ kg})}{\left(130\times10^{6}\text{ eV}\right)\left(\frac{1.60\times10^{-19}\text{ J}}{1\text{ eV}}\right)} = \boxed{9.2\times10^{8}}$$

---

11. ***REASONING*** The reaction given in the problem statement is written in the shorthand form: ${}^{17}_{8}\text{O}\,(\gamma, \alpha\text{n})\,{}^{12}_{6}\text{C}$. The first and last symbols represent the initial and final nuclei, respectively. The symbols inside the parentheses denote the incident particles or rays (left side of the comma) and the emitted particles or rays (right side of the comma).

SOLUTION Using the reasoning above and noting that an α particle is a helium nucleus, ${}^{4}_{2}\text{He}$, we have

$$\boxed{\gamma + {}^{17}_{8}\text{O} \rightarrow {}^{12}_{6}\text{C} + {}^{4}_{2}\text{He} + {}^{1}_{0}\text{n}}$$

---

15. ***REASONING*** Energy is released from this reaction. Consequently, the combined mass of the daughter nucleus ${}^{12}_{6}\text{C}$ and the α particle ${}^{4}_{2}\text{He}$ is less than the combined mass of the parent nucleus ${}^{14}_{7}\text{N}$ and ${}^{2}_{1}\text{H}$. The mass defect is equivalent to the energy released. We proceed by determining the difference in mass in atomic mass units and then use the fact that 1 u is equivalent to 931.5 MeV (see Section 31.3).

SOLUTION The reaction and the atomic masses are as follows:

$$\underbrace{{}^{2}_{1}\text{H}}_{2.014\,102\text{ u}} + \underbrace{{}^{14}_{7}\text{N}}_{14.003\,074\text{ u}} \rightarrow \underbrace{{}^{12}_{6}\text{C}}_{12.000\,000\text{ u}} + \underbrace{{}^{4}_{2}\text{He}}_{4.002\,603\text{ u}}$$

The mass defect Δm for this reaction is

$$\Delta m = 2.014\,102\text{ u} + 14.003\,074\text{ u} - 12.000\,000\text{ u} - 4.002\,603\text{ u} = 0.014\,573\text{ u}$$

Since 1 u = 931.5 MeV, the energy released is

$$(0.014\,573\text{ u})\left(\frac{931.5\text{ MeV}}{1\text{ u}}\right) = \boxed{13.6\text{ MeV}}$$

---

19. ***REASONING*** The rest energy of the uranium nucleus can be found by taking the atomic mass of the ${}^{235}_{92}\text{U}$ atom, subtracting the mass of the 92 electrons, and then using the fact that 1 u is equivalent to 931.5 MeV (see Section 31.3). According to Table 31.1, the mass of an electron is $5.485\,799 \times 10^{-4}$ u. Once the rest energy of the uranium nucleus is found, the desired ratio can be calculated.

SOLUTION The mass of ${}^{235}_{92}\text{U}$ is 235.043 924 u. Therefore, subtracting the mass of the 92 electrons, we have

$$\text{Mass of } {}^{235}_{92}\text{U nucleus} = 235.043\,924\text{ u} - 92(5.485\,799 \times 10^{-4}\text{u}) = 234.993\,455\text{ u}$$

The energy equivalent of this mass is

$$(234.993\,455\text{ u})\left(\frac{931.5\text{ MeV}}{1\text{ u}}\right) = 2.189 \times 10^{5}\text{ MeV}$$

Therefore, the ratio is

$$\frac{200 \text{ MeV}}{2.189 \times 10^5 \text{ MeV}} = \boxed{9.0 \times 10^{-4}}$$

---

23. ***REASONING*** We first determine the energy released by 1.0 kg of $^{235}_{92}\text{U}$. Using the data given in the problem statement, we can then determine the number of kilograms of coal that must be burned to produce the same energy.

SOLUTION The energy equivalent of one atomic mass unit is given in the text (see Section 31.3) as

$$1 \text{ u} = 1.4924 \times 10^{-10} \text{ J} = 931.5 \text{ MeV}$$

Therefore, the energy released in the fission of 1.0 kg of $^{235}_{92}\text{U}$ is

$$(1.0 \text{ kg of } ^{235}_{92}\text{U}) \left(\frac{1.0 \times 10^3 \text{ g/kg}}{235 \text{ g/mol}}\right) \left(\frac{6.022 \times 10^{23} \text{ nuclei}}{1.0 \text{ mol}}\right)$$

$$\times \left(\frac{2.0 \times 10^2 \text{ MeV}}{\text{nuclei}}\right) \left(\frac{1.4924 \times 10^{-10} \text{ J}}{931.5 \text{ MeV}}\right) = 8.2 \times 10^{13} \text{ J}$$

When 1.0 kg of coal is burned, about 3.0×10^7 J is released; therefore the number of kilograms of coal that must be burned to produce an energy of 8.2×10^{13} J is

$$m_{\text{coal}} = \left(8.2 \times 10^{13} \text{ J}\right)\left(\frac{1.0 \text{ kg}}{3.0 \times 10^7 \text{ J}}\right) = \boxed{2.7 \times 10^6 \text{ kg}}$$

---

27. ***REASONING*** We first determine the total power generated (used and wasted) by the plant. Energy is power times the time, according to Equation 6.10, and given the energy, we can determine how many kilograms of $^{235}_{92}\text{U}$ are fissioned to produce this energy.

SOLUTION Since the power plant produces energy at a rate of 8.0×10^8 W when operating at 25 % efficiency, the total power produced by the power plant is

$$\left(8.0 \times 10^8 \text{ W}\right)4 = 3.2 \times 10^9 \text{ W}$$

The energy equivalent of one atomic mass unit is given in the text (see Section 31.3) as

$$1 \text{ u} = 1.4924 \times 10^{-10} \text{ J} = 931.5 \text{ MeV}$$

Since each fission produces 2.0×10^2 MeV of energy, the total mass of ${}^{235}_{92}\text{U}$ required to generate 3.2×10^9 W for a year (3.156×10^7 s) is

$$\overbrace{\left(3.2\times10^{9}\ \text{J/s}\right)\left(3.156\times10^{7}\ \text{s}\right)}^{\text{Power times time gives energy in joules}}$$

$$\times \underbrace{\left(\frac{931.5\ \text{MeV}}{1.4924\times10^{-10}\ \text{J}}\right)}_{\text{Converts joules to MeV. See Sec. 31.3.}} \underbrace{\left(\frac{1.0\ {}^{235}_{92}\text{U nucleus}}{2.0\times10^{2}\ \text{MeV}}\right)}_{\text{Converts MeV to number of nuclei}} \underbrace{\left(\frac{0.235\ \text{kg}}{6.022\times10^{23}\ {}^{235}_{92}\text{U nuclei}}\right)}_{\text{Converts number of nuclei to kilograms}} = \boxed{1200\ \text{kg}}$$

---

31. ***REASONING*** To find the energy released per reaction, we follow the usual procedure of determining how much the mass has decreased because of the fusion process. Once the energy released per reaction is determined, we can determine the mass of lithium ${}^{6}_{3}\text{Li}$ needed to produce 3.8×10^{10} J.

SOLUTION The reaction and the masses are shown below:

$$\underbrace{{}^{2}_{1}\text{H}}_{2.014\ \text{u}} + \underbrace{{}^{6}_{3}\text{Li}}_{6.015\ \text{u}} \rightarrow \underbrace{2\ {}^{4}_{2}\text{He}}_{2(4.003\ \text{u})}$$

The mass defect is, therefore, 2.014 u + 6.015 u − 2(4.003 u) = 0.023 u. Since 1 u is equivalent to 931.5 MeV, the released energy is 21 MeV, or since the energy equivalent of one atomic mass unit is given in Section 31.3 as $1\ \text{u} = 1.4924\times10^{-10}\ \text{J} = 931.5\ \text{MeV}$,

$$(21\ \text{MeV})\left(\frac{1.4924\times10^{-10}\ \text{J}}{931.5\ \text{MeV}}\right) = 3.4\times10^{-12}\ \text{J}$$

In 1.0 kg of lithium ${}^{6}_{3}\text{Li}$, there are

$$(1.0\ \text{kg of}\ {}^{6}_{3}\text{Li})\left(\frac{1.0\times10^{3}\ \text{g}}{1.0\ \text{kg}}\right)\left(\frac{6.022\times10^{23}\ \text{nuclei/mol}}{6.015\ \text{g/mol}}\right) = 1.0\times10^{26}\ \text{nuclei}$$

Therefore, 1.0 kg of lithium ${}^{6}_{3}\text{Li}$ would produce an energy of

$$(3.4\times10^{-12}\ \text{J/nuclei})(1.0\times10^{26}\ \text{nuclei}) = 3.4\times10^{14}\ \text{J}$$

If the energy needs of one household for a year is estimated to be 3.8×10^{10} J, then the amount of lithium required is

$$\frac{3.8\times10^{10}\ \text{J}}{3.4\times10^{14}\ \text{J/kg}} = \boxed{1.1\times10^{-4}\ \text{kg}}$$

---

39. ***REASONING*** The momentum of a photon is given in the text as $p = E/c$ (see the discussion leading to Equation 29.6). This expression applies to any massless particle that travels at the speed of light. In particular, assuming that the neutrino has no mass and travels at the speed of light, it applies to the neutrino. Once the momentum of the neutrino is determined, the de Broglie wavelength can be calculated from Equation 29.6 ($p = h/\lambda$).

SOLUTION

a. The momentum of the neutrino is, therefore,

$$p = \frac{E}{c} = \left(\frac{35\ \text{MeV}}{3.00\times10^{8}\ \text{m/s}}\right)\left(\frac{1.4924\times10^{-10}\ \text{J}}{931.5\ \text{MeV}}\right) = \boxed{1.9\times10^{-20}\ \text{kg}\cdot\text{m/s}}$$

where we have used the fact that 1.4924×10^{-10} J = 931.5 MeV (see Section 31.3).

b. According to Equation 29.6, the de Broglie wavelength of the neutrino is

$$\lambda = \frac{h}{p} = \frac{6.63\times10^{-34}\ \text{J}\cdot\text{s}}{1.9\times10^{-20}\ \text{kg}\cdot\text{m/s}} = \boxed{3.5\times10^{-14}\ \text{m}}$$

---

41. ***REASONING AND SOLUTION*** The lambda particle contains three different quarks, one of which is the up quark u, and contains no antiquarks. Therefore, the remaining two quarks must be selected from the down quark d, the strange quark s, the charmed quark c, the top quark t, and the bottom quark b. Since the lambda particle has an electric charge of zero and since u has a charge of $+2e/3$, the charges of the remaining two quarks must add up to a total charge of $-2e/3$. This eliminates the quarks c and t as choices, because they each have a charge of $+2e/3$. We are left, then, with d, s, and b as choices for the remaining two quarks in the lambda particle. The three possibilities are as follows:

(1) u,d,s	(2) u,d,b	(3) u,s,b

---

45. ***REASONING*** To find the energy released per reaction, we follow the usual procedure of determining how much the mass has decreased because of the fusion process. Once the energy released per reaction is determined, we can determine the amount of gasoline that must be burned to produce the same amount of energy.

SOLUTION The reaction and the masses are shown below:

$$\underbrace{3\ {}^{2}_{1}\mathrm{H}}_{3(2.0141\ \mathrm{u})} \rightarrow \underbrace{{}^{4}_{2}\mathrm{He}}_{4.0026\ \mathrm{u}} + \underbrace{{}^{1}_{1}\mathrm{H}}_{1.0078\ \mathrm{u}} + \underbrace{{}^{1}_{0}\mathrm{n}}_{1.0087\ \mathrm{u}}$$

The mass defect is, therefore,

$$3(2.0141\ \mathrm{u}) - 4.0026\ \mathrm{u} - 1.0078\ \mathrm{u} - 1.0087\ \mathrm{u} = 0.0232\ \mathrm{u}$$

Since 1 u is equivalent to 931.5 MeV, the released energy is 21.6 MeV, or since it is shown in Section 31.3 that 931.5 MeV $= 1.4924 \times 10^{-10}$ J, the energy released per reaction is

$$(21.6\ \mathrm{MeV})\left(\frac{1.4924 \times 10^{-10}\ \mathrm{J}}{931.5\ \mathrm{MeV}}\right) = 3.46 \times 10^{-12}\ \mathrm{J}$$

To find the total energy released by all the deuterium fuel, we need to know the number of deuterium nuclei present. The number of deuterium nuclei is

$$(6.1 \times 10^{-3}\ \mathrm{g})\left(\frac{6.022 \times 10^{23}\ \mathrm{nuclei/mol}}{2.0141\ \mathrm{g/mol}}\right) = 1.8 \times 10^{21}\ \mathrm{nuclei}$$

Since each reaction consumes three deuterium nuclei, the total energy released by the deuterium fuel is

$$\frac{1}{3}\ (3.46 \times 10^{-12}\ \mathrm{J/nuclei})(1.8 \times 10^{21}\ \mathrm{nuclei}) = 2.1 \times 10^{9}\ \mathrm{J}$$

If one gallon of gasoline produces 2.1×10^{9} J of energy, then the number of gallons of gasoline that would have to be burned to equal the energy released by all the deuterium fuel is

$$\left(2.1 \times 10^{9}\ \mathrm{J}\right)\left(\frac{1.0\ \mathrm{gal}}{2.1 \times 10^{9}\ \mathrm{J}}\right) = \boxed{1.0\ \mathrm{gal}}$$